T0383135

5G New Radio Non-Orthogonal Multiple Access

This book provides detailed descriptions of downlink non-orthogonal multiple transmissions and uplink non-orthogonal multiple access (NOMA) from the aspects of majorly used 5G new radio scenarios and system performance.

For the downlink, the discussion focuses on the candidate schemes in 3GPP standards which are not only applicable to unicast services but also to broadcast/multicast scenarios. For the uplink, the main target scenario is massive machine-type communications where grant-free transmission can reduce signaling overhead, power consumption of devices and access delays. The design principles of several uplink NOMA schemes are discussed in-depth, together with the analysis of their performances and receiver complexities.

Devoted to the basic technologies of NOMA and its theoretical principles, data analysis, basic algorithms, evaluation methodology and simulation results, this book will be an essential read for researchers and students of digital communications, wireless communications engineers and those who are interested in mobile communications in general.

Yifei Yuan is Chief Expert of China Mobile Research Institute. Dr. Yuan graduated from Tsinghua University and Carnegie Mellon University. He specializes in the research and standardization of key air-interface technologies for 3G, 4G, 5G and 6G mobile networks. He has more than 20 years of experience at Bell Labs, ZTE and China Mobile.

Zhifeng Yuan is Senior Expert in the Algorithm Department of ZTE Corporation. With more than 15 years of experience at ZTE Corporation, he focuses on transmitter designs and advanced receiver algorithms of non-orthogonal multiple access (NOMA), channel coding, modulations and waveform for mobile communications.

5G New Radio Non-Orthogonal Multiple Access

Yifei Yuan and Zhifeng Yuan

CRC Press is an imprint of the
Taylor & Francis Group, an **informa** business

First edition published 2023
by CRC Press
6000 Broken Sound Parkway NW, Suite 300, Boca Raton, FL 33487-2742

and by CRC Press
4 Park Square, Milton Park, Abingdon, Oxon, OX14 4RN

CRC Press is an imprint of Taylor & Francis Group, LLC

Published through arrangement with the original publisher, Posts and Telecom Press Co., Ltd.

Library of Congress Cataloging-in-Publication Data
Names: Yuan, Yifei, editor. | Yuan, Zhifeng, editor.
Title: 5G new radio non-orthogonal multiple access / [edited by] Yifei Yuan and Zhifeng Yuan.
Description: First edition. | Boca Raton, FL : CRC Press, 2023. | Includes bibliographical references. |
Identifiers: LCCN 2022025363 (print) | LCCN 2022025364 (ebook) |
ISBN 9781032372754 (hbk) | ISBN 9781032372761 (pbk) |
ISBN 9781003336167 (ebk)
Subjects: LCSH: 5G mobile communication systems. | Multiple access protocols (Computer network protocols)
Classification: LCC TK5103.25 .A147 2023 (print) | LCC TK5103.25 (ebook) | DDC 621.3845/6--dc23/eng/20220812
LC record available at https://lccn.loc.gov/2022025363
LC ebook record available at https://lccn.loc.gov/2022025364

ISBN: 978-1-032-37275-4 (hbk)
ISBN: 978-1-032-37276-1 (pbk)
ISBN: 978-1-003-33616-7 (ebk)

DOI: 10.1201/9781003336167

Typeset in Minion
by codeMantra

Contents

Foreword

Not only considered as a strategic technology of industry that attracts significant attention from many countries, 5G has also triggered quite high expectations of human society to effectively support enhanced mobile broadband, ultra-reliable low-latency communications and massive machine-type communication. In the communications industry, greater interest is seen in the innovative upgrade of the systems by integrating advanced wireless and network technologies of 5G. In cellular communications, a key performance indicator is how many users and throughput can be supported simultaneously by a cell. Technologies that enable the multiplexing of multiple users are called multiple access technologies. The first four generations of mobile communications all feature the evolution of multiple access technologies: frequency division multiple access (FDMA) in 1G, time division multiple access (TDMA) primarily in 2G, code division multiple access (CDMA) in 3G and orthogonal frequency division multiple access (OFDMA) in 4G. Multiple access is also a key enabling technology in 5G. However, there are more than one multiple access technology in 5G. For enhanced mobile broadband services, while OFDM with cyclic prefix (CP-OFDM) is adopted for 5G physical channels, OFDM with spread discrete-time Fourier Transform (DFT-s-OFDM) can also be used for the uplink. For massive machine-type communication with a low data rate, the overhead of OFDMA is significant, which results in low efficiency of transmission. This prompts non-orthogonal multiple access (NOMA) in which multiple users share the same time and frequency resources and rely on interference cancelation techniques at the receivers to separate each user's signal. Several NOMA schemes have been proposed so far, including multi-user shared access (MUSA), sparse-code multiple access (SCMA), resource-spread multiple access (RSMA) and pattern-division multiple access (PDMA). The standardization work is ongoing.

This book is not merely to discuss various NOMA schemes. Rather, it begins with the fundamental theory and technology of NOMA, followed by the discussion of downlink unicast and multicast/broadcast and the comparison between direct symbol superposition, bit-flipping-based superposition and bit partition–based superposition. For the uplink NOMA, the book focuses on grant-free transmissions that include contention-free and contention-based designs and algorithms.

There are quite a number of published books on 5G wireless and networks, but very few are specifically about NOMA. This book is devoted to the basic technologies of NOMA and contains theoretical principles, data analysis, basic algorithms, evaluation methodology and simulation results. The book has clear structures and abundant information, written in plain language. It can either serve the general public or be used as a reference book for university courses in the field of digital communications. It is believed that the book can attract more engineers of communications industry to study NOMA, which is not only beneficial to the completion of 5G standards but also to the technological study of next-generation mobile communications.

Hequan Wu
Academician of Chinese Academy of Engineering

Preface

Non-orthogonal multiple access (NOMA) is one of the fundamental technologies of 5G physical layer. By superposition transmission of multiple users over the same time and frequency resources, the system capacity and the number of supported users can be significantly improved. While the multiple access of 5G systems is still primarily orthogonal, non-orthogonal can be a complement to orthogonal multiple access (OMA) and used for both scheduling-based transmission and grant-free/contention-based transmission. Its user scenarios span over massive machine-type communication (mMTC), enhanced mobile broadband (eMBB) small packet data and ultra-reliable low-latency communication (uRLLC). Therefore, NOMA has gained wide attention from both academia and the industry. In 2015–2016, 3GPP studied and specified downlink NOMA for Release 14. In 2017, a study item of NOMA based on 5G new air interface was approved, with the focus on uplink grant-free transmissions. Many companies had participated, where very diverse solutions were proposed and evaluated. The 3GPP NOMA study item was completed in December 2018, and part of the study was specified in the follow-up work item of two-step RACH.

Chinese companies began their study on NOMA very early and have accumulated a deep knowledge base. In the NOMA sub-group of the IMT-2020 (5G) Promotion Group of China, companies such as ZTE, Huawei and CATT proposed NOMA solutions like MUSA, SCMA and PDMA, respectively, in 2014, together with a large amount of forward-looking study. This spurred more Chinese and foreign companies to engage in the NOMA study and propelled the approval of the NOMA study item in 3GPP where ZTE became the rapporteur company to lead the effort.

In this book, NOMA technologies and solutions in 3GPP are discussed in a systematic and comprehensive manner. The contents cover multi-user superposition transmission (MUST) for downlink in Release 14 and uplink NOMA in 5G NR Release 16. The discussions usually start from theories in academia, and then to the practical design and engineering, finally to the corresponding standards specifications, accompanied by rich simulation results at the link and system levels. The target readers include wireless communications engineers and researchers/students in universities/institutes.

This book is the work of Yifei Yuan, Zhifeng Yuan, et al. Specifically, Chapter 1 was written by Yifei Yuan; Chapter 2 was written by Yifei Yuan, Zhifeng Yuan, Jianqiang Dai and Hong Tang; Chapter 3 was written by Yifei Yuan, Hong Tang and Weimin Li; Chapter 4 was written by Jianqiang Dai and Yifei Yuan; Chapter 5 was written by Yifei Yuan, Zhifeng Yuan, Nan Zhang, Weimin Li, Ziyang Li, Qiujin Guo and Jian Li; Chapter 6 was written by Yifei Yuan, Zhifeng Yuan, Li Tian, Chen Huang, Yuzhou Hu, Chunlin Yan and Ziyang Li; Chapter 7 was written by Ziyang Li, Qiujin Guo, Hong Tang, Weimin Li, Jian Li, Yifei Yuan, Chen Huang and Li Tian; Chapter 8 was written by Nan Zhang, Wei Cao, Zhifeng Yuan, Jianqiang Dai, Ziyang Li, Hong Tang, Weimin Li, Jian Li and Yihua Ma. The entire book was planned and managed by Yifei Yuan and Zhifeng Yuan. The authors sincerely thank Xinhui Wang, Zhongda Du, Liujun Hu, Guanghui Yu, Gang Bai, Peng Geng, Bo Sun and Wei Han for their strong support. They thank Professors Linglong Dai, Kewu Peng and Xin Su at Tsinghua University and Professor Wen Chen at Shanghai Jiaotong University for the illuminating discussions. They thank the member companies/institutes in IMT-2020 (5G) NOMA sub-group, including CAICT, Huawei, CATT, China Mobile, Qualcomm, Samsung, NTT DOCOMO, Nokia and Ericsson for their technical contributions. Finally, they thank Posts & Telecom Press for the synchronized collaboration and efficient work, which expedited the publication process so that the book can be unveiled sooner.

This book is based on the limited ken of the authors regarding the study and standardization of 5G non-orthogonal multiple access. Hence, imperfections are inevitable and hopefully can be excused by kind readers. Any suggestions are highly appreciated.

Authors

Authors

Yifei Yuan is Chief Expert of China Mobile Research Institute. Dr. Yuan graduated from Tsinghua University and Carnegie Mellon University. He specializes in the research and standardization of key air-interface technologies for 3G, 4G, 5G and 6G mobile networks. He has more than 20 years of experience at Bell Labs, ZTE and China Mobile.

Zhifeng Yuan is Senior Expert in the Algorithm Department of ZTE Corporation. With more than 15 years of experience at ZTE Corporation, he focuses on transmitter designs and advanced receiver algorithms of non-orthogonal multiple access (NOMA), channel coding, modulations and waveform for mobile communications.

Abbreviations

π/2-BPSK **pi/2-BPSK** **Pi/2-BPSK**	π/2-Binary-Phase Shift Keying
128QAM	128-Quadrature Amplitude Modulation
16QAM	16-Quadrature Amplitude Modulation
1G	First-Generation mobile communication
256QAM	256-Quadrature Amplitude Modulation
2G	Second-Generation mobile communication
32QAM	32-Quadrature Amplitude Modulation
3G	Third-Generation mobile communication
3GPP	Third-Generation Partnership Project
4G	Fourth-Generation mobile communication
5G **5G-NR**	Fifth-Generation mobile communication
64QAM	64-Quadrature Amplitude Modulation
AWGN	Additional White Gaussian Noise
BEC	Binary Erasure Channel
BER	Bit Error Rate
BICM	Bit Interleaver, Coding and Modulation
BLER	BLock Error Rate
BP	Belief Propagation
bps	bit per second
bps/Hz	bps per Hz
BPSK	Binary-Phase Shift Keying
BSC	Binary-Discrete Symmetric Channel

CDMA	Code Division Multiple Access
CQI	Channel Quality Indicator
CRC	Cyclic Redundancy Check
CSI	Channel State Information
CWIC	Code-Word-level Interference Cancellation
DCI	Downlink Control Information
DPC	Dirty Paper Coding
DRX	Discontinuous Reception
eMBB	enhanced Mobile BroadBand
EPA	Extended Pedestrian a model
ETU	Extended Typical Urban model
EVM	Error Vector Magnitude
EXIT chart	EXtrinsic Information Transition chart
FAR	False Alarm Rate
FDD	Frequency Division Duplex
FDMA	Frequency Division Multiple Access
FEC	Forward Error Correction
FER	Frame Error Rate
FFT	Fast Fourier Transform
FFT-BP	Belief Propagation with FFT
GSM	Global System of Mobile communications
HARQ	Hybrid Automatic Retransmission reQuest
HSDPA	High-Speed Downlink Packet Access
HSPA	High-Speed Packet Access
HSUPA	High-Speed Uplink Packet Access
IOT	Internet of Things
IR	Incremental Redundancy
IRC	Interference Rejection Combining
KPI	Key Performance Index
LDPC	Low-Density Parity Check
LLR	Log-Likelihood Ratio
Log-BP	Belief Propagation in Logarithm domain
Log-FFT-BP	Belief Propagation with FFT in Logarithm domain

Log-MAP	Maximum APP in Logarithm domain Maximum a Posteriori Probability in Logarithm domain
LTE	Long-Term Evolution
MAC	Media Access Control
Max-Log-MAP	Maximum-Maximum APP in Logarithm domain Maximum-Maximum a Posteriori Probability in Logarithm domain
ML	Maximum Likelihood
mMTC	massive Machine-Type Communication
MAP	Maximum APP Maximum a Posteriori Probability
MBB	Mobile BroadBand
MCS	Modulation and Coding Scheme
MIB	Mutual Information per transmitted Bit
MIMO	Multiple Input Multiple Output
min-sum	minimum-sum-product algorithm
MMSE	Minimum Mean-Squared Error
MU-MIMO	Multi-User Multi-Input Multi-Output
NR	New Radio access technology
OFDM	Orthogonal Frequency Division Multiplexing
OFDMA	Orthogonal Frequency Division Multiple Access
PBCH	Physical Broadcast CHannel
PDCCH	Physical Downlink Control CHannel
PDF	Probability Density Function
PDSCH	Physical Downlink Shared CHannel
PER	Package Error Rate
PRB	Physical Resource Block
PUCCH	Physical Uplink Control CHannel
PUSCH	Physical Uplink Shared CHannel
QAM	Quadrature Amplitude Modulation
QLDPC	q-array LDPC Non-binary LDPC
QoS	Quality of Service

QPSK	Quadrature Phase Shift Keying
RAN	Radio Access Network
RE	Resource Element
RV	Redundancy Version
SDU	Segment Data Unit
SIC	Successive Interference Cancellation
SINR	Signal-to-Interference-plus-Noise Ratio
SLIC	Symbol-Level Interference Cancellation
SNR	Signal-to-Noise Ratio
TB	Transport Block
TBS	Transport Block Size
TDMA	Time Division Multiple Access
TD-SCDMA	Time Division-Synchronous Code Division Multiple Access
THP	Tomlinson-Harashima Precoding
UCI	Uplink Control Information
UE	User Equipment
UMB	Ultra-Mobile Broadband
UMTS	Universal Mobile Telecommunication System
URLLC	Ultra-Reliable and Low-Latency Communication
VN	Variable Node
WCDMA	Wideband Code Division Multiple Access
WiMAX	Worldwide Interoperability for Microwave Access

CHAPTER 1

Introduction

Yifei Yuan

1.1 EVOLUTION OF MOBILE COMMUNICATIONS

Radio resources are limited, especially for the spectrum bands suitable for personal communications. The goal of mobile communications is always to improve spectrum utilization. In 1968, the cellular communications concept was proposed by AT&T Bell Laboratories, which is to connect multiple hexagonal cells similar to beehives and form a large network. Frequency resources can be reused between cells, which drastically increases the system capacity. Over the past several decades, there has been tremendous progress in cellular technologies, leading to multi-magnitude of improvements in spectral efficiency, user data rate and system capacities. Mobile communications have undergone four generations of evolution. In addition to the big jumps in peak data rates, system bandwidth and system throughput compared to the previous generation, each new generation is featured by its unique way of multiple access, as listed in Table 1.1.

The first generation (1G) of mobile communications is featured by frequency division multiple access (FDMA) where only voice services are supported. Each user is allocated a fixed frequency resource. Since analog modulation such as frequency modulation is used, there is no way to carry out the source coding and channel coding in order to compress the information and improve reliability. Transmit power is not well controlled either. All these lead to very poor resource utilization. For instance, in the widely deployed 1G network – Advanced Mobile Phone System (AMPS), each user is allocated with 30 kHz bandwidth and each cell can support no more than 30 users in 1 MHz system bandwidth. The analog circuits are very difficult to integrate: the first-generation mobile handsets are

DOI: 10.1201/9781003336167-1

TABLE 1.1 Multiple-Access Schemes for Previous Generations of Mobile Networks

Mobile Network Generation	First Generation (1G)	Second Generation (2G)	Third Generation (3G)	Fourth Generation (4G)
Multiple-access scheme	FDMA, fixed allocation	TDMA dominant, e.g., GSM; CDMA, e.g., IS-95	CDMA, primarily with TDMA in 3.5G for data services	OFDMA primarily

typically very bulky, power-thirsty and expensive, so out of reach for ordinary people.

The dominant multiple-access scheme of the second generation (2G) of mobile communications is time division multiple access (TDMA). Its basic service is voice service. The most successful 2G standard is the Global System of Mobile Communications (GSM) developed mainly by European companies. In GSM, a frequency band is first divided into a number of narrow bands, each having a 200 kHz bandwidth. In each narrow band, users are allocated in different time slots in a round-bin fashion. In order to reduce the interference from neighboring cells and maintain good voice quality at cell edges, 7 or 11 adjacent cells are formed into a cluster. No frequency reuse is allowed within a cluster. Air interfaces of 2G cellular networks are digitized where the analog voice is sampled and quantized, and then compressed via a voice encoder. Channel coding is applied to correct the bit errors in noise, fading and interference-prone wireless environment. Via digital modulation and power control, the spectral efficiency per link is significantly increased and so is the system capacity. In the later stage of 2G, the IS-95 system, developed by Qualcomm, Inc., began to be deployed in part of North America. IS-95 is a code division multiple access (CDMA) technology with a direct spread spectrum. It can be regarded as a harbinger of 3G.

CDMA is widely used in the third generation (3G) of mobile communications. The systems' capability of interference mitigation is significantly increased in 3G so that the neighboring cells can fully reuse the frequency, therefore bringing a huge gain in system capacity. CDMA2000 and Universal Mobile Telecommunication System (UMTS) are two major mobile standards in 3G. CDMA2000 was mainly deployed in North America, Korea, China, etc. The system bandwidth of CDMA2000 is 1.2 MHz per carrier. Its standards development body is 3GPP2. UMTS/high-speed packet access (HSPA) was developed by 3GPP, the most influential wireless standards development body in the world. European operators and system vendors played important roles in UMTS which has been deployed worldwide. The system bandwidth of UMTS is 5 MHz per carrier. Hence, it is also called wideband CDMA (WCDMA). In order to support a higher rate for data services, CDMA and UMTS evolved independently along their own paths, resulting in the evolution data optimized (EV-DO) and HSPA, respectively. Both EV-DO and HSPA incorporate TDMA into CDMA where a shorter slot duration is used to facilitate faster link adaptation. There is another 3G standard called Time Division Synchronous CDMA (TD-SCDMA) which was developed mainly by Chinese and some European companies. TD-SCDMA is part of 3GPP standards and is widely deployed in China.

The fourth generation (4G) of mobile communications is featured by orthogonal frequency division multiple access (OFDMA). There are several reasons for this choice. Firstly, the system bandwidth of 4G systems is at least 20 MHz, much greater than that of 3G. Wider bandwidth means a faster time-domain sampling rate and more resolvable multipaths. If CDMA is used, the receiver would suffer from severe multipath interference. Although such interference can be suppressed by using advanced receivers, the processing complexity would be prohibitive. By contrast, if the system bandwidth can be divided into multiple mutually orthogonal subcarriers and each subcarrier is narrow enough to be considered as flat, no complicated equalization or interference suppression would be needed, which drastically reduces the implementation cost of receivers. This is especially crucial for multi-input-multi-output antenna (MIMO) receivers. In some sense, it is orthogonal frequency division multiplexing (OFDM) that promotes the wide usage of MIMO in 4G, thus improving the link and system capacity. It should be pointed out that the idea of TDMA is also used in 4G where the slot duration is even shorter than that of 3G. Note that 4G is not purely OFDM, e.g., some control channels and reference signals of 4G systems still rely on CDMA for differentiation.

In the early stage of 4G standards development, there were three major standards: Ultra Mobile Broadband (UMB), World Interoperability for Microwave Access (WiMAX) and Long-Term Evolution (LTE). The core technology of UMB was from Qualcomm's research in IEEE 802.20. Later on, with more companies participating, for instance, Lucent Technologies, Nortel and Samsung, 3GPP2 began to specify the technical details. The specification work of UMB was completed near the end of 2007. However, due to the lack of interest from several big operators like Verizon, the work on UMB was suspended after 2008. WiMAX can be considered as an extension of Wi-Fi to support wider area and mobility scenarios.

The standardization of WiMAX was completed in 2007. WiMAX was in Sprint's deployment plan at the beginning. As the business operation of Sprint deteriorated, plus the very loose industry alliance and lack of concrete business case, few countries or areas have widely adopted WiMAX in their 4G networks.

The first release of LTE in 3GPP is Release 8, frozen in 2008. With the suspension of UMB and the marginalization of WiMAX, LTE became the mainstream global 4G mobile standard. Since 2009, 3GPP began to specify LTE-Advanced. As a major enhancement, its version is dubbed as Release 10 which can fully meet the key performance requirements of IMT-Advanced.

In addition to OFDM and MIMO, a number of new features have been introduced to LTE-Advanced, for instance, carrier aggregation, intercell interference cancellation, wireless relay, enhanced downlink control channel and device-to-device communications. All these technologies help to improve the average spectral efficiency, the peak rate, the network throughput, coverage, etc. The features are not only applicable to macro cells but also to heterogeneous networks made up of macro nodes and low-power nodes.

1.2 SYSTEM REQUIREMENTS FOR 5G MOBILE COMMUNICATIONS

Different from previous generations, the applications of 5G are very diverse. Peak data rate and average spectral efficiency of cells are no longer the only key performance indicators (KPIs). Quite a number of other KPIs are introduced as part of the system design requirements, such as user experienced rate, number of connections, very low latency, very high reliability and power efficiency. Apart from the wide coverage scenario, the deployment scenarios of 5G extend to dense hotspots, machine-type communications (MTC), vehicle-to-vehicle (V2V) communications, large outdoor events, subways, etc. All these indicate that technologies of 5G will be diversified, not to be featured in just one technology per generation as before. Among the technologies of 5G, non-orthogonal multiple access (NOMA) is regarded as very promising [1–2].

1.2.1 Major Use Scenarios and Deployment Scenarios

For mobile broadband users, the goal of 5G communications is to achieve a high-speed data experience similar to optical fiber communications. For internet of things (IoT), 5G systems should support myriad types of applications, for instance, transportations, medical services, agriculture, finance, construction, power grid and environment protection, all with a common character – massive connection. Figure 1.1 shows a few major applications for 5G mobile broadband and the IoT [3].

In IoT networks, data collection services include low data rate services such as meter reading, and high data rate services such as video monitoring. Meter reading is characterized by massive connection, low-cost terminal devices, low power consumption and small data packets. Video monitoring requires not only high throughput but also a high density of deployment. For the control type of services, some are time-sensitive like vehicle-to-vehicle (V2) and some are not time-sensitive such as appliances at family homes.

There are three major use scenarios of 5G: enhanced mobile broadband (eMBB), ultra-reliable and low latency communications (uRLLC) and massive MTC (mMTC). For eMBB type of services, a high data rate is crucial for data stream applications which can tolerate 50~100 ms delay. The latency requirement is about 5~10 ms for interactive types of applications and about tens of milliseconds for virtual reality and online games. After 2020, it is expected that the cloud memory will store 30% of digital information, meaning that the data rate of the mobile network that connects the cloud and the terminals should be comparable to that of optical fiber communications. For uRLLC type of services, the latency requirement is very challenging for time-sensitive control systems. mMTC types of services encompass low-speed data collection, high-speed data collection, time-insensitive control, etc.

There are many deployment scenarios for eMBB. A few major scenarios are indoor hotspots (InH), dense urban, rural and urban macro [4].

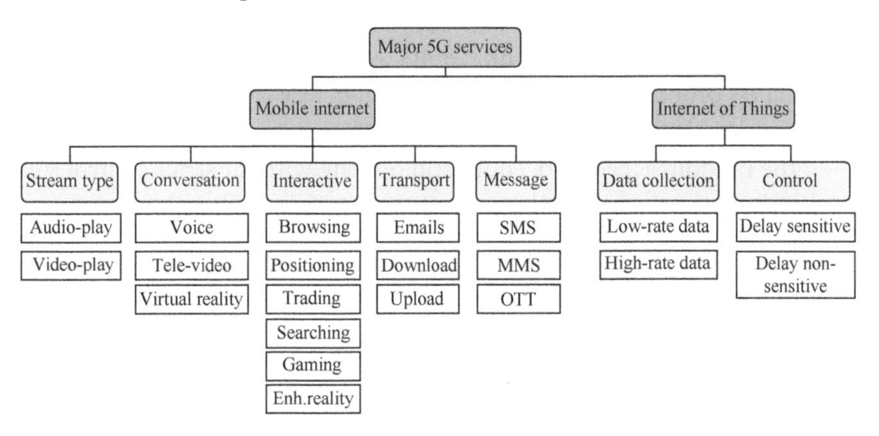

FIGURE 1.1 Major services in 5G.

For indoor hotspots, the coverage of each node is relatively small. Very dense population is expected, and the KPIs are focused on high data throughput and good experience rates. For dense urban deployment, the network can be homogeneous (e.g., single layer) or heterogeneous (e.g., two layers). The target deployment areas are downtowns and crowded street blocks. The traffic load will be very high, and the systems should well support indoor and outdoor users. In Table 1.2, major deployment scenarios are listed together with some system parameters.

1.2.2 Key Performance Indicators

5G KPIs include peak data rate, peak spectral efficiency, system bandwidth, control plane latency, user plane latency, latency for infrequent small packet transmission, interruption time for mobility, interoperability between systems, reliability, coverage, battery life, energy efficiency at terminal devices, spectral efficiency per cell/node, traffic capacity over unit area, user experience rate and connection density. Among them, there are a few KPIs related to multiple access [5]:

- Control plane latency refers to the time duration to switch from the idle state to the start of data transmission in a connected state. The required latency is 10 ms. More specifically, control plane latency assumes that a terminal is not in discontinued reception mode, e.g., the terminal would regularly monitor the paging from the network. The latency is calculated as the time lapse between layer 2/3 segmented data unit transmission to a successful reception. For uRLLC, the user plane latency should not exceed 0.5 ms for both downlink and uplink. For eMBB, the user plane latency requirement is 4 ms for both downlink and uplink. In 4G systems, the multi-step random access procedure should be carried out before the grant-based transmission. It is expected in 5G that NOMA can significantly simplify the random access procedure and operate without strict control signaling, thus reducing the control plane and user plane latency.
- Battery life is defined as the power-on time without recharging. For massive IoT, the battery life depends on the coverage situation, the number of bits to be uploaded per day, the number of bits to be downloaded per day and the capacity of the battery. The most important factor is the time required for each random access and the subsequent data transmission. For orthogonal multiple-access systems like

TABLE 1.2 Major Deployment Scenarios of 5G and Key System Parameters

Deployment Scenario	Indoor Hotspot	Dense Urban	Rural	Urban Macro
Carrier frequency	30， 70, or 4 GHz	4 GHz + 30 GHz (two layers)	700 MHz, 4 GHz, or 2 GHz	2, 4, or 30 GHz
Total bandwidth (after carrier aggregation)	At 70 GHz: up to 1 GHz (uplink + downlink) At 4 GHz: up to 200 MHz (uplink + downlink)	At 30 GHz: up to 1 GHz (uplink + downlink) At 4 GHz: up to 200 MHz (uplink + downlink)	At 700 MHz: up to 20 MHz (uplink + downlink) At 4 GHz: up to 200 MHz (uplink + downlink)	At 4 GHz: up to 200 MHz (uplink + downlink) At 30 GHz: up to 1 GHz (uplink + downlink)
Deployment	Single-floor, indoor, open-space office area	Two layers. Macro layer has hexagonal grids as the first layer. Randomly distributed low-power nodes as the second layer	Single layer of hexagonal grids	Single layer of hexagonal grids
Site-to-site distance	20 m Total 12 nodes per 120 * 50 m^2 area	20 m for the macro layer; 3 low-power nodes (outdoor) per macro cell	1732 m or 5000 m	500 m
Number of antenna elements at the base station	Up to 256 Tx or Rx	Up to 256 Tx or Rx	4 GHz: up to 256 Tx or Rx 700 MHz: up to 64Tx or Rx	Up to 256 Tx or Rx
Number of antenna elements at the terminal	30 GHz or 70 GHz: up to 32 Tx or Rx 4 GHz: up to 8 Tx or Rx	30 GHz: up to 32 Tx or Rx 4 GHz: up to 8 Tx or Rx	4 GHz: up to 8 Tx or Rx 700 MHz: up to 4 Tx or Rx	30 GHz: up to 32 Tx or Rx 4 GHz: up to 8 Tx or Rx
User distribution and mobile speed	100% indoor with 3 km/h ten users per node	Macro layer: uniform distributed, 80% indoor users with 3 km/h; 20% outdoor users with 30 km/h ten users per macro cell	50% outdoor vehicles with 120 km/h, 50% indoor users with 3 km/h ten vehicles/ users per base station	20% outdoor vehicles with 30 km/h; 80% indoor users with 3km/h ten vehicles/ users per base station

4G, the entire random access procedure would take a long time, causing high power consumption. In this sense, NOMA has an advantage compared to orthogonal multiple access if the target is to reduce the power consumption at terminals.

- In eMBB scenarios under the full-buffer traffic assumption, the required spectral efficiency per cell for 5G systems is about 3 times the requirement for 4G systems, both for average spectral efficiency and for cell edge spectral efficiency. It is well known that orthogonal multiple-access systems cannot approach the system performance bounds. NOMA can further improve the system throughput.
- Connection density refers to the number of total machine-type devices that can be supported within a certain area, under certain quality of service (QoS) requirements. QoS should be considered in conjunction with the traffic arrival rate, the time duration for transmission, the bit error rate, etc. For urban deployment, the required connection density is 1 million terminal devices per square kilometer. In orthogonal multiple access-based 4G, the design is primarily tailored for eMBB with a limited number of simultaneously transmitting users and each with high data throughput. Such a design is not suitable for massive connection scenarios. NOMA has the potential to efficiently support a large number of low-date-rate devices.

1.2.3 General Methodology for Performance Evaluation

For KPIs such as user plane latency, control plane latency, latency of infrequent small packet transmission and battery life, analytical methods may be used for performance evaluation. Whereas for KPIs such as system spectral efficiency and connection density, system-level simulations are generally required for the performance evaluation. Regarding NOMA, the urban macro is a suitable deployment scenario due to: (1) macro base stations typically refer to a single-layer network, e.g., no low-power nodes to be deployed to improve the system capacity. Hence, NOMA would be an important technology to increase the throughput, especially when the number of antennas at the macro station is limited; (2) in macro cell deployment, the number of active users per cell is relatively large. This would facilitate the user-pairing and scheduling to improve the performance; (3) the coverage of a macro cell is large, leading to a significant near-far effect which is beneficial to improve the system throughput and reduce the receiver complexity.

In addition to the indispensable system-level simulations, the study on NOMA involves a lot of link-level simulations. Traditional link-level simulation assumes a single user, where the multiple-access scheme is primarily orthogonal. However, in the case of NOMA, multiuser link-level simulation is crucial in order to precisely model the inter-user interference and the effect of interference cancelation or suppression. By doing so, a more accurate link to system mapping (or physical layer abstraction) can be developed so that each link's behavior can be more precisely modeled in system-level simulations.

1.3 MAJOR TYPES OF SCHEMES FOR DOWNLINK NOMA

The primary use of non-orthogonal multiple in downlink is eMBB, with the target of improving the system capacity. Several types of downlink NOMA are listed below [6]:

1. Direct superposition: direct linear superposition of modulation symbols of multiple users that share the same time and frequency resources. More complex receivers are generally required.
2. Flipping and superposition: some user bits are flipped before being superimposed, to maintain a Gray mapping property in the constituent constellation. Less complex receivers can be used.
3. Bit partition: to maintain regular Quadrature Amplitude Modulation (QAM) constellation after the superposition of modulation symbols. Gray mapping property is satisfied, and the constellation points are equal-distanced. Less complex receivers can be used.

1.4 MAJOR TYPES OF SCHEMES FOR UPLINK NOMA

While NOMA can also be used for eMBB uplink, the related issues are often implementation specific and transparent to the air interface standards. Hence, the focused scenario of uplink NOMA in this book is mMTC. The main goal is to support massive connections. Major types of schemes are listed below [7]:

1. Short spreading code: to use non-orthogonal spreading codes that are typically short. This can reduce the blind decoding complexity. Orthogonal spreading can be used as complementary to improve coverage. Devices can automatically choose the short sequences.

2. Bit-interleaver based: different bit interleavers assigned to different users. Iterative detections and decoding are required at the receivers.
3. Structured spreading matrix based: different spreading codes applied to different bits of multiple users. Modulation and spreading codes are jointly designed. Spreading codes may have sparsity property to reduce the complexity of the maximum-likelihood type of receivers.

REFERENCES

1. L. Dai, B. Wang, Y. Yuan, C. I. S. Han, Z. Wang, "Non-orthogonal multiple access for 5G: Solutions, challenges, opportunities, and future research trends," *IEEE Communications Magazine*, Vol. 53, No. 9, 2015, pp. 74–81.
2. M. Vaezi, Z. Ding, H. V. Poor, *Multiple Access Technologies for 5G Wireless Networks and Beyond*. Springer: Berlin, Germany, 2019.
3. Y. Yuan, L. Zhu, "Application scenarios and enabling technologies of 5G," *China Communications*, Vol. 11, No. 11, 2014, pp. 69–79.
4. 3GPP, TR 38.802, Study on new radio access technology, Physical layer aspects.
5. 3GPP, TR 38.913, Study on scenarios and requirements for next generation access technologies.
6. 3GPP, TR 36.859, Study on downlink multiuser superposition transmission (MUST) for LTE (Release 13).
7. 3GPP, TR 38.812, Study on non-orthogonal multiple access (NOMA) for NR.

CHAPTER 2

Basics of Downlink Multiple Access

Yifei Yuan, Zhifeng Yuan,
Jianqiang Dai and Hong Tang

2.1 PRINCIPLE OF DOWNLINK MULTIPLE ACCESS

As mentioned in Chapter 1, one important application scenario of downlink non-orthogonal multiple access (NOMA) is eMBB. The transmission eMBB is usually scheduling-grant-based and the design goal is to maximize the spectral efficiency of the system, under a certain user fairness criterion. In downlink NOMA, the serving base station sends user-specific data to different users that share the same time and frequency resources. In the field of information theory, downlink NOMA is often called the "broadcast channel" which carries user-specific data traffic. It is different from the primary broadcast channel in air-interface standards that carries the system information common to all the potential users in a cell, a type of control information for the initial access and mobility management of user terminals. The broadcast channel in information theory can be illustrated in Figure 2.1.

Take the case of two users as an example, the data provided by the source to the terminals contains three parts: M_0 is the data common to User 1 and User 2. M_1 and M_2 are the individual data for User 1 and User 2, respectively. After the joint encoding, the composite-coded sequence X^n is generated. The broadcast channel can be represented using the joint probability function $p(y_1, y_2|x)$. At the receiver, User 1 and User 2 perform decoding for the received signal, p_2 and y_2^n, respectively. Note that in Figure 2.1, the joint channel coding has a general sense and would include channel coding,

DOI: 10.1201/9781003336167-2

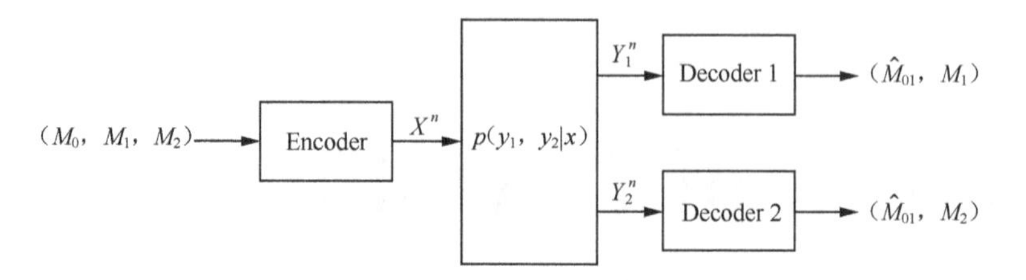

FIGURE 2.1 An example of a "broadcast channel" in information theory.

MIMO (multi-input multi-output) precoding, modulation, radio resource mapping, etc. To optimize the performance, the channel-state information (CSI) for User 1 and User 2 should be considered in the generalized joint encoding. Such CSI may be the full information about the channel, like $p(y_1, y_2|x)$, or other reduced/compressed information of the channels. The receiver in Figure 2.1 is also in a general sense and can be a signal detector, symbol demodulator, channel decoder, etc.

For downlink transmissions, parallel joint decoding is rarely used in practical systems. This is not only due to the very complex processing required at the terminal receiver, but also the security concern. However, it does not mean that only single-user detection should be used. In fact, advanced receivers with reasonable complexities can be considered in both academia and the industry.

For the general form of the broadcast channels, as shown in Figure 2.1, the bound analysis of its system capacity is a difficult problem in academia. Nevertheless, there are a few types of broadcast channels widely used in practical systems whose capacities can be more easily analyzed or calculated, for instance:

1. For the downlink, a base station is usually equipped with multiple transmit antennas to support multiuser MIMO (MU-MIMO). The optimal precoder for MU-MIMO is generally nonlinear [1], for instance, dirty paper coding. However, in practical engineering, linear precoder is often applied that relies on complex-domain signal processing whose performance can be quite close to that of dirty paper coding in many situations. When the antenna spacing is small and the same polarization is assumed, the spatial correlation between antennas would be high. In this case, low-rank beamforming is preferable because it can effectively differentiate users by the angles of the departure. The relevant CSI in this case is the large-scale fading which is typically slow-varying and reciprocal between

down and uplink so that the open-loop MIMO can be employed. When the antenna spacing is large or cross-polarization is assumed, the spatial correlation between antennas would be low where instantaneous "co-phasing" should be used. In such a case, the relevant CSI primarily depends on the small-scale fading which is fast varying. In such antenna settings, downlink and uplink channels are no longer reciprocal if operating in a frequency division duplex (FDD). It means that the CSI measured for the uplink cannot be used to infer the CSI for the downlink. Hence, CSI feedback is required. In general, MU-MIMO does not require very advanced interference cancelation algorithms in the receivers. Quite often, symbol-level MMSE receivers can provide enough good performance. Note that there are plenty of academic papers/books and technical contributions to standards development organizations (SDOs) addressing spatial multiplexing which is not to be further elaborated on here. It should be pointed out that due to two, four or eight transmit antennas typically deployed at a base station, the discussion and the simulations about downlink NOMA would inevitably touch upon the MU-MIMO. This aspect will be elaborated on in Chapter 4. Despite the complex coupling between NOMA and MU-MIMO, when discussing the principle of NOMA, we in this chapter focus on downlink NOMA where NOMA paired users share the same spatial precoder.

2. When the channel in Figure 2.1 satisfies $p(y_1, y_2|x) = p(y_2|x) \cdot p(y_2|y_1)$, that is, $X \rightarrow Y_1 \rightarrow Y_2$ can form a Markov Chain, the general-meaning "broadcast channel" can be written in a degraded form in which the channel quality of User 1 is always better than User 2. For degraded "broadcast channel", it can be proved that by using superposition coding and successive interference cancelation (SIC), the system performance bound can be achieved in theory [2]. Superposition coding can be linear or nonlinear. Apart from SIC receivers, less advanced receivers such as modulation symbol-level interference cancelation can also be used. Hence, its requirements for CSI feedback to support NOMA are not high, e.g., only amplitude type of information, like the channel quality indication (CQI) would be needed. Superposition coding is the key point in Chapters 2 and 4.

3. For the broadcast channel with a single transmit antenna, one type of nonlinear precoding, Tomlinson-Harashima precoding (THP) can be considered. The design principle of THP is to put the burden

of interference cancelation on the transmitter, thus allowing less advanced receivers which do not have the capability of interference. While the requirements for the receiver are lower than other schemes, THP relies on high-accurate CSI feedback, not only the amplitude but also the phase information of the fading channel. THP will be discussed in Section 2.7.

It should be emphasized that the above three listed aspects may not be independent of each other. Due to the wide deployment of multiple transmit antennas and the support of CQI in air-interface standards, MU-MIMO and superposition coding would often work together. This would make the resource scheduling, the choice of precoding and user pairing more complex. In those MU-MIMO+NOMA systems, the aim of the dynamic resource scheduler is to maximize the system throughput, as well as to maintain certain fairness between different users.

In theory, the number of superposed users can be many. As the number of users increases, the system throughput tends to grow, under the proportional fairness criterion, as shown in Figure 2.2. Here the users are asymmetric, their signal-to-noise ratios (SNRs) are not the same, reflecting certain near and far effects. Note, for pure power domain NOMA transmission (when users share the same precoding matrix/vector of MIMO), NOMA would only show the performance gain over orthogonal multiple

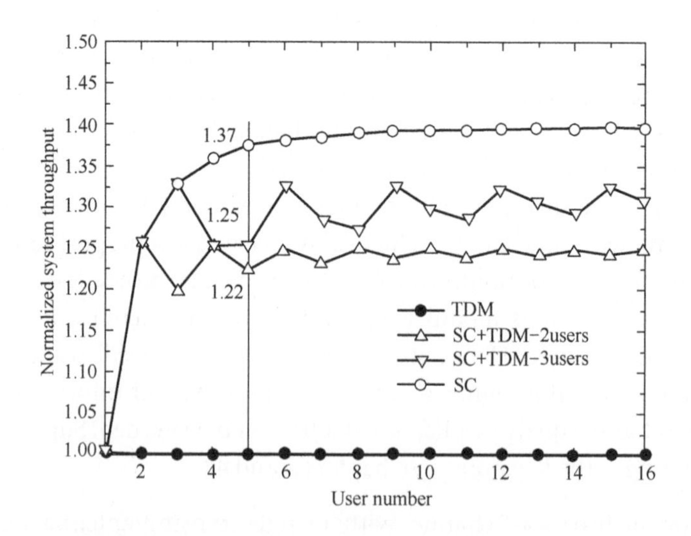

FIGURE 2.2 Relationship between downlink NOMA system performance and the number of superposed users.

access (OMA) in asymmetric user case. In Figure 2.2, the distribution of SNR is uniform between [–3, 21] dB. The receiver is assumed ideal, e.g., maximum likelihood detection and perfect channel coding. It is observed that when two users are superposed, the gain of NOMA over OMA is about 25%. As the number of users is increased to 5, the NOMA gain is about 37%. Beyond that point, the performance becomes saturated when the number of users further increases.

For eMBB services which often carry big data, the number of active users per cell is not expected to be large. Hence, the chance that many users share the same time and frequency resources would not be high. In addition, if we consider the complexity of resource scheduling, the receiver implementation and the overall system performance, superposing two users would glean most of the potential gain of NOMA and yet significantly reduce the challenges of system deployment and terminal implementation. Hence, the focus of downlink NOMA would be a two-user superposition.

A two-user degraded broadcast channel with a single transmit antenna can be illustrated in Figure 2.3 [3] where the sum rates between NOMA and OMA are compared. Here UE1 and UE2 refer to the user far away from the base station (BS) and close to the BS, respectively. The transmit powers (at the BS) for UE1 and UE2 are denoted as P_1 and P_2, respectively. The channel gains of BS-UE1 and BS-UE2 are $|h_1|^2$ 和 $|h_2|^2$, respectively. Here the channel gain includes path loss, large-scale shadow fading and small-scale fading. For simplicity, it is also assumed that the channel

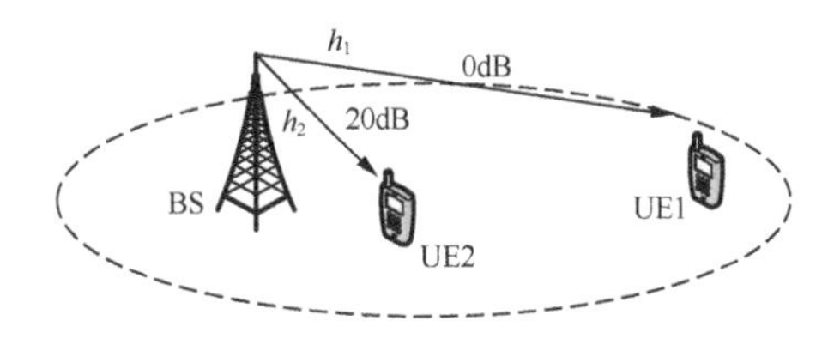

Rate	UE2(strong)	UE1(weak)
NOMA(A)	3 bit/(s·Hz)	0.9 bit/(s·Hz)
OMA(B)	3 bit/(s·Hz)	0.64 bit/(s·Hz)
OMA(C)	1 bit/(s·Hz)	0.9 bit/(s·Hz)

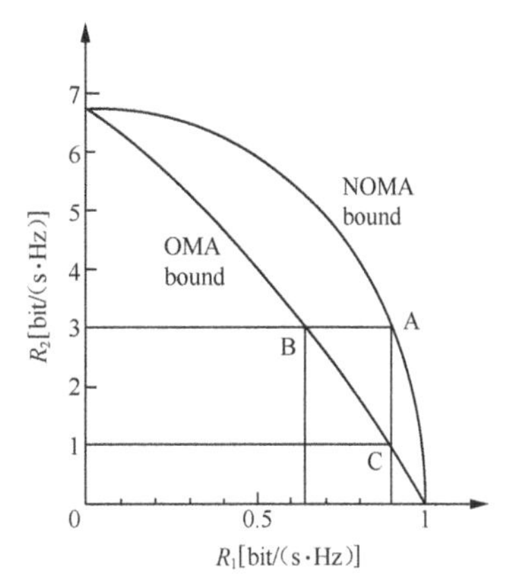

FIGURE 2.3 Sum-rate regions for NOMA and OMA for a two-user downlink transmission.

remains constant during a data packet transmission, and there is no delay spread, e.g., flat channel. The total transmit power ($P_{1+}P_2$) should be constant, while the ratio between P_1 and P_2 can be adjusted.

In the case of OMA, UE1 occupies *a* portion of the entire frequency resource while UE2 occupies the rest, e.g., $1-a$. Therefore, the SNRs of UE1 and UE2 are $P_1|h_1|^2/\alpha N_0$ and $P_1|h_1|^2/\alpha N_0$, respectively. Here N_0 is the power spectral density of additive white Gaussian noise (AWGN) that includes thermal noise and other-cell interference (assumed white). According to the Shannon channel capacity formula, the achievable rates of UE1 and UE2 for OMA can be written as

$$\begin{aligned} R_1 &< \alpha \cdot \log\left(1+\frac{P_1|h_1|^2}{\alpha N_0}\right) \\ R_2 &< (1-\alpha)\cdot \log\left(1+\frac{P_2|h_2|^2}{(1-\alpha)N_0}\right) \end{aligned} \tag{2.1}$$

In order to obtain the rate bounds, e.g., $\{R1, R_2\}$, in Figure 2.3 for OMA, for each pair of $\{P_1, P_2\}$, multiple ratios of frequency allocations need to be tried to get the envelope.

For NOMA transmission, UE1 and UE2 share the same time-frequency resources. At its receiver, UE1 would see the interference from the transmission to UE2 as well as the thermal noise/other-cell interference $P_2|h_1|^2+N_0$. For UE2, if the interference due to the transmission to UE1 can be completely canceled, the post-processing SNR can be written as $P_2|h_2|^2/N_0$. Here it is assumed that (1) UE1 is far away from BS and the modulation coding set (MCS) for UE1 is relatively low and (2) more power is allocated for the transmission to UE1, e.g., P_1 is significantly higher than P_2. Hence, UE2 which is close to the BS would receive a very strong signal for UE1, and then be able to detect/decode the data for UE1, and cancel it. According to the Shannon formula, the achievable rates for UE1 and UE2 in the NOMA case can be represented as

$$\begin{aligned} R_1 &< \log\left(1+\frac{P_1|h_1|^2}{P_2|h_1|^2+N_0}\right) \\ R_2 &< \log\left(1+\frac{P_2|h_2|^2}{N_0}\right) \end{aligned} \tag{2.2}$$

For NOMA, for each pair of power allocation $\{P_1, P_2\}$, there would be one corresponding point in the sum-rate bound. For example in Figure 2.3, the wideband downlink SNR of UE1 is 0 dB, e.g., $\frac{P_1+P_2}{N_0}|h_1|^2=1$. The wideband DL SNR of UE2 is 20 dB, e.g., $\frac{P_1+P_2}{N_0}|h_2|^2=100$. As the figure shows, regardless of the transmit power ratio, the sum-rate bound of NOMA is always higher than that of OMA. More specifically, at point A in NOMA bound, UE1 has a spectral efficiency of 0.9 bps/Hz, while the spectral efficiency of UE2 is 3 bps/Hz. At point B in OMA bound, UE1 has the spectral efficiency of 0.64 bps/Hz, much lower than the NOMA case, while the spectral efficiency of UE2 is still 3 bps/Hz, similar to the case of OMA. Let us check also point C in OMA, the spectral efficiency of UE1 is 0.9 bps/Hz, similar to that of point A in NOMA. However, the spectral efficiency of point C is only 1 bps/Hz, much lower than the NOMA case (3 bps/Hz). From the above, it is observed that for OMA, if we want to increase the rate for cell-edge users, the rate for cell-center users would suffer significantly. By contrast, NOMA can drastically improve the cell-edge performance without the significant degradation of cell-center performance. Hence, under the proportional fairness criterion, NOMA can increase the system performance compared to OMA.

It should be pointed out that in practical systems, no matter whether OMA or NOMA is to be used, the system would normally operate in the region between point A, point B and point C due to the following reasons:

1. If the operating point lies too toward the left, it means that a large portion of transmission power would be allocated for the close-to-BS users, e.g., UE2, which does not sound reasonable given the small path loss of UE2. This would lead to very high spectral efficiency for UE2, e.g., 6 bps/Hz. At the same time, the spectral efficiency of UE1 would be only 0.1 bps/Hz, due to the small portion of transmit power for UE1 and severe path loss. Such a big disparity in data rates between users would be quite unfair;

2. On the other hand, if the power allocation for UE1 is set too low, it would be difficult for UE2 to successfully decode UE1's data. More advanced receivers may be required in this case.

The sum-rate analysis discussed above under the constraint of proportional fairness can provide certain guidance for the practical design of

NOMA. Apart from the sum-rate analysis, several engineering aspects should also be considered:

- Compared to the system equipment, mobile terminals are more sensitive to hardware cost and power consumption. Even if advanced receivers would be required, their complexity should be kept low. Hence, the performance of practical downlink NOMA systems can be well below the sum-rate analysis. While the receiver algorithms are implementation specific and transparent to air-interface standards, certain types of receivers would be suitable for certain specific NOMA transmission schemes, thus indirectly affecting the choice of transmitter-side solutions in standards.
- Modulations for a single user are typically quadrature phase-shift keying (QPSK) or quadrature altitude modulation (QAM). After superposition coding, the composite constellation would no longer have regular patterns such as the equally spaced QAM points. If the constellation points are too arbitrary, it would impose a heavy burden on RF hardware implementation, for instance, to meet error vector magnitude (EVM) requirements.
- Flexibility of transmit power allocation. The transmit power ratio needs to be informed to users, otherwise, blind detection needs to be carried out by terminals to determine which ratio has been used. This would significantly increase the terminal implementation complexity. Too flexible power allocation would make the constellations too arbitrary. On the other hand, if too much constraint is imposed on the power allocation, only a small number of operating points are allowed in the sum-rate curve. This would significantly constrain the resource scheduling for NOMA and negatively impact the system performance.
- Realistic channel estimation and its impact on interference cancelation. A key assumption of Eq. (2.2) is that the interference due to the transmission to far users can be completely canceled. However, in practical systems, symbol-level interference cancelation is often used where the symbol error rate of hard detection cannot be ignored. Combined with the channel estimation error, the reconstructed interference signal may significantly deviate from the actual interfering signal, causing noticeable residual interference. Even when

code-word-level interference cancelation is used, there is still a channel estimation error that makes it difficult to cancel the interference completely.

- Compatibility to legacy terminals. It is desirable to be able to support legacy terminals, e.g., OMA users, in the systems. Such compatibility is more business-viable and allows smooth migration of the networks from OMA to NOMA. This design aspect will be discussed in detail in Chapter 4.
- The total downlink control signaling. In addition to the aforementioned transmit power ratio indication, downlink control signaling also includes the potential scheduling information, which specifically supports NOMA transmission, on top of those already defined for OMA transmission. These additional scheduling signaling would be potentially used, depending on the NOMA transmission schemes and receiver algorithms.

2.2 SIMULATION EVALUATION METHODOLOGY

Simulations for downlink NOMA include link-level simulations and system-level simulations. Different from the studies on other technologies, link-level simulations of NOMA usually involve multiple users, in order to accurately represent the cross-user interference. In downlink NOMA, the signals targeting the far-away (e.g., cell-edge) user and the nearby (e.g., cell-center) user would experience the same channel. Due to the fast fading, the actual behavior and the result of the interference cancelation would be time-varying. In general, system-level simulations are needed to provide comprehensive performance evaluations for NOMA.

2.2.1 Parameters and Metrics for Link-Level Simulations

Commonly-used parameters for NOMA link-level simulations are listed in Table 2.1 [4]. It is noticed that there are multiple configurations related to antennas, for example, reference signals, spatial correlations between antennas, the number of transmitting/receiving antennas and transmission modes. This means that there is a certain relation between MIMO and NOMA, which is to be further elaborated. Several other parameters are related to the link adaptation, for instance, the delay of CSI reporting, the periodicity and granularity of CSI reporting, and the number of hybrid automatic retransmission request (HARQ) transmissions. These parameters

TABLE 2.1 Link-Level Simulation Parameters

Parameter	Value
Carrier frequency	2 or 3.5 GHz
System bandwidth	10 MHz
Allocated bandwidth	5 MHz
Reference signal (pilot)	Cell common reference signal (CRS) or user-specific demodulation reference signal (DMRS)
Channel and mobility speed	AWGN， EPA/ETU/EVA, 3 or 60 km/h
Spatial correlation between antennas	Low
(num Tx antennas, num Rx antennas)	(2, 2), (4, 2), or (4, 4), Note: (4, 4) is optional
Transmission mode	Two transmit antennas: CRS based Four transmit antennas: DMRS
Delay of CSI reporting	5 ms
Periodicity of CSI reporting	5 ms
Granularity of CSI reporting	Wideband or sub-band
Link adaptation	Dynamic, or with fixed MCS
EVM requirement (Tx side, Rx side)	Ideal in AWGN, or (8%, 4%) for other channels
HARQ	At most 4 HARQ transmissions

are only enabled when the link adaptation is turned on. EVM refers to error magnitude measurement and represents the quality of modulation signal. The lower the EVM is, the smaller the difference is between the amplitude/phase of the actual modulation signal and the amplitude/phase of the ideal signal. Due to the signal superposition in NOMA transmission, the composite constellation would be more complicated than the legacy modulations of a single user, in terms of the number of constellation points, the spacing between the points and the shape of the constellations. Hence, EVM is an important measure that affects not only the transmitter but also the receiver. Quite often, due to the strong nonlinear effect introduced by the power amplifier at the transmitter, the requirement for EVM at the transmitter is more stringent in order to minimize the waveform distortion.

For downlink NOMA link-level simulations, the following three metrics are often used:

- The first metric is the traditionally used block error rate (BLER) vs. SNR curves. Normally, the BLER would decrease as SNR increases, just like a water-fall. Each power allocation would require a bunch of BLER vs. SNR curves, depending on the number of superimposed users. If multiple power allocations are considered, multiple link curves are needed, each would contain one curve for the far-away (e.g., cell-edge) user and the other curve for the nearby (e.g., cell-center) user.

- The second metric is the sum-rate vs. SNR curves. Generally speaking, the sum rate would increase monotonically as SNR increases and saturate after SNR reaches a certain point. Compared to BLER vs. SNR curves, sum rate vs. SNR considers link adaptation and does not need to differentiate the rate for far-away users and the rate for nearby users.
- The third metric is the capacity vs. power allocation, e.g., rate region. Normally, when a user's rate decreases when less power is allocated to this user. At the same time, the other user's rate increases. The entire curve would form a closed region and the curve is usually convex, like the one in Figure 2.3. The rate region metric is especially useful for NOMA transmission, under certain BLER targets. Each curve corresponds to a pair of channel gains of the cell-edge user and cell-center user. Since different power allocations can be reflected on the same curve, the metric of capacity vs. power allocation is quite illustrative for NOMA link-level evaluations.

2.2.2 Link to System Mapping

Link to system mapping is typically used to model the key signal processing blocks in the receivers. Different types of receivers may have different links to system mapping models. Several models are listed in [5], mainly for two types of receivers: one is the maximum likelihood (ML)/reduced ML receiver, the other can be code-word-level interference cancelation (CWIC), symbol-level interference cancelation (SLIC) and minimum mean squared error-inference rejection and cancelation (MMSE-IRC).

2.2.2.1 Algorithm and Link to System Mapping for ML Receivers

The link to the system mapping method for ML receivers is based on information theory, with certain adjustments by empirical data. The mapping itself does not model the details of signal processing in the receivers. The basic idea is to calculate the mutual information per transmitted bit (MIB) for each resource element (RE). Then, the MIBs of all the REs in a subframe are averaged to get the average mutual information. In the end, the effective signal to interference and noise ratio, $SINR_{eff}$ is calculated from the average mutual information and looked up in the SNR-BLER table to figure out the target BLER.

The MIB on each RE in the ML receiver is denoted as MIB_{ML} and can be empirically represented as

$$\text{MIB}_{\text{ML}} = \beta \cdot C_{\text{BICM}} \tag{2.3}$$

where C_{BICM} is the normalized spectral efficiency of bit interleaved coded modulation (BICM). β is a weighing factor and can be obtained by comparing with the BLER curves of actual receivers. The smaller the value of β is, the bigger the performance difference is between the actual receiver and the ideal receiver that is based on mutual information calculation. The BICM-normalized spectral efficiency of the nearby (e.g., cell-center) user can be written as

$$C_{\text{BICM,near}} = 1 - \frac{1}{m_{\text{near}}} \sum_{i=0}^{m_{\text{near}}-1} E_{t,z} \left[\log_2 \frac{\sum_{s_{\text{target}} \in M} p(z \mid s_{\text{target}})}{\sum_{s_{\text{target}} \in M_{\text{near}}(i,t)} p(z \mid s_{\text{target}})} \right] \tag{2.4}$$

The normalized spectral efficiency of BICM of the far-away (e.g., cell-edge) user can be written as

$$C_{\text{BICM,far}} = 1 - \frac{1}{m_{\text{far}}} \sum_{i=0}^{m_{\text{far}}-1} E_{t,z} \left[\log_2 \frac{\sum_{s_{\text{target}} \in M} p(z \mid s_{\text{target}})}{\sum_{s_{\text{target}} \in M_{\text{far}}(i,t)} p(z \mid s_{\text{target}})} \right] \tag{2.5}$$

where S_{target} is the superposed signal. m_{near} and m_{far} are the number of bits in the constellation for cell-center (e.g., near) and cell-edge (e.g., far) users, respectively. For example, QPSK corresponds to 2 bits, whereas 16-QAM corresponds to 4 bits. M is the set corresponding to the points in the S_{target} constellation. $M_{\text{near}}(i,t)$ and $M_{\text{far}}(i,t)$ are the constellation points of the near user and the far user, respectively, when the i-th bit has the value t where $t \in \{0, 1\}$; z is the post-MMSE signal. $p(z \mid s_{\text{target}})$ is the transition probability density function, depending primarily on the distance between z and S_{target}. Figure 2.4 is an example showing how the log-likelihood ratio (LLR) of the third bit in a 16-QAM constellation is calculated. The solid triangle represents the post-MMSE signal z. The left figure shows the Euclidean distances between z and all the constellation points of which the third bit is 0. The right figure shows the Euclidean distances between z and all the constellation points of which the third bit is 1. It is obvious that the distances in the left figure are smaller than in the right figure, meaning that the third bit of z is more likely to be 0.

Considering the case of two transmit antennas, AWGN channel and QPSK modulation for the near (cell-center) user, we can calculate the weight factor β for mutual information, under non-Gray mapping and Gray mapping, respectively. As shown in Table 2.2, β values for Gray mapping are higher than those of non-Gray mapping, meaning that the

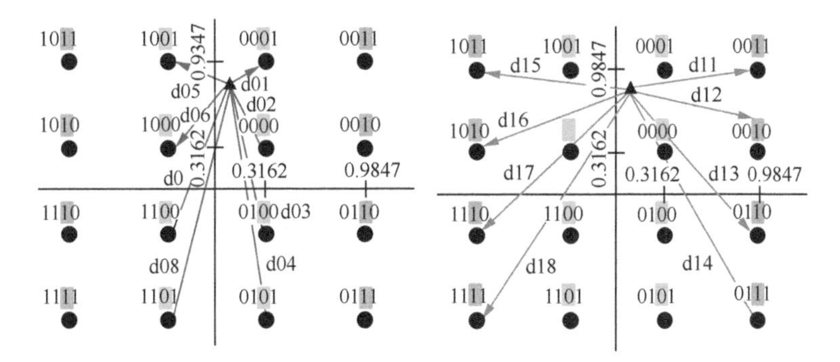

FIGURE 2.4 An example of LLR calculation.

TABLE 2.2 Weight Factor β for Mutual Information under Gray Mapping and Non-Gray Mapping

Power Ratio for Far UE		0.7153		0.9091		0.9617		0.9844	
MCS Level	**Mod**	**Non-Gray**	**Gray**	**Non-Gray**	**Gray**	**Non-Gray**	**Gray**	**Non-Gray**	**Gray**
MCS1	QPSK	0.0133	0.0429	0.4000	0.3952	0.8881	0.8779	1.0296	1.0431
MCS2	QPSK	0.0802	0.1034	0.5961	0.5840	0.9322	0.9205	0.9538	0.9658
MCS3	QPSK	0.1293	0.1536	0.8041	0.7927	0.9693	0.9583	0.9640	0.9740
MCS4	QPSK	0.2581	0.2814	0.9209	0.9203	0.9791	0.9692	0.9677	0.9765
MCS5	QPSK	0.4229	0.4454	1.0185	1.0229	1.0354	1.0277	1.0217	1.0315
MCS6	QPSK	0.5752	0.5934	1.0186	1.0259	1.0250	1.0204	1.0140	1.0236
MCS7	16QAM	0.7053	0.7135	0.9918	0.9975	0.9969	0.9928	0.9929	0.9980
MCS8	16QAM	0.8084	0.8135	0.9996	1.0031	1.0031	0.9999	1.0006	1.0029
MCS9	16QAM	0.8616	0.8653	0.9939	0.9970	0.9971	0.9949	0.9947	0.9965
MCS10	64QAM	0.8528	0.8561	0.9689	0.9706	0.9730	0.9716	0.9711	0.9746
MCS11	64QAM	0.8956	0.8981	0.9787	0.9795	0.9818	0.9811	0.9800	0.9828
MCS12	64QAM	0.9427	0.9449	1.0052	1.0061	1.0076	1.0073	1.0058	1.0077
MCS13	64QAM	0.9519	0.9540	1.0024	1.0035	1.0043	1.0043	1.0027	1.0039
MCS14	64QAM	0.9557	0.9575	0.9993	0.9999	1.0006	1.0010	0.9998	1.0001
MCS15	64QAM	0.9715	0.9727	1.0049	1.0051	1.0051	1.0052	1.0046	1.0049

capacity of a Gray-mapped constellation is higher than a non-Gray-mapped constellation.

After averaging over multiple Res, avg(MIB_{ML}) is obtained. Then, with the mapping from SINR to MIB, the effective SINR can be calculated as

$$\mathrm{SINR}_{\mathrm{eff}} = f^{-1}(\mathrm{avg}(\mathrm{MIB}_{\mathrm{ML}})) \tag{2.6}$$

where $f(.)$ is the mapping from SINR to MIB. This mapping can be determined in advance, for instance, to use the mapping in IEEE 802.16 m.

2.2.2.2 Link to System Mapping for CWIC, SLIC and MMSE-IRC Receivers

In this mapping method, the interference cancelation mechanism in the receivers is explicitly modeled. Assuming that N layers of data are superposed on the same time-frequency resources, the received signal on the r-th REs can be represented as

$$\mathbf{y}_{\mathrm{rx}}(r) = \sum_{k=1}^{N} H_k(r)\mathbf{x}_k(r) + \mathbf{e}_{\mathrm{IN}}(r) + \mathbf{e}_{\mathrm{EVM}}(r) \tag{2.7}$$

where $H_k(r)$ represents the effective channel of the k-th layer that also includes the transmit power, e.g., $H_k(r) = \sqrt{p_k}h_k$. $x_k(r)$ is the data of the k-th layer, constrained by $\|\mathbf{x}_k(r)\| = 1$. $e_{\mathrm{IN}}(r)$ denotes the other-cell interference and thermal noise. $e_{\mathrm{EVM}}(r)$ represents the effect of EVM that can be modeled as complex Gaussian. Its standard deviation on the i-th receive antenna can be represented as $\sigma_{\mathrm{EVM},i} = \sqrt{\mathrm{EVM}_{\mathrm{rx}}^2 + \mathrm{EVM}_{\mathrm{tx}}^2}\,\|\mathbf{h}_{k,i}\|_F$, where $\mathrm{EVM}_{\mathrm{rx}}$ is the EVM at the receiver, e.g., $\mathrm{EVM}_{\mathrm{rx}} = 0.04$. $\mathrm{EVM}_{\mathrm{tx}}$ is the EVM at the transmitter, e.g., $\mathrm{EVM}_{\mathrm{tx}} = 0.08$. $h_{k,i}$ is the effective channel on the i-th receive antenna for the k-th data layer.

Taking the near (cell-center) user as an example, the MMSE weight vector for the detection of the n-th data layer can be written as

$$\begin{aligned} w_n(r) &= H_n^H(r) * R_{yy}^{-1} \\ &= \sqrt{P_{\mathrm{near}}}h_n^H(r) * \Big(\big(\sqrt{P_{\mathrm{near}}}h_n(r)\big)\big(\sqrt{P_{\mathrm{near}}}h_n(r)^H\big) \\ &\quad + \big(\sqrt{P_{\mathrm{far}}}h_n(r)\big)\big(\sqrt{P_{\mathrm{far}}}h_n(r)^H\big) + \sum_{k \neq n} H_k(r)H_k^H(r) + \sigma^2 I \Big)^{-1} \end{aligned} \tag{2.8}$$

where $H_n = \sqrt{P_{\text{near}}} h_n$ is the effective channel of the near user, $\left(\sqrt{P_{far}} h_n(r)\right)\left(\sqrt{P_{far}} h_n(r)^H\right)$ is the interference from far (cell-edge) user in the same cell. $\sum_{k \neq n} H_k(r) H_k^H(r)$ is the interference from the other cells. P_{near} is the transmit power allocated to the near (cell-center) user. P_{far} is the transmit power allocated to the far (cell-edge) user. They should satisfy: $P_{\text{near}} + P_{\text{far}} = 1$.

According to Eq. (2.2) and the MMSE criterion, the SINR for the near (cell-center) user can be written as

$$\text{SINR}_n^{\text{MMSE}}(r) = \frac{P_{\text{near}} \left\|\mathbf{w}_n(r)\mathbf{h}_n(r)\right\|^2}{P_{\text{far}} \left\|\mathbf{w}_n(r)\mathbf{h}_n(r)\right\|^2 + \sum_{k=1, k \neq m, k \neq n}^{N} \left\|\mathbf{w}_n(r)\mathbf{h}_k(r)\right\|^2 + \mathbf{w}_n(r)\mathbf{R}_{ee}\mathbf{w}_n^H(r)} \tag{2.9}$$

where $\mathbf{R}_{ee} = E\left\{\mathbf{e}_T \mathbf{e}_T^H\right\}$ is the covariance matrix of the interference and the noise (including EVM). $\mathbf{e}_T = \mathbf{e}_{\text{IN}} + \mathbf{e}_{\text{EVM}}$

In the case of the CWIC receiver, considering the channel estimation error, there would be some residual inference that cannot be completely canceled. The SINR of the n-th data stream can be represented as:

$$\text{SINR}_n^{\text{CWIC}}(r) = \frac{P_{\text{near}} \left\|\mathbf{w}_n(r)\mathbf{h}_n(r)\right\|^2}{P_{\text{far}} \sigma_{\tilde{H}}^2 + \sum_{k=1, k \neq m, k \neq n}^{N} \left\|\mathbf{w}_n(r)\mathbf{h}_k(r)\right\|^2 + \mathbf{w}_n(r)\mathbf{R}_{ee}\mathbf{w}_n^H(r)} \tag{2.10}$$

where $\sigma_{\tilde{H}}^2 = tr(\mathbf{R}_{ee}) / (G_{\text{CE}} N_{\text{rx}})$ is the variance of the channel estimation, e.g., $\tilde{\mathbf{H}} = \mathbf{H} - \hat{\mathbf{H}}$. N_{rx} denotes the number of receiver antennas. G_{CE} is the processing gain of the channel estimator. According to Ref. [5], $G_{\text{CE}} = 10$.

To simplify the analysis, SLIC can be considered as a weaker version of CWIC, e.g., its SINR degraded by a fixed value compared to that of CWIC. That is:

$$\text{SINR}_n^{\text{SLIC}}(r) = \frac{\Delta_{\text{SLIC}} P_{\text{near}} \left\|\mathbf{w}_n(r)\mathbf{h}_n(r)\right\|^2}{P_{\text{far}} \sigma_{\tilde{H}}^2 + \sum_{k=1, k \neq m, k \neq n}^{N} \left\|\mathbf{w}_n(r)\mathbf{h}_k(r)\right\|^2 + \mathbf{w}_n(r)\mathbf{R}_{ee}\mathbf{w}_n^H(r)} \tag{2.11}$$

where according to the simulations, $\Delta_{\text{SLIC}} = 10^{-0.07}$. Regarding MMSE-IRC, similar equations can be used, except that due to its inferior performance to SLIC, the value of $\Delta_{\text{MMSE-IRC}}$ is less than Δ_{SLIC}

Once the $\text{SINR}_n(r)$ is obtained by averaging multiple REs, the effective SINR can be calculated as

$$\text{SINR}_n^{\text{eff}} = \varphi^{-1}\left(\frac{1}{M}\sum_{r=1}^{M}\varphi\left(\text{SINR}_n(r)\right)\right) \tag{2.12}$$

The target BLER can be determined by looking up SNR-BLER curves for the AWGN channel. The SINR for the far (cell-edge) user can be obtained in a similar way.

2.2.3 Parameters for System-Level Simulations

Configurations of system-level simulation parameters include three aspects: (1) deployment scenarios and cell layout, (2) traffic model and (3) detailed parameter setting.

2.2.3.1 Deployment Scenarios and Cell Topology

In the third generation of mobile networks and the early stage of the fourth generation, only homogeneous deployment is simulated where the networks consist only of macro base stations, uniformly spaced with the same transmit power and antenna gains. The main purpose of this homogenous network is to support wide coverage, as illustrated in Figure 2.5.

Since the later stage of the fourth generation, low-power nodes were introduced on top of the macro homogeneous layer. The transmit power, the antenna gains, and the antenna heights of these low-power nodes are

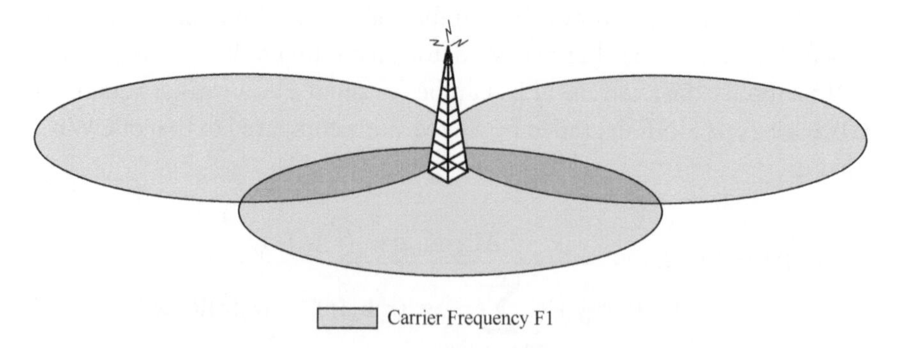

FIGURE 2.5 Wide coverage by homogeneous deployment.

much lower than those of macro stations. Since their coverage is typically below a hundred meters, multiple low-power nodes would be deployed in a macro cell, and their locations can be more flexible, clustered or uniformly distributed. Macro base stations and low-power nodes form heterogeneous networks. Low-power nodes may operate in the same or different carriers from that of the macro layer, as illustrated in Figure 2.6.

For downlink NOMA, the prevailing view is that NOMA is more suitable for homogeneous networks of wide coverage. The reasons are as follows:

- Compared to the orthogonal transmission, the gain of NOMA is more pronounced when far and near users share the same time-frequency resources. The more significant the near-far effect is, the more motivation is for NOMA. The near-far effect would be more common in homogeneous wide-coverage cases, rather than in heterogeneous deployment where the cell edge becomes blurred.
- More number of active users per cell is expected in wide-coverage homogeneous networks, compared to the case of low-power nodes. This makes it easier for the scheduler to pair the near users and the far users for NOMA transmissions and reduce the negative impact

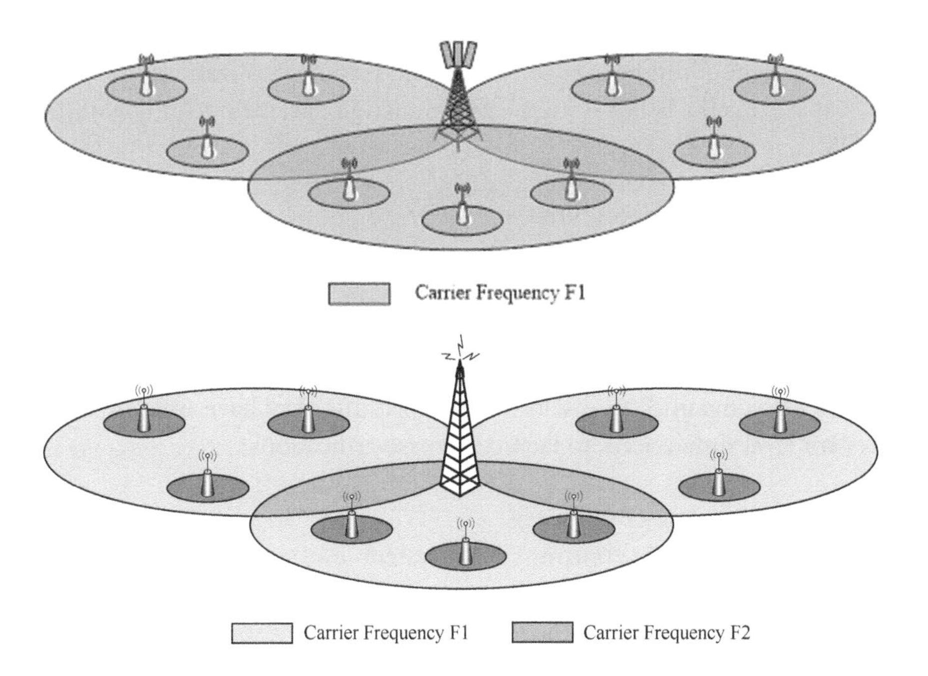

FIGURE 2.6 Heterogeneous networks.

on the other schemes, e.g., MU-MIMO, that coexist with NOMA. Essentially, both NOMA and MU-MIMO would compete against each other in selecting suitable users within the limited pool of users. If the number of active users is small, the pool of users is small, which would exacerbate the mutual destruction between technologies.

2.2.3.2 Traffic Models and Metrics

For data services, the simplest traffic model is a full buffer, e.g., a user always has data to transmit in the buffer, and the rate of the buffer clearance would never outpace the traffic arrival rate. A full buffer represents the situation when a system is fully loaded. In realistic networks, a fully loaded case rarely occurs. A more practical traffic model is File Transfer Protocol (FTP). By adjusting the packet arrival rate and the packet size, the system loading can be controlled. System loading is often reflected in resource utilization. In general, the higher the resource utilization, the higher the system loading, or in other words, more active users are supported in a cell. This means that NOMA would show more gains when the system loading is high.

Regarding the evaluation metrics for system-level simulations assuming the full-buffer traffic, average cell throughput, 5% cell-edge throughput, or throughput cumulative distribution function (CDF) can be used. For FTP traffic, average throughput, 5% throughput, 50% throughput, and 95% throughput can be used, under certain resource utilization.

Table 2.3 is the list of system-level simulation parameters for downlink NOMA.

2.2.4 Scheduling Algorithms

Scheduling algorithms for base stations are typically implementation specific, and would not be standardized. However, for NOMA, the scheduling algorithm would be much more complicated than OMA and can have a big impact on the system performance. Therefore, it is encouraged in SDOs for companies to disclose the algorithms they have used in their system-level simulations, to facilitate cross-verifications.

2.2.4.1 Criterion for User Pairing

The essence of NOMA transmission in system-level simulation is to enable resource sharing among multiple users as long as certain requirements are fulfilled. In order to reduce the scheduler's complexity, some initial screen-out should be carried out if some basic requirements cannot be met. For instance, if the precoding matrices of the two users are very similar to each

TABLE 2.3 System-Level Simulation Parameters for Downlink NOMA

Parameter	Value
Layout	Hexagonal, 19 macro BSs, each having three cells
Inter-site distance	500 m
System bandwidth	10 MHz
Carrier frequency	2.0 GHz
BS transmit power (per cell)	46 dBm
Pathloss model	ITU UMa, with the 3D distance between an eNB and a UE applied
Penetration loss	Outdoor: 0 dB Indoor: $(20+0.5d_{in})$ dB (d_{in}: independent uniform random value between [0, 25] for each link)
Shadow fading	ITU Uma
Fast fading model	ITU UMa
Antenna pattern at the base station	3D
Antenna height at the base station	25 m
Antenna gain + coupling loss	17 dBi
Antenna height at the terminal	1.5 m
Antenna gain at the terminal	0 dBi
Antenna configuration	Base station: 2 Tx, cross-pol 4 Tx, cross-pol, half-lambda spacing between antenna clusters 8 Tx, cross-pol, half-lambda spacing between antenna clusters Terminal: 2 Rx, cross-pol 4 Rx, cross-pol, half-lambda spacing between antenna clusters Mandatory: 2Tx/2Rx, 4Tx/2Rx Optional: 4Tx/4Rx, 8Tx/2Rx
Traffic model	FTP traffic model 1 with For RUs of 60%, 80% and 90%, packet size = 0.1 Mbytes For a RU of 60%, packet size = 0.5 Mbytes Full-buffer (optional)
User dropping method	20% users outdoor; 80% users indoor
Minimum distance between the BS and the terminal	35 m
Number of superposed users	2

(*Continued*)

TABLE 2.3 (*Continued*) System-Level Simulation Parameters for Downlink NOMA

Parameter	Value
Noise figure at the terminal	9 dB
Mobility speed	Outdoor: 3, 60 km/h; Indoor: 3 km/h
Cell selection criterion	RSRP for intra-frequency
Cell selection threshold	3 dB
Control signaling overhead	3 OFDM symbols per slot used for downlink control, CRS ports and DM-RS with 12/24 REs per PRB depending on the assumed number of MIMO layers and TM
CSI feedback	Non-ideal channel estimation and interference estimation Feedback period: 5 ms Feedback delay: 5 ms
EVM	Tx EVM: 8%, FFS smaller values UE Rx EVM: 4%

other, or if no significant near-far effect is observed, these two users are not suitable to be paired for NOMA transmission. The example below provides a better illustration,

Assuming the precoding matrix of UE1 is W_1 with rank r_1, the precoding matrix of UE2 is W_2 with rank r_2 as follows:

$$W_1 = \left[v_{1,1} \; v_{1,2} \cdots v_{1,r_1} \right] \tag{2.13}$$

$$W_2 = \left[v_{2,1} \; v_{2,2} \cdots v_{2,r_2} \right] \tag{2.14}$$

$$[v_{i,j}] = \frac{v_{i,j}}{\sqrt{v_{i,j} v_{i,j}}} \tag{2.15}$$

where $[v_{i,j}]$ is the normalized version of $v_{i,j}$, $i = 1,2, j = 1,2,\ldots,r_i$ The SINR of UE1 is denoted as s_1 and the SINR of UE2 is denoted as s_2, both in dB scale. Here the SINR can be of large scale or like CQI reported in CSI feedback. The pairing of UE1 and UE2 should fulfill the following:

- A vector in the vector set $\{[v_{1,1}],[v_{1,1}],\ldots,[v_{1,r_1}]\}$ is equal to a vector in a set of $\{[v_{2,1}],[v_{2,1}],\ldots,[v_{2,r_2}]\}$.
- $|s_1 - s_2|$ satisfies a certain criterion, for instance, $|s_1 - s_2| \geq 10$ dB where $|\cdot|$ denotes the absolute value. Assuming the above criterion is met, if $s_1 > s_2$, UE1 is considered as the near user, UE2 is considered as the far user and vice versa.

2.2.4.2 Transmit Power Allocation

For those user pairs that meet the user pairing criterion, the BS needs to allocate the transmit power between the candidate paired users in order to maximize the sum throughput. Assuming UE1 is the near user with power p_1, UE2 is the far user with power p_2. Here p_1 and p_2 are normalized and constrained by

$$p_1 + p_2 = 1 \tag{2.16}$$

$$p_1 < p_2 \tag{2.17}$$

There are many combinations of (p_1, p_2) to be tried, for example, $(0.5-\Delta p, 0.5+\Delta p)$, $(0.5-2\Delta p, 0.5+2\Delta p)$,…,$(p_{\min}, p_{\max})$, where Δp is the step size of power adjustment, $0 < \Delta p < 0.5$ For each power combination, we need to calculate the sum rate. In the end, the power ratio that can deliver the maximum sum throughput and fulfill the proportional fairness criterion will be chosen.

2.2.4.3 Calculation of SINR for NOMA

After the BS allocates the power for near and far users, it is time to calculate the SINRs of the candidate paired users. These SINRs would be used to derive the capacity and the metric for proportional fairness, etc.

Assuming that UE1 and UE2 are paired users, the SINR reported by UE1 is s_1 and its allocated power is p_1, the SINR reported by UE2 is s_1 and its allocated power is P_2, and the SINRs of UE1 and UE2 after the pairing becomes

$$\overline{s_1} = s_1 + 10\log_{10}(p_1) \tag{2.18}$$

$$\overline{s_2} = s_2 + 10\log_{10}(p_2) \tag{2.19}$$

where both $\overline{s_1}$ and $\overline{s_2}$ are dB scale SINR after the pairing.

2.2.4.4 Calculation of the PF Metric

Proportional fairness (PF) not only targets performance maximization but also considers the fairness which is captured in the special treatment of the history of scheduling. Essentially, whether the channel is in a good fade or in bad fade would depend on the instantaneous rate normalized by the average rate over a long period of time, which is defined as

$$M_k = \frac{R_k(t)}{T_k(t)} \tag{2.20}$$

where M_k is the PF metric of the k-th user at the moment of t. $R_k(t)$ is the data rate of the k-th user at the moment of t. $T_k(t)$ is the average rate of the k-th user from moment 1 to moment t, which can be updated as

$$T_k(t) = \left(1 - \frac{1}{\Delta t}\right) T_k(t-1) + \frac{1}{\Delta t} R_k(t) \tag{2.21}$$

where Δt is the length of the exponential window, and $T_k(t-1)$ is the average rate of the k-th user from moment 1 to moment $t - 1$.

2.2.4.5 Procedure of Scheduling

The resource scheduling would be based on the above important steps: user pairing criterion, power allocation, SINR calculation and PF metric calculation. Here is an example where a sub-band-level proportional fairness scheduling is summarized:

Step 1: For each sub-band, calculate the PF metric assuming orthogonal transmission. Select the user whose PF metric is the highest.

Step 2: Loop through all the combinations of user pairs and transmit power allocations, and calculate the PF metric for each combination. To select the user pair whose PF metric is the highest for non-orthogonal transmission.

Step 3: Compare the PF metrics between orthogonal transmission (OMA) and non-orthogonal transmission (NOMA) of this cell at this time instant. If the PF metric of NOMA is higher, there should be non-orthogonal transmission in this sub-band; otherwise, orthogonal transmission should be carried out. If the current sub-band is for near users, other sub-bands of this user should not be considered for far users.

Steps 1–3 are repeated over all sub-bands.

2.3 DIRECT SUPERPOSITION OF SYMBOLS

In 3GPP Release 13 Study Item of multiuser superposition transmission (MUST), direct superposition is called MUST Category 1 where the modulations of each user are superposed in vector form without any bit flipping or transformation. As Figure 2.7 shows, two QPSK signals x_1 and

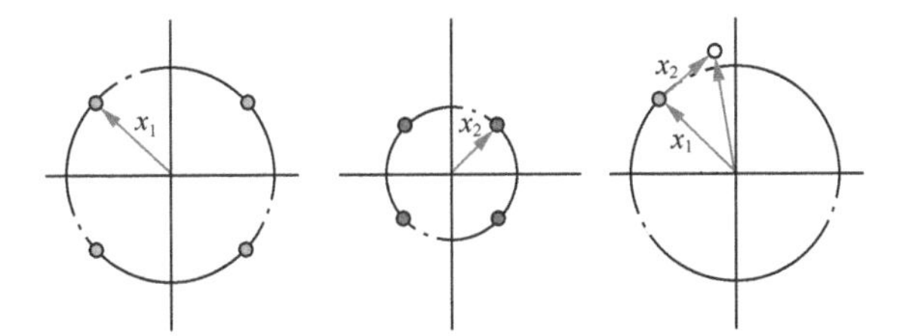

FIGURE 2.7 Vector sum of two constellation points.

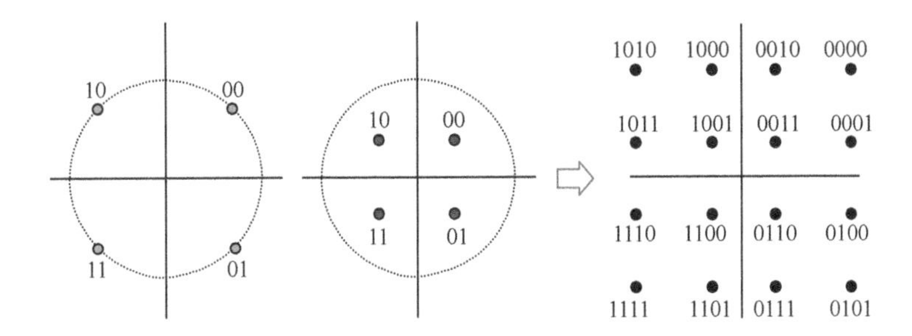

FIGURE 2.8 An example of superposition of two QPSK constellations with a power ratio of 4:1.

x_2 correspond to two vectors in the constellations. The summed vector x contains the information of both x_1 and x_2.

Figure 2.7 is an example of the superposition of two vectors in the constellations. In the following, we show all possible cases of the superposition. Figure 2.8 is an example of the superposition of two QPSK constellations with the power ratio of 4:1, to get a 16-point constellation. It is observed that although the QPSK constellation of each user has a Gray mapping property, the composite constellation may not always have a Gray mapping property, in the case of direct superposition. It is well known that the performance of the non-Gray-mapped constellation is inferior to Gray-mapped. Direct superposition leads to rather arbitrary bit mapping in the composite constellation which depends on the power ratio and may be difficult to optimize.

2.3.1 Transmitter-Side Processing

The transmitter-side processing for direct superposition is illustrated in Figure 2.9. The data blocks of the two users denoted as TB1 and TB2, respectively, are separately encoded, rate-matched, bit scrambling and modulated using legacy modulation. The modulated symbols are power scaled

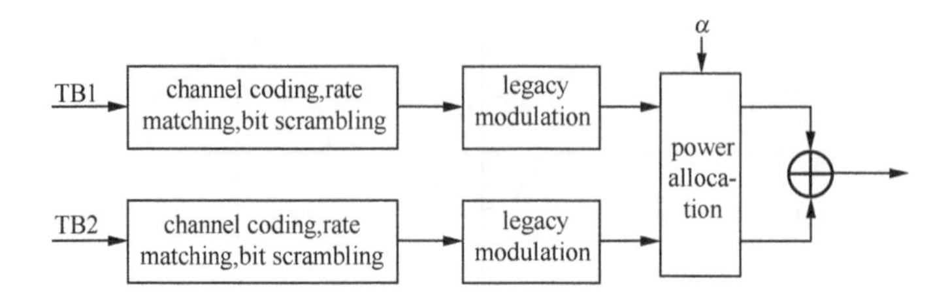

FIGURE 2.9 Transmitter-side processing for direct superposition.

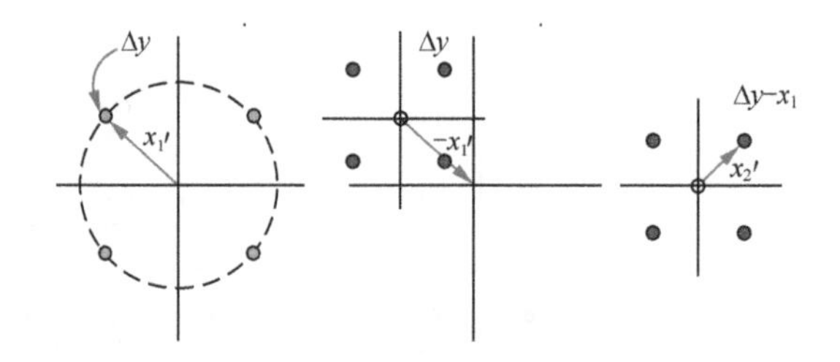

FIGURE 2.10 SIC-type demodulation.

according to the ratio α, and then directly superposed. The left two blocks, channel coding, etc. and modulation are exactly the same as LTE or NR.

2.3.2 Receiver Algorithm

Since the composite constellations of direct superposition are in general not Gray-mapped, more advanced receivers would be required. A relatively simple receiver such as SLIC may not have enough capability of canceling the interference. For near (cell-center) users, a more complicated CWIC may be necessary in order to deliver enough good performance.

SIC is a widely used CWIC. It is very suitable for the case when users have different SNRs, e.g., one close to the BS and the other at the cell edge. Figure 2.10 depicts the SIC-based demodulation:

Step 1: to detect the signal for the far user x_1' based on the received signal y

Step 2: the signal for the near user may be one of the points surrounding x_1', to subtract x_1' from y

Step 3: to determine the signal for the near user x_2'

Figure 2.11 is the block diagram of the CWIC receiver. Not only the core algorithms of CWIC can be described but also basic physical layer

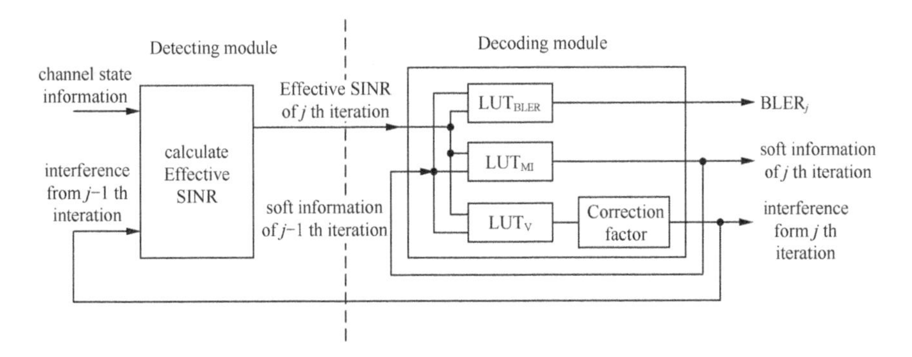

FIGURE 2.11 Block diagram of CWIC.

processing can be modeled in this figure. One essence of CWIC is that after one user's data is successfully decoded, the receiver would regenerate the clean received signal for this particular user according to the information bits, modulation order, code rate and CSI. As long as the channel can be estimated accurately and there is no cyclic redundancy check (CRC) missing detection, this user's interference with other NOMA paired users can be almost entirely removed.

For the downlink, a user is normally not required to decode other users' data. However, in the case of NOMA, if the CWIC is used, the target user may need to decode other users' data in order to perform interference cancelation. Even though the physical layer decoding does not mean that the user would be able to interpret the higher layer data of other users, still, compared to orthogonal transmission, this is an additional processing and would increase the power consumption at the terminals.

CWIC requires more signaling support for the operation. In order to reconstruct the interference signal, apart from the power allocation information, the target user also needs to know the modulation order, the resource allocation, etc. Certainly, when the time and frequency resources of the target user and the interfering users are completely overlapped, the resource allocation of the interfering users need not be signaled to the target user. However, the complete overlap of resources is a scheduling constraint and may have a negative impact on the system throughput. When CWIC is used together with HARQ, the hardware processing and control signaling design would be much more complicated than that of less advanced receivers.

It is noted that advanced receivers would only be used for near (cell-center) users. The transmit power for far users (as the interferer to near users) is relatively strong, but the modulation order of far users is typically low. Hence, it is not challenging for near users to decode the signal of far users and subtract them. For far users, the signal for the near user would be very weak. It is very challenging for the far user to decode the near user's signal also because of the higher-order modulation used for the near user. Hence, linear receivers such as MMSE-IRC are usually used for far users.

2.4 GRAY MAPPING WITH FLEXIBLE POWER RATIOS

The superposition with Gray mapping and flexible power ratio is called MUST Category 2 in 3GPP. As discussed earlier, the requirement for receivers can be relaxed when the composite constellation has a Gray mapping property. Figure 2.12 is an example of two QPSK superpositions with a power ratio of 4:1. The composite 16-point constellation is Gray-mapped.

2.4.1 Transmitter-Side Processing

There are many ways to ensure Gray mapping in the composite constellation. In the following, two typical methods are discussed.

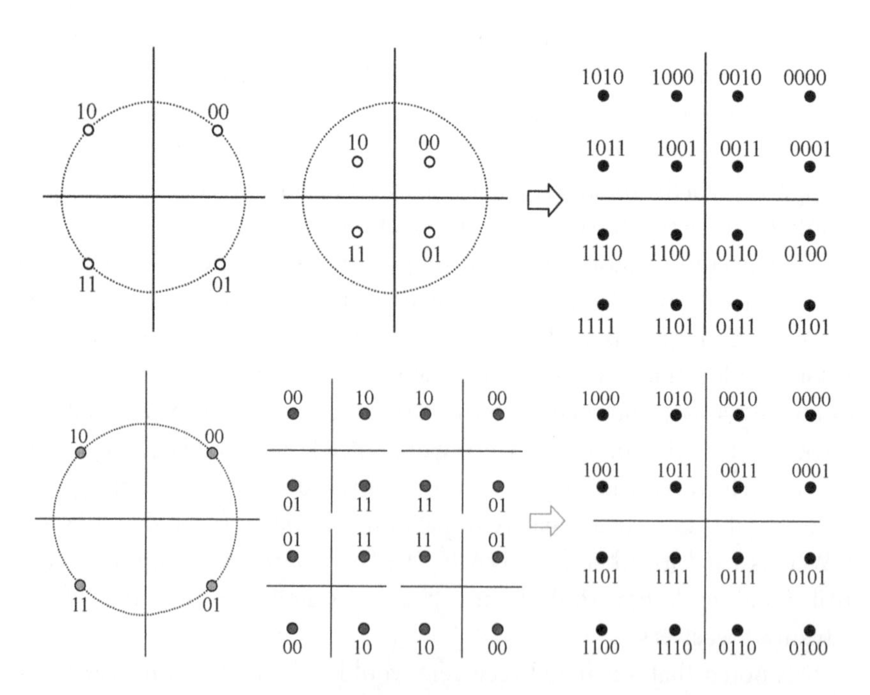

FIGURE 2.12 Composite constellation with Gray mapping property.

2.4.1.1 Superposition with Mirror Transformation

For the user allocated with less power (usually the cell-center user, corresponding to the lowest two digits), the location of this user's constellation point in the composite constellation depends on the constellation point of the user allocated with more power (usually the cell-edge user, corresponding to the highest two digits). In the mirror transformation-based superposition, if the constellation point of the far user is in the first quadrant, the four constellation points in the upper-right corner of the composite constellation would remain the same, similar to the direct superposition. However, if the constellation point of the far user is in the second quadrant, the four constellation points in the upper-left corner of the composite constellation should be mirrored horizontally. If the constellation point of the far user is in the third quadrant, the four constellation points in the lower-left corner of the composite constellation should be mirrored both horizontally and vertically. If the constellation point of the far user is in the fourth quadrant, the four constellation points in the lower-right corner of the composite constellation should be mirrored vertically [6].

As illustrated in Figure 2.13, assuming the allocated power for the far user is P_1. When QPSK is used, its constellation point S_1 in complex

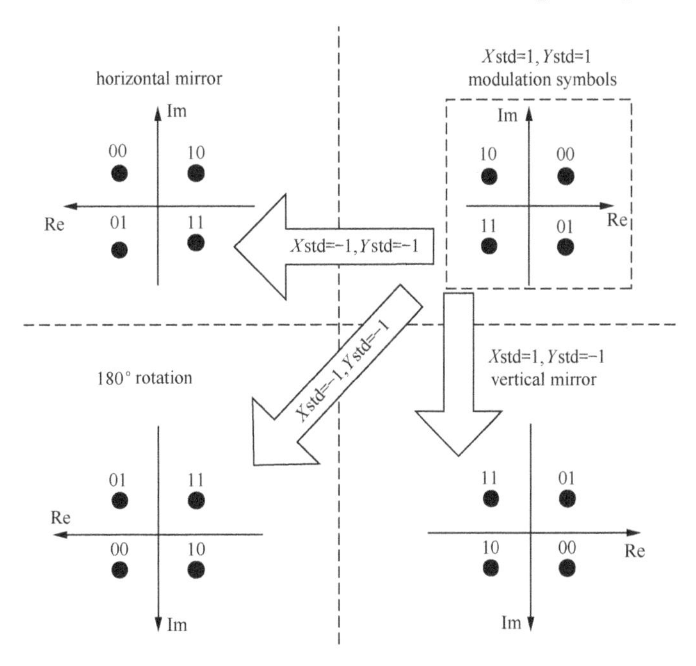

FIGURE 2.13 An example showing the principle of mirror transformation-based superposition.

form can be represented as $\sqrt{P_1}\cdot(x_{1+}y_1\cdot i)$. Assuming the allocated power for the near user is P_2. When QPSK is used, its constellation point S_2 in complex form can be represented as $\sqrt{P_2}\cdot(x_2+y_2\cdot i)$.$P_1$ corresponds to unnormalized integer-grid constellation as S_{std}为$X_{\text{std}}+Y_{\text{std}}\cdot i$ where the coordinate $(X_{\text{std}}, Y_{\text{std}})$ can be {(1, 1), (–1, 1), (–1, –1), (1, –1)}. The composite constellation points can be written as $(S_1+\Delta S)$. When the symbol S_{std} is $(1+i)$, that is $X_{\text{std}}=1$, $Y_{\text{std}}=1$, ΔS would be $\sqrt{P_2}\cdot(x_2+y_2\cdot i)$, similar to S_2. When the symbol S_{std} is $(-1+i)$, that is $X_{\text{std}}=-1$, $Y_{\text{std}}=1$, ΔS would be $\sqrt{P_2}\cdot(-x_2+y_2\cdot i)$, equivalent to the horizontal mirror transformation of S_2. When the symbol S_{std} is $(1-i)$, that is $X_{\text{std}}=1$, $Y_{\text{std}}=-1$, ΔS should be $\sqrt{P_2}\cdot(x_2-y_2\cdot i)$, equivalent to the vertical mirror transformation of S_2. When the symbol S_{std} is $(-1-i)$, that is $X_{\text{std}}=-1$, $Y_{\text{std}}=-1$, ΔS should be $\sqrt{P_2}\cdot(-x_2-y_2\cdot i)$, equivalent to both horizontal and vertical mirror transformation or 180° rotation of S_2.

2.4.1.2 Inclusive OR of Bits

Figure 2.14 shows how to ensure Gray mapping property by inclusive-OR logic. The coded bits of the near user and the far user are first input to a bit conversion table to output $c_1, c_2, \ldots, c_n$ and $d_1, d_2, \ldots, d_m$. Here n and m represent the numbers of bits carried by each modulation symbol of the near user and the far user, respectively. The bit conversion can be represented as

$$\begin{aligned} c_1 &= a_1 \odot (b_1 \odot b_3 \ldots \odot b_{m-1}) \\ c_2 &= a_2 \odot (b_2 \odot b_4 \ldots \odot b_m) \\ c_i &= a_i (i = 3,4,\ldots,n) \end{aligned} \tag{2.22}$$

When the far user is modulated by QPSK and its power allocation is higher than the near user, the above bit conversion can be simplified as

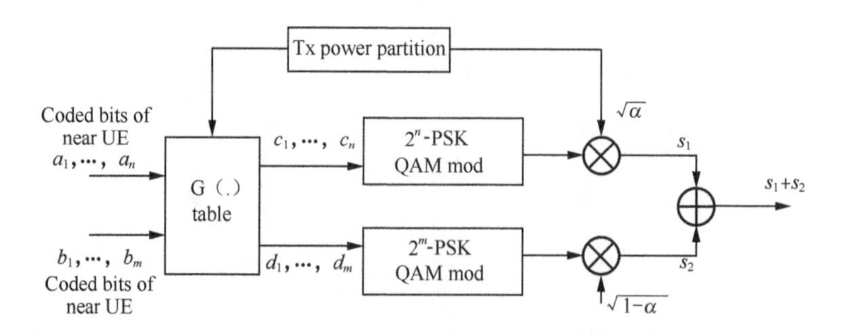

FIGURE 2.14 Inclusive OR method.

$$
\begin{aligned}
c_1 &= a_1 \odot b_1 \\
c_2 &= a_2 \odot b_2 \\
d_i &= b_i, \text{for } i = 1,2
\end{aligned}
\tag{2.23}
$$

It is seen from Eq. (2.23) that after the bit conversion, the two highest bits of the near user are the inclusive OR of the original two highest bits between the near user and the far user. After the bit conversion, the rest of the processing is quite similar to that of direct superposition, e.g., each going through the legacy modulation, power scaled and then superposed. Note that here the inclusive OR can also be replaced all by the exclusive-OR which shares the same principle.

The following example shows how to use the bit conversion table. Assuming that c_1 has two bits as shown in Figure 2.15. When c_1 is "10", it corresponds to the solid dots in Figure 2.15a, whereas other constellation points are hollow dots. Assuming that c_2 has two bits, as shown in Figure 2.15b, when c_2 is "10", it corresponds to the solid dots in Figure 2.15b.

As shown in Figure 2.16, c_1 is "10" and c_2 is "10", by using Eq. (2.23) or going through the bit processing blocks, we can obtain the concatenated bit streams "1011".

It should be pointed out that although the power allocation can be quite flexible, from the receiver implementation point of view, the composite constellation does not only need to have Gray mapping property but also should avoid overlapping of constellation clusters which can happen when the near user is allocated more power. Figure 2.17 shows an example of

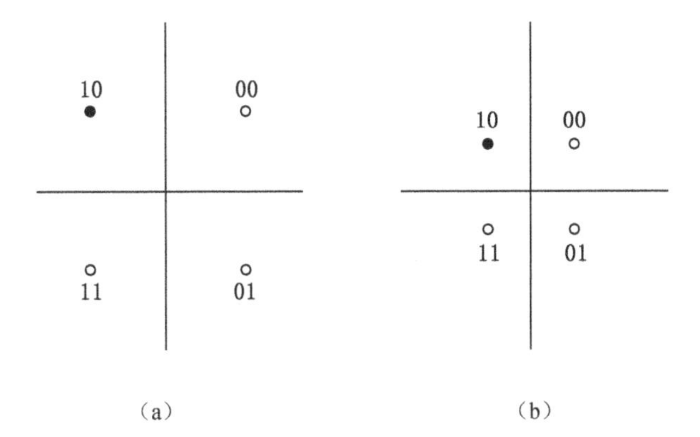

FIGURE 2.15 Input bits. (a) 2-bit constellation for C1 and (b) 2-bit constellation for C2.

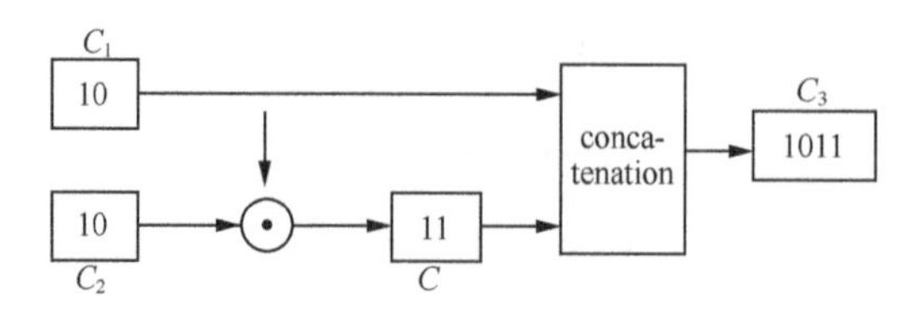

FIGURE 2.16 An example of bit transformation.

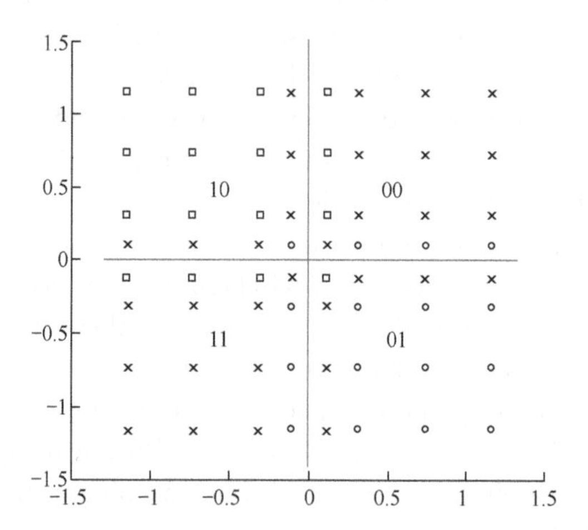

FIGURE 2.17 Overlapping in the composite constellation (QPSK for the far user + 16 QAM for the near user) [7].

TABLE 2.4 Range of Power Ratios to Avoid Overlapping in the Composite Constellation

Modulation Pair: Near + Far	Value of P_1 (far)
QPSK+QPSK	$P_1 \geq 0.5$
16QAM+QPSK	$P_1 \geq 0.6429$
64QAM+QPSK	$P_1 \geq 0.7$

such overlapping. Since the four centers of the clusters (corresponding to the far user) are not well separated and the constellation points within each cluster (corresponding to the near user) are quite well spaced, some of the constellation points spill to the neighboring quadrants. While such overlapping can be handled by using more advanced receivers such as ML detectors, it would be quite challenging for less advanced receivers such as SLIC, leading to noticeable performance loss. In order to avoid overlapping, the power ratio between the far user and the near user should have a certain range, depending on the modulation pairing as shown in Table 2.4.

2.4.2 Receiver Algorithms

Since the mirror transformation method can ensure Gray mapping property in the composite constellation, less advanced receivers such as SLIC can be used. As opposed to the CWIC, SLIC does not need to decode the information bits of the interfering user. Instead, it can infer the modulation symbols once the interference signal has been detected. Then the receiver can reconstruct the interfering signal with the knowledge of the power ratio.

There are two types of SLIC, explicit and implicit. In the explicit interference cancelation, the modulation symbols with more allocated power would be reconstructed and canceled. Since there is no channel coding involved, there would be some detection errors that may propagate and degrade the performance. In the implicit interference cancelation, joint detection would be carried, with the transmitted signal as the high order modulation. LLR for each bit is calculated. The near user and the far user's LLRs are then input to the channel decoder, respectively. During this process, there is no interference cancelation either at the modulation symbol level or code-word level. There is no error propagation either.

Note that the far user can use a linear MMSE receiver without significant performance degradation.

2.5 BIT PARTITION

Bit partition for downlink NOMA is called MUST Category 3 in 3GPP MUST Study Item. There are two types of bit partitions. The first type can be considered as a special case of mirror transformation-based superposition, that is, by further restricting the power ratio between the far and near users, the composite constellation not only has Gray mapping property but also corresponds to one of the legacy constellations already specified for LTE. In another word, the composite constellation should be regular in shape and the constellation points would be equally spaced [8]. Table 2.5 shows several examples. For instance, when the far user of QPSK is allocated 80% power, leaving 20% power to the near user of QPSK, the composite constellation is a classic 16-QAM.

The name "bit partition" comes from the concept that the composite constellation can be considered as a very big regular constellation. The constellation of a user can be considered as a sub-constellation under a big regular constellation. As shown in the second row of Table 2.5, a legacy 64-QAM constellation can carry six coded bits. These six bits can

TABLE 2.5 Power Rations and Modulation Pairs in the First Type of the Bit Partition Method

Legacy Modulation Order	Modulation of Far User	Modulation of Near User	Power Percentage of Far User
16QAM	QPSK	QPSK	0.8
64QAM	QPSK	16QAM	0.762
64QAM	16QAM	QPSK	0.952
256QAM	QPSK	64QAM	0.753
256QAM	16QAM	16QAM	0.941...
256QAM	64QAM	QPSK	0.988...

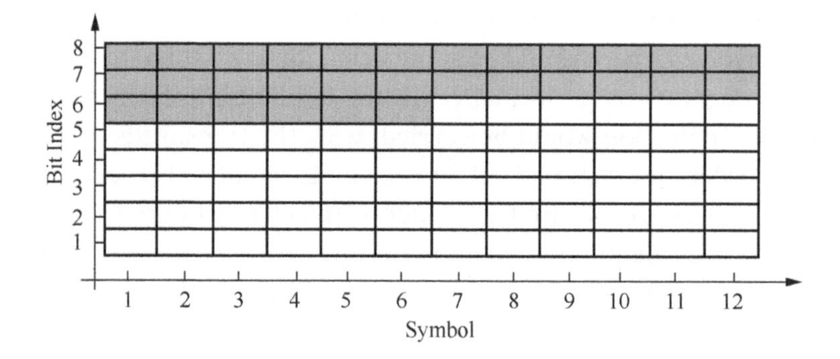

FIGURE 2.18 An example of the second type of the bit partition method.

be partitioned into two bits and four bits, corresponding to the far and the near users, respectively, with the power ratio of 0.762:0.238, one with QPSK and the other with 16-QAM. Since the composite constellation is the legacy QAM, the specification work of bit partition-based NOMA is rather simple, e.g., just to define a table containing the power ratios for different pairings of modulation orders.

Since the values for the power ratio are restricted, the bit partition method puts more constraint on user pairing and resource allocation in the scheduler implementation, compared to the mirror transformation method. Hence, its performance is a little inferior.

The above example is the first type of bit partition method where the partition is the same across all the modulation symbols of a code block. If the bit partition can vary with respect to the modulated symbols, this would be called the second type of bit partition method. Figure 2.18 shows an example where the composite constellation is 256 QAM, each carrying eight bits. It is seen that the bit partition ratio is 3:5 for the first six modulation symbols, whereas the bit partition ratio is changed to 6:2 for the later six modulation symbols.

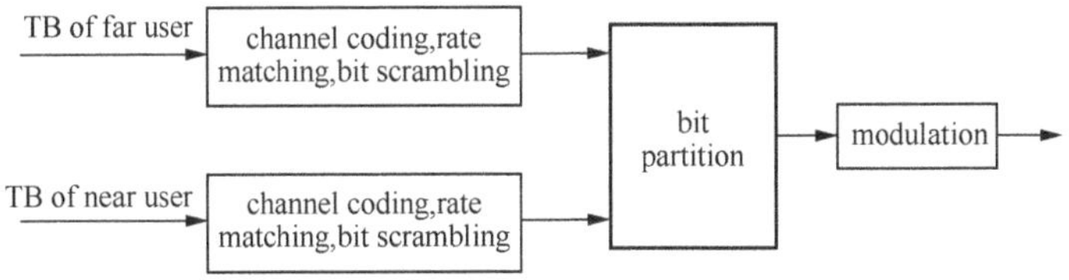

FIGURE 2.19 Block diagram of transmitter-side processing for the bit partition method.

In the second partition method, due to the flexible adjustment of the partition ratio down to the granularity of modulated symbol, in theory, it can match the CSI more precisely. However, it has a big impact on LTE standards, e.g., requiring significant changes in CQI reporting and transport block size (TBS) tables [9]. The legacy CQI table is designed assuming the same modulation order for all the bits in a code block. Modulation symbol level of flexible bit partition would inevitably need a much more complicated calculation of CQI. Legacy TBS tables are designed with the legacy CQI. Hence any big changes in the CQI definition would affect the TBS tables.

2.5.1 Transmitter-Side Processing

Figure 2.19 shows the basic processing units at the transmitter side for the bit partition method. Transport blocks (containing the information bits) of the far user and the near user are first encoded, rate-matched and bit scrambling. As mentioned above, due to the potential support of the second type of bit partition, the detailed calculation of the rate matching may be different from the legacy LTE. After the bit scrambling, coded bits are bit partitioned, possibly modulated symbol-dependent, and then mapped to the corresponding legacy constitute constellations.

2.5.2 Receiver Algorithms

Bit partition method not only can ensure Gray mapping property in the composite constellation but also the equal-distance property between adjacent constellation points. Hence, less advanced receivers, e.g., SLIC, can be used. Note that for the far users, a linear MMSE-IRC receiver should be enough.

2.6 PERFORMANCE EVALUATION

2.6.1 Link-Level Performance

To verify the rate regions in Section 2.1, the AWGN channel is simulated where the SNRs of the near user (denoted as UE1) and the far user (UE2) are 20 and 0 dB, respectively. The far user is QPSK modulated while

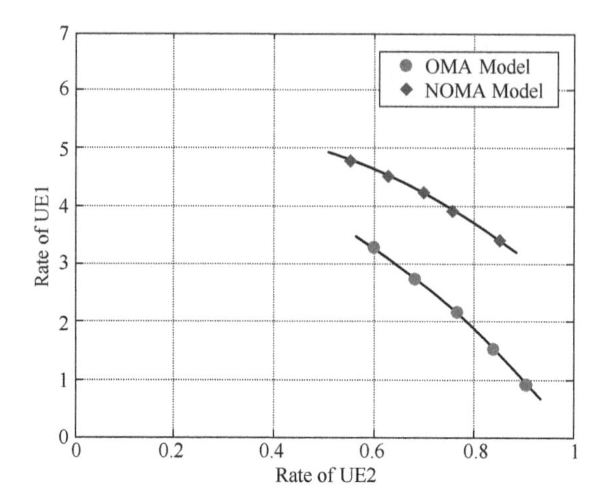

FIGURE 2.20 Rate region comparison via AWGN channel simulations.

the near user can be QPSK, 16-QAM, or 64-QAM. The power ratios are {0.7:0.3}, {0.75:0.25}, {0.8:0.2}, {0.85:0.15}, and {0.9:0.1}. For OMA, the ratio of the degree of freedom ranges from 0.1 to 0.9.

Select a few rate pairs that are close to the theoretical rate region, the rate region of AWGN channel simulation is shown in Figure 2.20. It is observed that in the interest of practical systems, e.g., the cell-edge spectral efficiency is within 0.5–0.9 bps/Hz/cell, the sum rate of NOMA is significantly higher than that of OMA.

The performance of (Gray-mapped+SLIC receiver) vs. (non-Gray-mapped+SLIC receiver) can be compared via BLER curves. More specifically, by adjusting the power ratio between the near and the far users, first to align the performance of the far user. Then, the power ratio is adjusted to compare the performance of the near user. It is seen in Figure 2.21 that the solution that ensures the Gray property in the composite constellation outperforms the direct superposition solution.

$$\text{QPSK}(\text{near}) + \text{QPSK}(\text{far})$$

$$\text{16-QAM}(\text{near}) + \text{QPSK}(\text{far})$$

Rate regions are simulated with a more realistic setting, and the results are shown in Figure 2.22. UE1 denotes the near user with an SNR of 20 dB. UE2 denotes the far user with an SNR of 8 dB. Compared with Figure 2.20, simulation assumptions for Figure 2.22 are fading channels, with more practical receivers and realistic channel estimations.

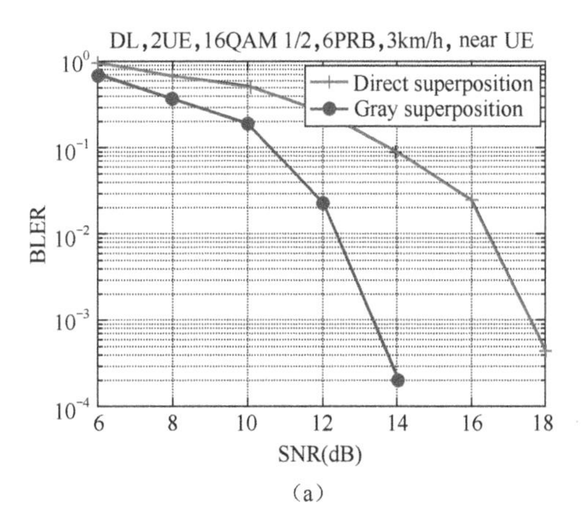

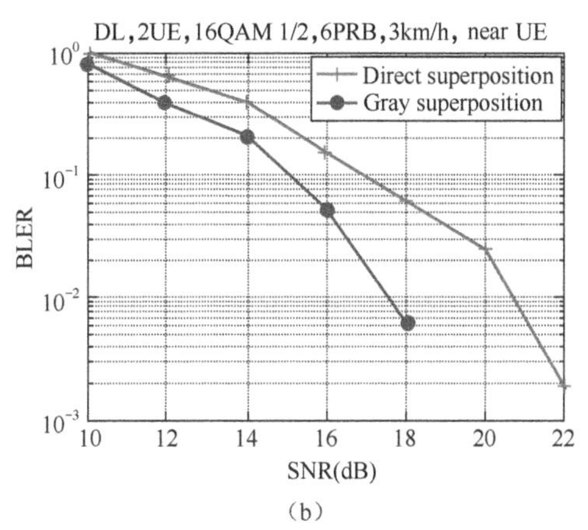

FIGURE 2.21 BLER performance comparison between Gray-mapped and non-Gray-mapped. (a) QPSK (near)+QPSK (far) and (b) 16-QAM (near)+QPSK (far).

In Figure 2.22, the triangle dots correspond to non-Gray mapping+CWIC. The diamond dots correspond to the Gray-mapped composite constellation+SLIC. It is observed that the rate regions of these two combinations are overlapped. This means that if Gray mapping cannot be ensured, more advanced receivers such as CWIC are required in order to match the performance of Gray-mapped. If neither Gray mapping nor CWIC-like receivers are available, the rate region would correspond to the square dots in Figure 2.22. Poor performance especially occurs when the near user is allocated more power. This leads to the relatively weak signal of the far user at the near user, and therefore a higher chance that the far user's data cannot be decoded correctly with the less advanced receiver, e.g., SLIC. As the far user is allocated more power, it is

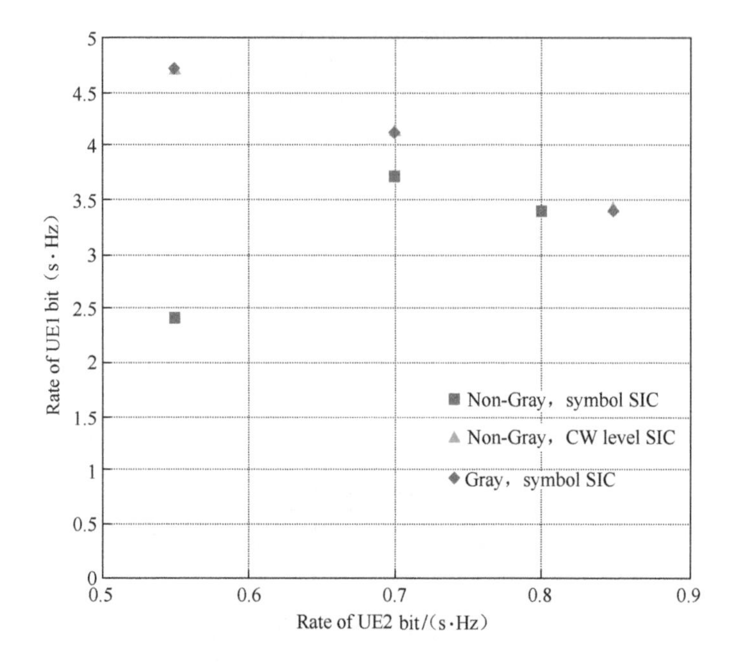

FIGURE 2.22 Rate regions via simulations with a more realistic setting.

TABLE 2.6 System Performance Comparison between NOMA and OMA for Full-Buffer Traffic, 2Tx and 2Rx Antennas

Wideband Scheduling	Cell Average Spectral Efficiency (bps/Hz)	Cell-Edge Spectral Efficiency (bps/Hz)
SU-MIMO	1.3108	0.206
DL NOMA	1.4320	0.269
Gain	9.25%	30.70%

easier for the near user to decode the far user's data using less advanced receiver, and then cancel it.

2.6.2 System Performance

2.6.2.1 Full-Buffer Traffic and Wideband Scheduling

Simulation parameters can be based on Table 2.3, in particular two transmit antennas and two receive antennas of cross-polarization and mobile speed of 3 km/h, MMSE+CWIC receiver and wideband scheduling. The simulated system performances of SU-MIMO and NOMA are compared in Table 2.6. It is observed that in terms of cell-edge spectral efficiency, the gain of NOMA is about 26%. For the average cell throughput, the gain of NOMA is about 8%.

In 3GPP MUST study, many companies simulated NOMA system performance with wideband scheduling for full-buffer traffic. The gain of

TABLE 2.7 Impact of Different Precoding, Four Tx Antennas, and Full-Buffer Traffic

	Baseline Spectral Efficiency (SU-MIMO) (bps/Hz)	Spectral Efficiency with Potentially Different MIMO Precoders (bps/Hz)	Relative Gain Over the Baseline (%)	Spectral Efficiency with the Same MIMO Precoder (bps/Hz)	Relative Gain Over the Baseline (%)
Cell average	1.5722	2.1235	35.06	1.6091	2.35
Cell edge	0.0337	0.0358	6.24	0.0394	17.07

NOMA over OMA ranges from 4.4% to 12.9% for average spectral efficiency, and 13%–31% for cell-edge spectral efficiency.

As mentioned earlier, both MU-MIMO technology and NOMA technology aim to improve the system capacity by scheduling multiple users at the same time and frequency resources. Since the number of active users in a cell is limited, the scheduler needs to balance between MU-MIMO and NOMA. When both technologies are used, each gain will be degraded, e.g., the total gain would not be the sum of the gain of each. Note that the user pairing criteria of the two technologies are not the same. For MU-MIMO, users with low spatial correlation and similar SNR would be paired, whereas for NOMA, users with very different SNR, e.g., one close to the BS and the other at the cell edge, would be paired.

Table 2.7 shows the system performance of four Tx antennas. There are ten active users per cell. It is seen that when the paired users use different spatial precoders, the average spectral efficiency per cell for NOMA+MU-MIMO is about 35% higher than that of OMA. If the same precoder is used, the gain of NOMA in terms of average spectral efficiency is about 2%. This indicates that by deploying MU-MIMO and NOMA together, the average system throughput can be significantly improved over OMA+SU-MIMO. It is noticed that by using NOMA+MU-MIMO, the cell-edge spectral efficiency does not improve. In fact, there is slight degradation (17% to 6%) from NOMA to NOMA+MU-MIMO.

2.6.2.2 FTP Traffic, Two Transmit Antennas, Wideband Scheduling

Table 2.8 shows the performance comparison between NOMA and SU-MIMO (e.g., OMA) when the file size is 0.1 Mbytes with high loading. The performance metric is measured in user-perceived throughput (UPT). It is seen that for resource utilization at about 0.7, in terms of average throughput, the gain of NOMA over OMA is about 10.7%. In terms of cell-edge performance, the gain of NOMA is about 17.8%. When the resource utilization (RU) is about 0.8, the gain in average throughput

TABLE 2.8 Performances of Downlink NOMA and SU-MIMO for FTP1 Traffic with Wideband Scheduling

Wideband Scheduling	Target RU	Actual RU	Mean UPT (Mbps)	5% UPT (Mbps)	50% UPT (Mbps)	95% UPT (Mbps)
SU-MIMO (OMA)		0.7325	7.068	0.8827	4.572	21.68
DL NOMA	0.7	0.696	7.825	1.039	5.362	22.57
Gain		-	10.71%	17.80%	17.27%	4.11%
SU-MIMO (OMA)		0.8715	4.458	0.6529	2.542	15.59
DL MIMO	0.8	0.8402	5.151	0.7107	3.132	16.99
Gain		-	15.54%	8.85%	23.21%	9.00%

of NOMA is about 15.5%, and the gain in cell-edge throughput is about 8.85%. The degradation for resource RU=0.8 compared to RU=0.7 is due to more packets not being able to transmitted within the pre-defined time duration and therefore dropped.

The trend of system performance can be explained by the statistics of the number of active users and scheduled users, as shown in Figure 2.23. When the RU is 0.7, the maximum number of active users per cell is 21. When the RU is increased to 0.8, the maximum number of active users per cell becomes 24. This means that when more users are active, there are more users being scheduled.

The distributions of the number of scheduled users are shown in Figure 2.24. When RU is about 0.7, about 57% of users are scheduled with OMA, whereas 13% of users are scheduled with NOMA. When RU is increased to 0.8, about 59% of users are with OMA and 25% of users are with NOMA. This means that as the loading gets higher, the probability of successful pairing is increased, which explains why NOMA demonstrates significant gains in average UPT, 50% UPT and 95% UPT.

The performance gain of NOMA over OMA depends on the system loading. The closer the RU is to 1, the closer the performance is to full-buffer traffic. In 3GPP MUST study, many companies simulated the system performance of NOMA for FTP with wideband scheduling. When RU is about 60% and 80%, the system performance gains of NOMA over OMA are in the following ranges:

- For RU around 60%, the average throughput gain is from –9% to 8%, and the cell-edge throughput gain is from –13% to 15.9;
- For RU around 80%, the average throughput gain is 1% to 20%, and the cell-edge throughput gain is from 4.4% to 25.4%;

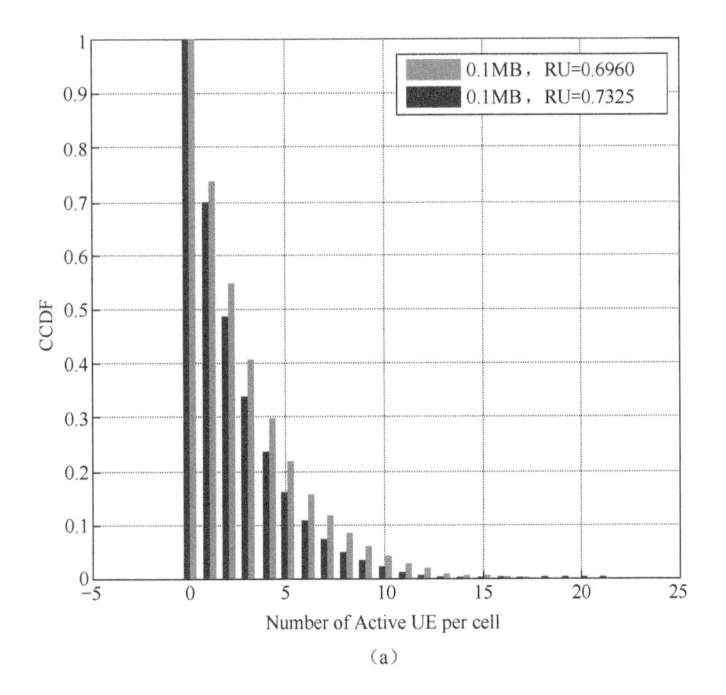

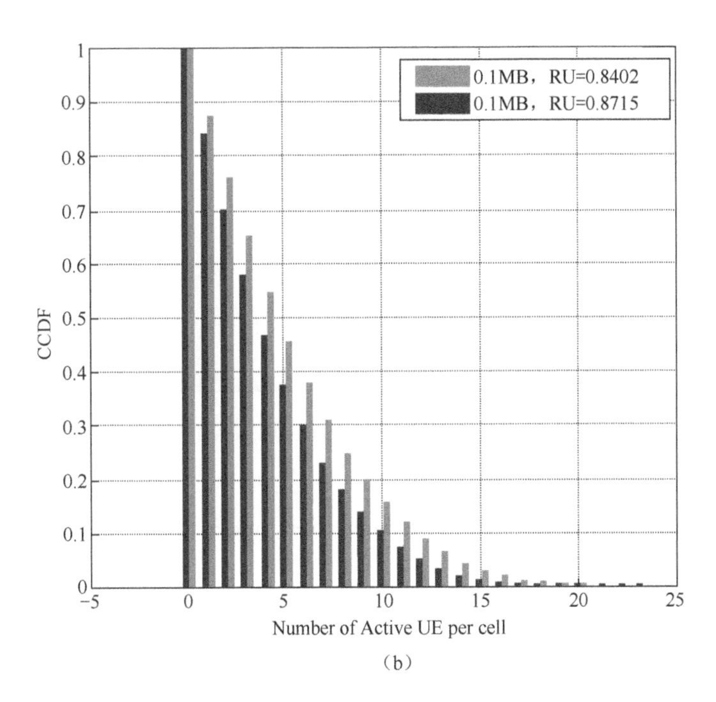

FIGURE 2.23 Distribution of the number of active users for different RUs. (a) RU is about 0.7 and (b) RU is about 0.8.

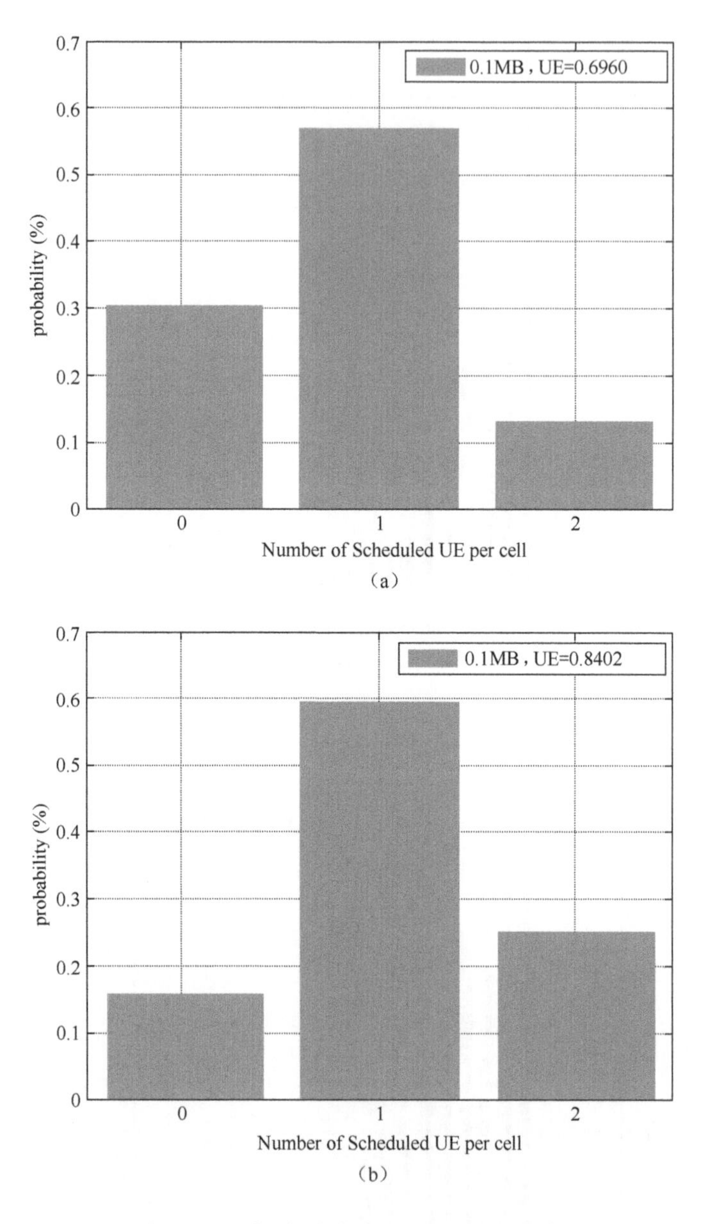

FIGURE 2.24 Distributions of scheduled users under different RUs. (a) RU is about 0.7 and (b) RU is about 0.8.

2.6.2.3 FTP Traffic, Two Transmit Antennas, and Sub-Band Scheduling

Figure 2.25 shows the distribution of the number of simultaneously sub-band scheduled users when the RU is around 76%. This distribution depends quite on the scheduling algorithm. It is observed that around 27% of the time, there is only one user scheduled within a cell.

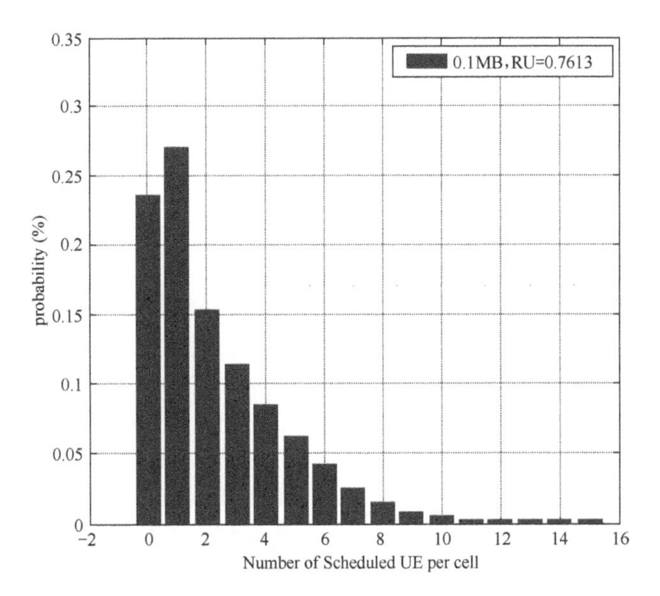

FIGURE 2.25 Distribution of number of users simultaneously sub-band scheduled.

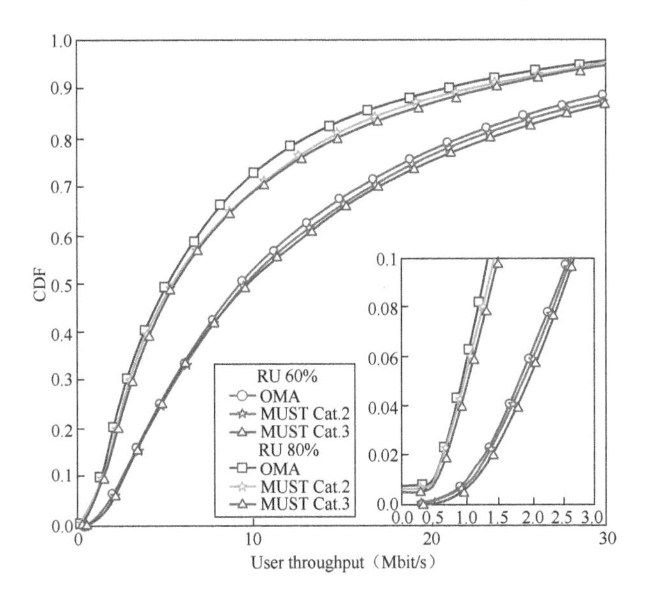

FIGURE 2.26 Distribution of UPT.

The distributions of UPT are illustrated in Figure 2.26 where the orthogonal single-user MIMO, mirror transformation-based NMA (e.g., MUST Category 2) and bit partition-based NOMA (e.g., MUST Category 3) are compared. Two transmit antennas are assumed for each BS. Two receive antennas are assumed for each terminal. Each user can have a single layer

TABLE 2.9 Gains of NOMA in UPT

		Category 2		Category 3	
Throughput (Mpbs)	**OMA**	**NOMA**	**Gain (%)**	**NOMA**	**Gain (%)**
Average	7.00	7/73	10.43	7.78	11.09
95%	24.24	25.81	6.45	26.67	10.00
50%	3.98	4.55	14.20	4.35	9.24
5%	0.76	0.86	13.61	0.79	3.77
Average of last 5%	0.42	0.55	29.35	0.46	8.88
RU (%)	88.23%	86.99%	-	87.87%	-

or two layers of data. The precoding matrix of two paired NOMA users is the same. There are four power ratios, α=0.14, 0.17, 0.23, 0.36, in the mirror transformation-based NOMA. Schedulers can dynamically switch between OMA and NOMA. The inner loop and outer loop link adaptation are simulated to compensate for the inaccuracy of CQI estimation and signaling delays.

Table 2.9 shows the UPT at different percentiles when the RU is around 85%. It is seen that Category 2 MUST outperforms Category 3 MUST, especially in terms of the middle and cell-edge rate, with better fairness.

2.7 OTHER TECHNIQUES

2.7.1 Tomlinson-Harashima Precoding

Considering a simple point-to-point channel, using x to denote the transmitted symbol and y as the receiver symbol, and w as the AWGN ~$N(0, \sigma^2)$, that is,

$$y = x + s + w \tag{2.24}$$

The interfering signal s is the signal the BS is sending at the same time and frequency resources. This interfering signal is known to the transmitter, but not known to the receiver. The power of the user signal x should satisfy the power constraint. This type of channel can be illustrated in Figure 2.27 [10]. Using u to represent the constellation to be transmitted. One straightforward way is $x = u - s$. Then at the receiver, the signal would be $y = u - s + s + w = u + w$. The issue with this method is that the power of the transmitted signal would increase with the power of the interference $|s|^2$. This violates the power constraint for the transmitter.

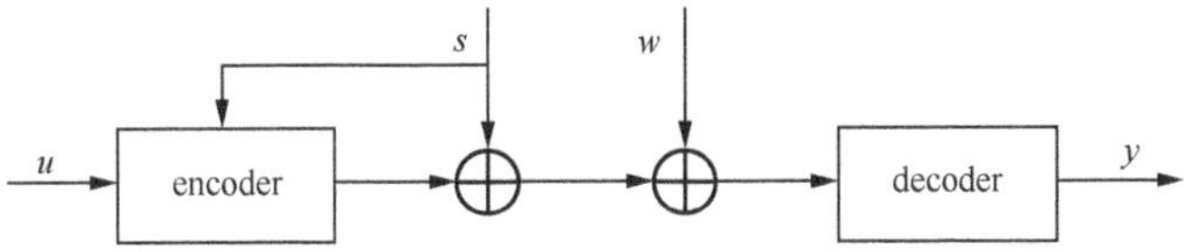

FIGURE 2.27 The channel whose interference is known to the transmitter [10].

The key idea of THP is to duplicate the constellation of the desired signal to be transmitted, and then form an extended constellation, as shown in Figure 2.28a. Within each duplicated small constellation, the relative positions of the points in the constellation remain unchanged. When selecting the constellation point to be transmitted p, we need to search for the constellation point that is the closest to the interfering signal s. As shown in Figure 2.28b, to transmit only $x=p-s$. When the interference power is relatively high, the power of the actual symbol would not be very high. Mathematically, assuming that the transmit signal u is a QAM modulated symbol, that is $\left\{a_I + ja_Q \middle| a_I, a_Q \in \left\{\pm 1, \pm 3, \ldots, \pm\sqrt{M}-1\right\}\right\}$. Then, to perform modulo operation to get $p \in \left\{2\sqrt{M}\left(p_I + jp_Q\right) \middle| p_I, p_Q \in Z\right\}$ where Z is an integer. At the receiver, the detector searches for the closest constellation point to the received symbol in the extended constellation, as illustrated in Figure 2.28d.

Let us consider a more general case with two-user transmission using THP. Assuming that the BS has multiple transmit antennas and each terminal has only one receive antenna. The received signals can be represented as

$$y_1 = \mathbf{h}_1^H\left(\mathbf{u}_1 x_1 + \mathbf{u}_2 x_2\right) + w_1 \tag{2.25}$$

$$y_2 = \mathbf{h}_2^H\left(\mathbf{u}_1 x_1 + \mathbf{u}_2 x_2\right) + w_2 \tag{2.26}$$

where x_1 and x_2 are the transmitted signals of UE1 and UE2, respectively. $\mathbf{u}_1$ and $\mathbf{u}_2$ are spatial precoders of UE1 and UE2, respectively. $\mathbf{h}_1$ and $\mathbf{h}_2$ are spatial channels from the BS to UE1 and UE2, respectively. Here, UE1 is assumed to be the near user, e.g., $\|\mathbf{h}_1\|^2 \geq \|\mathbf{h}_2\|^2$. If x_2 is treated as the known interference to be transmitted and the two users have the same spatial precoder, e.g., $\mathbf{u}_1 = \mathbf{u}_2 = \mathbf{u}$, then to perform THP to UE1,

$$y_1 = \mathbf{h}_1^H \mathbf{u}(p_1 - x_2) + \mathbf{h}_1^H \mathbf{u} x_2 + w_1 = \mathbf{h}_1^H \mathbf{u} p_1 + w_1 \tag{2.27}$$

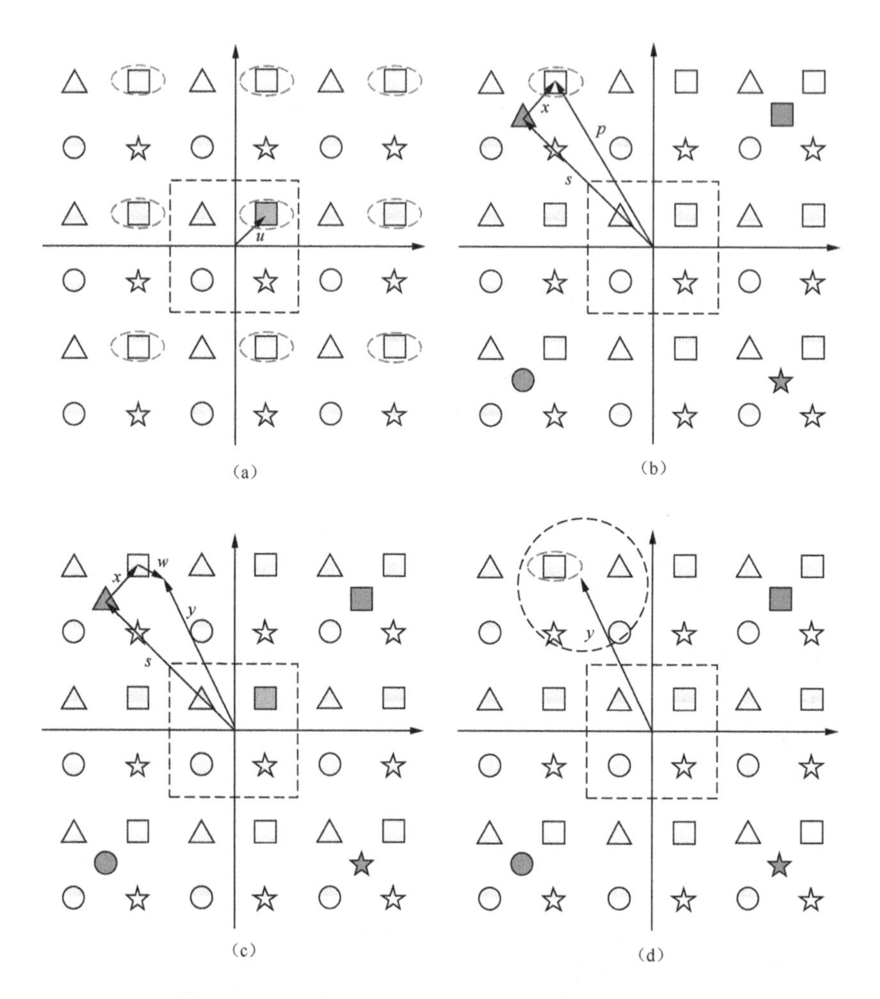

FIGURE 2.28 Illustration of THP precoding and decoding [10]. (a) replicating the constellation, (b) choosing that representation **p** in its equivalence class which is closest to the interference **s**, and transmitting the difference **x**=**p** − **s**, (c) the received signal: **y**=**x**+**s**+**w**, and (d) finding the point in the extended constellation that is closest to y and decodes the information bits corresponding to its equivalence class.

Here, p_1 which corresponds to the constellation point of UE1 in the extended constellation is the closest to the interference x_2. It is seen that for UE1 (near user), due to the interference cancelation implemented at the transmitter, the interference observed at the receiver is just AWGN, that is, no interference from UE2.

For UE2, since it is a far user, its modulation order is not high, e.g., QPSK. Hence, no specific treatment is needed at the transmitter side. Equation (2.26) can be written as

$$y_2 = \mathbf{h}_2^H \mathbf{u} x_2 + \left[\mathbf{h}_2^H \mathbf{u}(p_1 - x_2) + w_2 \right] \tag{2.28}$$

The receiver of UE2 can treat UE1's signal as noise. As seen in Figure 2.28, for s (which is x_2 of UE2), if the spatial precoder fully matches the spatial channel, e.g., $\mathbf{h}_2^H \mathbf{u} = 1$, its clean signal corresponds to the green triangle in the second quadrant. The noise observed by UE2 is the summation of UE1's x_1 and the AWGN w_2.

Equations (2.25) and (2.26) are suitable for two users with different precoders. In other words, THP can also be used for MU-MIMO. In this case, the BS needs to know the spatial channel of each user, e.g., $\mathbf{h}_1$ and $\mathbf{h}_2$, in order to accurately perform interference cancelation at the transmitter.

REFERENCES

1. B. M. Hochwald, C. B. Peel, and A. L. Swindlehurst, "A vector-perturbation technique for near-capacity multi-antenna multiuser communication - Part II: Perturbation," *IEEE Transactions on Communications*, Vol. 53, No. 3, 2005, pp. 537–544.
2. D. Tse, and P. Viswanath, *Fundamentals of Wireless Communication*. Cambridge University Press: Cambridge, 2005.
3. Y. Yuan, et al. "Non-orthogonal transmission technology in LTE evolution," *IEEE Communications Magazine*, Vol. 54, No. 7, 2016, pp. 68–74.
4. 3GPP, TR 36.859, Study on Downlink Multiuser Superposition Transmission (MUST) for LTE (Release 13).
5. 3GPP TR 36.866, Study on Network-Assisted Interference Cancellation and Suppression (NAICS) for LTE (Release 12).
6. 3GPP, R1-157609, Description of MUST Category 2, ZTE, RAN1#83, November 2015, Anaheim, USA.
7. 3GPP, R1-154454, Multiuser superposition transmission scheme for LTE, MediaTek, RAN1#82, August 2015, Beijing, China.
8. 3GPP, R1-152806, Multiuser superposition schemes, Qualcomm, RAN1#81, May 2015, Fukuoka, Japan.
9. 3GPP, R1-152493, Candidate schemes for superposition transmission, Huawei, RAN1#81, May 2015, Fukuoka, Japan.
10. 3GPP, R1-154701, Candidate schemes for superposition transmission based on dirty paper coding, Xinwei, RAN1#82, August 2015, Beijing, China.

CHAPTER 3

Non-Orthogonal Transmission for Downlink Broadcast/Multicast

Yifei Yuan, Hong Tang and Weimin Li

3.1 APPLICATION SCENARIOS

Downlink broadcast/multicast has been widely used in TV services, emergency communications, vehicle-to-vehicle communications, machine-type communications, etc.

- Multi-media TV and entertainment: TV programs of super high resolution, virtual reality, 360° (panorama), push-to-talk or push-to-video
- Vehicle-to-vehicle communications: automatic driving, warning for driving information, safety driving, transportation/traffic signals or indications
- Machine-type communications: software upgrade, public control information
- Emergency communications: warning for natural disasters (earthquake, tsunami and hurricane), amber warning and warning for leakage of chemical or radioactive materials

DOI: 10.1201/9781003336167-3

In many applications, the target consumers of broadcast/multicast services are all the subscribers within a coverage area. Different from the unicast services, the physical layers of broadcast/multicast normally would not support dynamic feedback of channel-state information (CSI). Hence, the transmitter is not able to apply appropriate precoding and link adaptation or hybrid automatic retransmission request (HARQ). In addition, the data sent to each broadcast or multicast group is the same. The performance of broadcast/multicast systems is typically measured as the coverage percentage for a specific data rate, instead of the data throughput for each cell. It is well known that the poor coverage area is often at the cell edge for two reasons: (1) cell edge users are far away from their serving base station and suffer from severe path loss, and (2) cell edge users are not far away from neighboring base stations and would experience strong interference from those stations.

Broadcast/multicast can be within a single cell, for instance, single-cell point-to-multipoint, where different contents are transmitted in different cells. However, in many scenarios, in order to fill the coverage holes, a single-frequency network (SFN) is deployed where neighboring base stations are precisely synchronized and send the same information using the same modulation order in the same time and frequency resources. At the terminal receiver, signals from multiple neighboring base stations are identical, except that they experience different path loss, shadow fading, small-scale fading and propagation delays. In SFN, if code-division multiple access is used, a highly sophisticated receiver is required in order to cancel the inter-symbol interference due to the propagation delay difference. In 4G cellular systems, orthogonal frequency-division multiplexing (OFDM) is used in the downlink, which is very beneficial to SFN. As long as the delay propagation delay is within the cyclic prefix, in theory, no interference will be introduced because the signals are coherently superposed. Using mathematical equations, assuming that terminals can receive broadcast/multicast signals from N base stations where the large-scale fading (including path loss) for the i-th base station is denoted as L_i, the small-scale fading is $H_i[k]$ and the propagation delay is τ_i. The combined channel in the frequency domain can be represented as

$$H_{\text{comb}}[k] = \sum_{i=1}^{N} L_i \cdot H_i[k] e^{j2\pi k \tau_i} \tag{3.1}$$

Assuming that there are enough number of cells participating in SFN transmission, the interference would be ignored, with only thermal noise left. This is analogous to achieving all-angle-like "illumination" as in the surgical rooms. The signal-to-power ratio (SNR) of the k-th subcarrier at the receiver can be written as

$$\mathrm{SNR}[k] = \frac{P_T \left| \sum_{i=1}^{N} L_i \cdot H_i[k] e^{j2\pi k \tau_i} \right|}{N_{\mathrm{thermal}}} \tag{3.2}$$

In Eq. (3.2), P_T is the transmit power on each subcarrier and N_{thermal} is the power of thermal noise on each subcarrier. Since in theory there is no inter-cell interference, user equipment (UE) geometry depends only on the inter-site distance (ISD). As ISD increases, the cumulative distribution function (CDF) of UE geometry moves to the left and the coverage is degraded.

3.2 BRIEF INTRODUCTION OF PHYSICAL MULTICAST CHANNEL (PMCH) IN LTE

The design principle of the broadcast/multicast channel in long-term evolution (LTE) Rel-8 is to reuse the channel design for unicast services as much as possible and at the same time sufficiently consider the characteristics of broadcast/multicast services. In order to support both unicast service and multicast service, a radio frame (10 ms) can have up to six subframes (each of 1 ms duration) configured as multicast/broadcast SFN (MBSFN) subframe. In frequency-division duplex, subframes #0, #4, #5 and #9 cannot be configured as MBSFN subframes. In Time Division Duplexing (TDD), subframes #0, #1, #5 and #6 cannot be configured as MBSFN. These subframes would carry system information and cell synchronization channels for unicast and multicast services.

The first consideration in the MBSFN subframe is to ensure the orthogonality between signals coming from multiple neighboring base stations. The most straightforward way is to extend the cyclic prefix to overcome different propagation delays. If the overhead of the cyclic prefix is increased and we still want to keep the subframe length unchanged, the number of OFDM symbols per subframe needs to be reduced. Two deployments are supported for PMCH: (1) unicast and multicast services share the same carrier, e.g., mixed carrier, and (2) only multicast services, e.g., dedicated carrier. In the mixed carrier case, there are 12 OFDM symbols within a broadcast/multicast MBSFN subframe. The length of the cyclic prefix is

16.6 μs. Its subcarrier spacing is the same as LTE unicast which is 15 kHz. In the dedicated carrier case, there are six OFDM symbols within a broadcast/multicast MBSFN subframe. The length of the cyclic prefix is 33.3 μs. The subcarrier spacing is 7.5 kHz. The length of an OFDM symbol (excluding the cyclic prefix) is increased to 0.133 ms.

PMCH was enhanced in LTE Rel-14 where a new MBSFN subframe structure was introduced. Such a subframe structure is used for dedicated carrier of broadcast/multicast and can support wider coverage and larger areas of SFN combination. More specifically, the subcarrier spacing is reduced to 1.25 kHz; hence, there are 144 subcarriers and 2 OFDM symbols in a physical resource block. The cyclic prefix is increased to 200 μs. Rel-14 enhancement of MBSFN also includes using subframe #0 that carries broadcast/multicast system information and the synchronization signal, therefore completely cutting off the reliance on unicast systems so that it can independently carry out the cell synchronization signal, acquire the system information and reduce the overhead of control signaling.

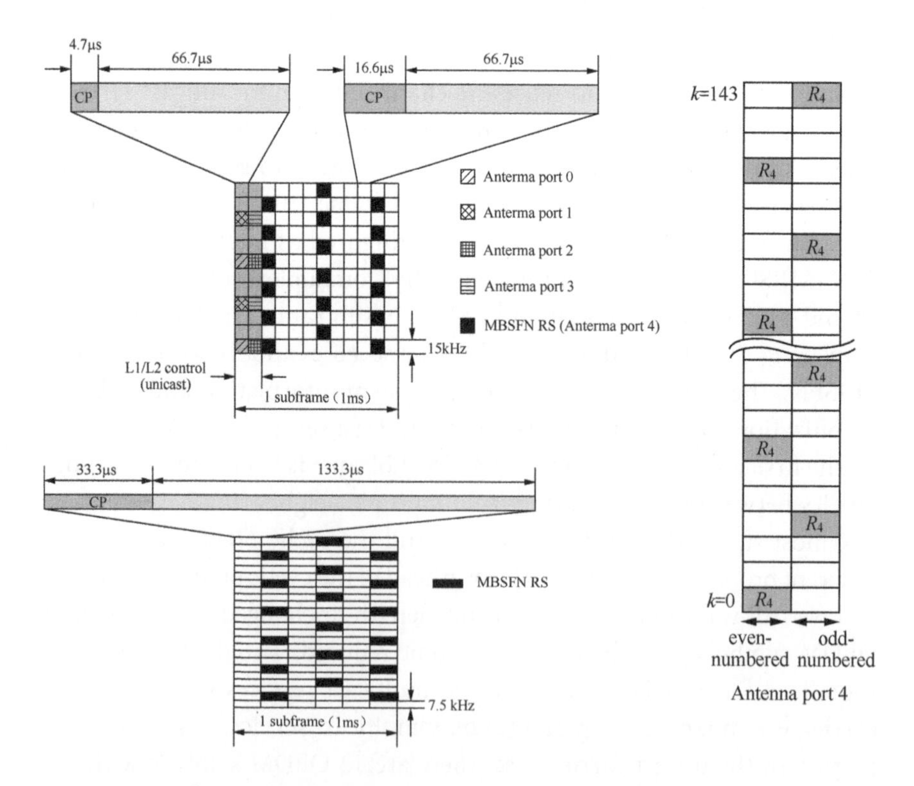

FIGURE 3.1 Three subframe structures of LTE multicast/broadcast.

As seen in Eq. (3.1), due to the different propagation delays and the nature of coherent combination, the frequency selectivity in the superposed channel is increased, meaning that the density of the demodulation reference signal should be increased in the frequency domain, as illustrated in Figure 3.1.

Figure 3.2 shows the relationship between the supported spectral efficiency (under a certain coverage percentage) and the site-to-site distance [1]. It is assumed that the system bandwidth is 10 MHz, and the spectral efficiency calculation considers various overheads such as extended cyclic prefix, multicast reference signal, physical downlink control channel, etc. It is observed that when the site-to-site distance is 500 m, 3.6 bps/Hz spectral efficiency can be supported over 95% of areas. When the site-to-site distance is increased to 1,732 m, only 1 bps/Hz of spectral efficiency can be supported over 95% of areas. This means that if the transmit power of the base stations (eNB) is kept the same, the density of eNB deployment should be increased in order to support a higher data rate for broadcast/multicast services.

Certainly, the data rate or the coverage percentage of MBSFN can be improved by increasing the transmit power of the base stations. However,

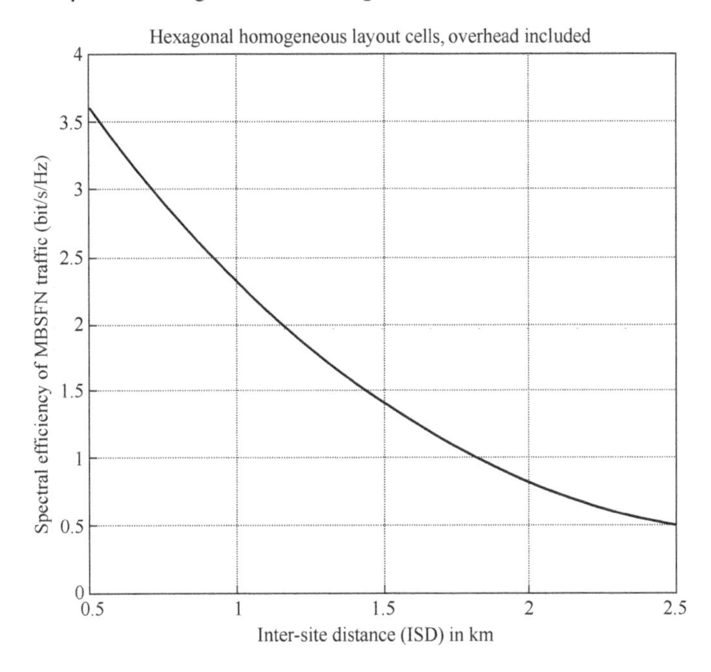

FIGURE 3.2 Supported spectral efficiency of MBSFN service (under 95% coverage) as a function of site-to-site distance.

in many situations, this may not be feasible, either from the perspective of equipment cost, power consumption and regulatory requirement or from the consideration of business models for operators. After all, MBSFN is a type of cellular service which is different from TV broadcasting services. A TV tower is usually much higher than the height of a typical base station. The transmit power of a TV station is also a magnitude higher than a base station. The spectrum of TV broadcasting is typically 700 MHz that can easily penetrate the building walls. Hence, one or two TV towers would be enough for the coverage of an entire city. This fits the pure broadcasting services. MBSFN may not follow the same operating model.

3.3 NON-ORTHOGONAL TRANSMISSION FOR BROADCAST/MULTICAST SERVICES

The principle of non-orthogonal transmission can also be applied to broadcast/multicast services where services of different data rates can share the same time and frequency resources in the same spatial dimension. Each service with a specific data rate targets a certain number of users whose UE geometry (wideband long-term SNR) is within a certain range. As Figure 3.3 shows, the basic service of lower rate targets general users and can reach the cell edge, whereas the enhanced service of higher rate targets those users that are close to the base stations.

The basic service can be audio-based programs or image-based or video-based programs with low pixel resolutions. Its subscribing fee is low and it has a large user pool. It needs to cover most of the users in the serving area. The enhanced service can be images or video-based programs with high resolutions. Its subscription fee is higher and its user pool may be small; thus, there is no need to cover most of the users in the service area. The enhanced service can also be found in event-driven hotspots such as sports stadiums, outdoor concerts and big gatherings. In this case, the coverage

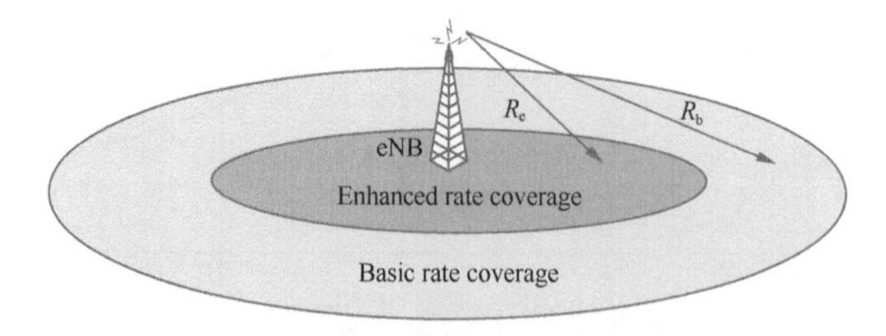

FIGURE 3.3 An example of two-layer non-orthogonal transmission for broadcast/multicast services.

area of the enhanced service is limited to those temporary spots with the purpose of providing high-quality broadcast/multicast services.

3.4 PERFORMANCE EVALUATION VIA SIMULATION

The physical channel of broadcast/multicast services does not support CSI feedback, nor the ACK/NACK feedback. Hence, link adaptation or HARQ retransmission is not possible. As far as the single link is concerned, its transmission rate in fading channel is far less than the ergodic rate using the Shannon formula for each fading realization. Hence, it is not very meaningful to use channel capacity to quantify the performance of broadcast/multicast services. In practice, outage probability is often used, that is the block error rate given an averaged SNR. This averaged SNR reflects the average gain of the fast fading channel over a long duration of time (not very short) and thus of long-term sense.

Broadcast/multicast services are one-to-many transmissions. The potential number of users is unlimited, and hence there is no clear notion of system capacity. The system performance of broadcast/multicast is often measured in the percentage of coverage for a service of a particular rate. In cellular systems, broadcast/multicast services are typically deployed in homogeneous networks. The base stations participating in MBSFN broadcast/multicast form a cluster whose size is configurable [2]. As Figure 3.4 shows, the cluster on the left is composed of a central base station, surrounded by a ring of six base stations, with a total of seven base stations. The cluster on the right is

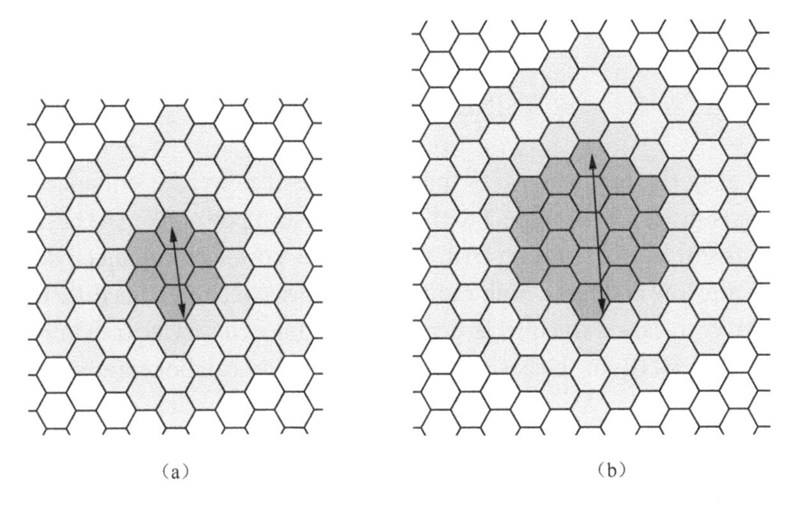

FIGURE 3.4 Clusters of base stations with different sizes to support MBSFN. (a) 7 base stations and (b) 19 base stations.

composed of a central base station, surrounded by two rings of base stations. The inner ring has six base stations, and the outer ring has 12 base stations. Hence, there are a total of 19 base stations in this cluster. The larger the cluster is, the less edge effect would be observed. Hence, the SNR would be higher and the coverage would be better. However, larger cluster deployment puts more stringent requirements for system operations, for instance, precise time synchronization across a large number of base stations and less propagation delays in the backhaul network. Nevertheless, the layout in Figure 3.4b with wrapped around is assumed when evaluating the performance of broadcast/multicast services below.

With the lack of CSI feedback, transmitter-side spatial domain precoding cannot be used to support closed-loop multiple-input multiple-output (MIMO). No rank information can be fed back from the terminal, and thus open-loop MIMO is not possible either. Hence, a single-antenna port or two antenna points are assumed for broadcast/multicast services with rank=1. Link-to-system mapping is based on block error rate (BLER) vs. long-term average SNR, that is, in order to satisfy the 1% block error rate, what is the required average SNR. The link-to-system mapping considers the impact of frequency selectivity of MBSFN transmission on the performance. For the transmission of enhanced service, it is assumed that the interference due to basic service transmission can be completely canceled at the receiver, the average signal-to-interference and noise ratio (SINR) for the enhanced service transmission can be represented as

$$\mathrm{SINR}_{\mathrm{enh}} = \frac{\alpha P_t \sum_{i=1}^{N} L_i^2}{N_{\mathrm{thermal}}} \tag{3.3}$$

In Eq. (3.3), the parameter α is the percentage of power for enhanced service transmission with respect to the total transmit power. For the basic service transmission, assuming that the transmit power of enhanced service is typically low and has a high modulation order/code rate, it is difficult for its receiver to cancel the interference due to enhanced service transmission. The average SINR for the basic service transmission can be represented as

$$\mathrm{SINR}_{\mathrm{base}} = \frac{(1-\alpha) \sum_{i=1}^{N} L_i^2}{N_{\mathrm{thermal}} + \alpha \sum_{i=1}^{N} L_i^2} \tag{3.4}$$

TABLE 3.1 Simulation Parameters for Broadcast/Multicast Services

Parameter		Value
Cell layout		Homogeneous hexagonal, 19 wrap-around
Carrier frequency		2 GHz
System bandwidth		10 MHz
Site-to-site distance		500 m
Pathloss model		ITU Uma
Indoor user ratio		80%
Transmit power of the base station		46 dBm
Average number of users		10, uniformly distributed
Antenna configuration		Number of transmitting antennas at the base station: 2 Number of receiving antennas at the terminal: 2
Antenna height at the base station		25 m
Antenna height at the terminal		1.5 m
Base station antenna gain + cable loss		14 dBi
Terminal anenna gain		0 dBi
Noise factor		9 dB
Shadow fading correlation	Between base station	0.5
	Between cells	1.0
Distance correlation of shadow fading		50 m
Mini distance between the base station and the terminal		25 m

Table 3.1 lists the simulation parameters for broadcast/multicast channels [3].

Two cases are simulated: (1) there is only a single-layer transmission for broadcast/multicast, e.g., all the transmit power allocated to the basic service, and (2) two-layer non-orthogonal transmission, one for the enhanced service with a higher rate and the other for the basic service with a lower rate. Five transmit power ratios are considered: [10%:90%], [20%:80%], [30%:70%], [40%:60%] and [50%:50%]. The CDF curves of wideband SINR for single-layer transmission of broadcast/multicast are shown in Figure 3.5. It is seen that most users would see wideband SINR higher than 15 dB since the signal only suffers from thermal noise.

Figure 3.6 shows the CDFs of the wideband SINR for the enhanced-layer transmission at different power partitions. The observation aligns with the intuition: the CDFs move to the right as more transmit power is allocated to the enhanced layer.

Figure 3.7 shows the CDFs of the wideband SINR for basic layer transmissions at different ratios of power. It is observed that when the power allocated to the basic layer is decreased from 90% to 50%, the CDFs of its SINR move quickly to the left.

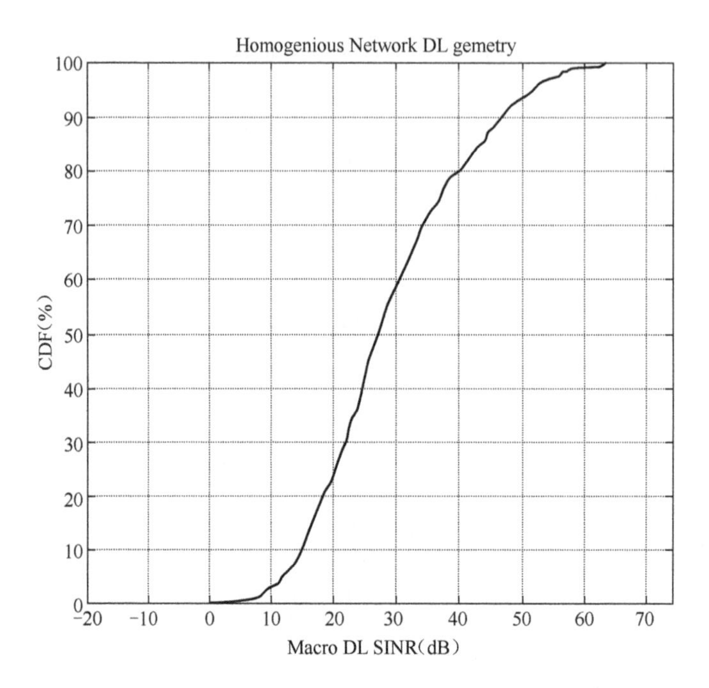

FIGURE 3.5 CDFs of wideband downlink SINR for single-layer broadcast/multicast service.

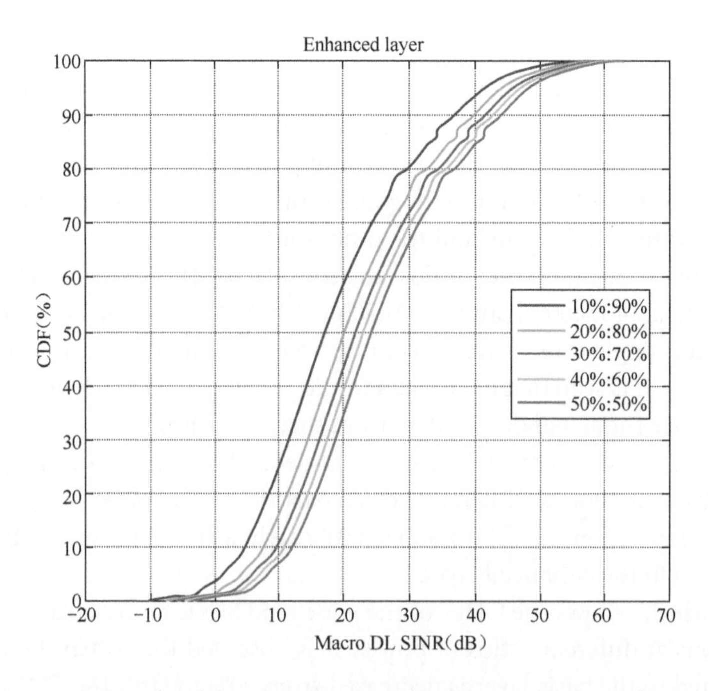

FIGURE 3.6 CDFs of wideband SINR when the enhanced layer is allocated with different powers.

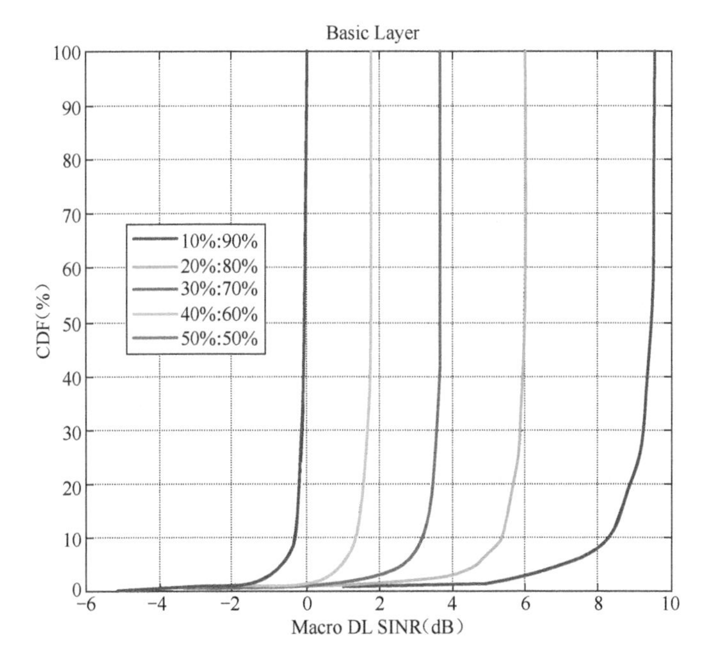

FIGURE 3.7 CDFs of wideband SINR for the basic layer at different power ratios.

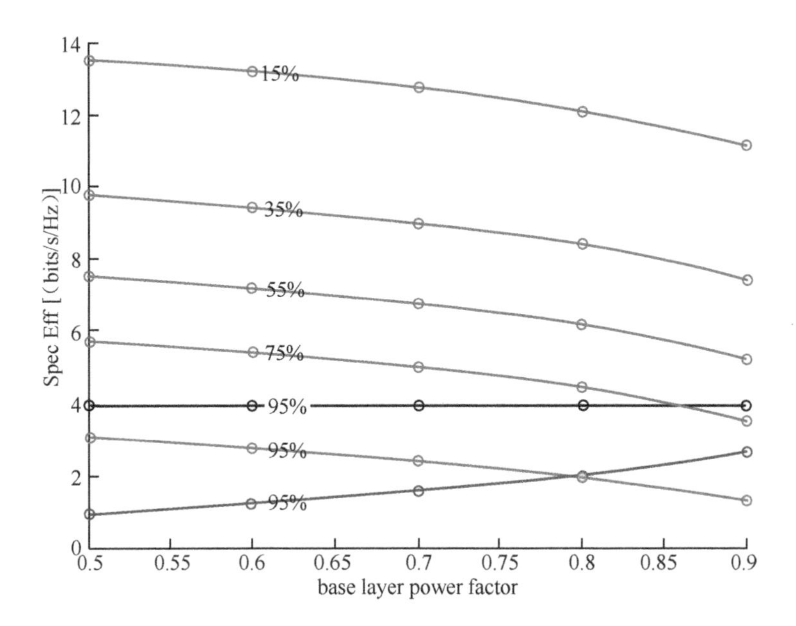

FIGURE 3.8 Spectral efficiencies and coverage percentage under different power partitions.

Several curves of spectral efficiency vs. the power allocation of the basic layer are shown in Figure 3.8. The curves correspond to three cases. The black curve is for single-layer transmission. Since the transmit power is entirely allocated to the basic layer, the curve remains constant in Figure 3.8 which is 4 bps/Hz with 95% coverage.

Blue curves in Figure 3.8 are for basic layer transmission, whereas red curves are for enhanced layer. When 50% power is allocated to the basic layer, its spectral efficiency is only about 0.8 bps/Hz in order to support 95% coverage. At this time, the spectral efficiency of the enhanced layer is 3.1 bps/Hz to support 95% coverage, and 13.5 bps/Hz if 15% of areas need to be covered. When more power is allocated to the basic layer, e.g., 80%, the spectral efficiency of the basic layer is increased to 2.1 bps/Hz with 95% coverage. At the same time, 12.2 bps/Hz can be achieved for the enhanced layer, with 15% coverage. This reflects that when the coverage requirements for the basic service and the enhanced service are different, the near-far effect is created by superposition transmission of the basic layer to the far users and the enhanced layer to the near users. Hence, the total spectral efficiency is increased. Figure 3.8 provides a reference to the operators to balance the coverage and the service quality in downlink non-orthogonal multiple access for broadcast/multicast services.

REFERENCES

1. Y. Yuan, *Key Technologies and System Performance of LTE-Advanced.* People Telecommunications Press: China, 2013.
2. 3GPP, RP-150860, Motivation on the study of PMCH using MUST, MediaTek, RAN#68, June 2015, Malmo, Sweden.
3. 3GPP, RP-150979, Multi-rate superposition transmission of PMCH, ZTE, RAN#68, June 2015, Malmo, Sweden.

CHAPTER 4

Standardization of Downlink Superposition Transmission

Jianqiang Dai and Yifei Yuan

In 3GPP, long-term evolution (LTE) multiuser superposition transmission (MUST) study item was completed in December 2015 [1], followed by the standardization work in Rel-14 [2] with the deployment scenarios extended to

Case 1: The same precoding matrix in the spatial domain for paired users, suitable for scenarios with a small number of transmit antennas

Case 2: Transmit diversity for paired users, suitable for scenarios with two transmitting antennas

Case 3: Different precoding matrices can be used for paired users, suitable for scenarios with many transmitting antennas

Among the three cases, Case 1 was the most important scenario in the study item of Rel-13 MUST. Case 2 refers to transmission mode (TM) with spatial domain diversity, e.g., LTE TM3 in 3GPP terminology. Compared to Case 1, Case 2 represents just another TM while still maintaining the same Gray mapping property as in Case 1. Hence, Case 1 and Case 2 are grouped in the same family. Note that Gray mapping is only meaningful in Case 1 and Case 2. There are two key specification aspects for Case 1

DOI: 10.1201/9781003336167-4

and Case 2: (1) the transmitter-side processing to ensure Gray mapping property in the composite constellation; (2) downlink physical layer control signaling to efficiently support MUST.

Case 3 was made explicit and added to the scope during the work item stage of Rel-14 MUST. Due to the different rotations of each user's constellation by different spatial domain precoders, the shape and the bit mapping of the composite constellation can be arbitrary. This type of superposition is quite different from the original non-orthogonal multiple access (NOMA) principle in Case 1. The essence of specifying Case 3 is to enhance multiuser multi-input multi-output (MU-MIMO) with the focus on control signaling enhancements of MU-MIMO. This reflects the strong coupling between MU-MIMO and NOMA mentioned in Chapter 2. Since Case 3 has nothing to do with Gray mapping, its specification is quite different from the specification of Case 1 and Case 2, which is to be further elaborated on in this chapter.

In the study item of MUST, there was a discussion on channel state information (CSI) feedback enhancement in order to more accurately reflect the signal-to-interference plus noise ratio (SINR) in the presence of interference between near and far users. However, since CSI feedback is often discussed and specified in the MIMO agenda item, CSI feedback-related issues were of low priority and not considered in the work item stage.

4.1 MERGED SOLUTION OF DOWNLINK NOMA

4.1.1 Unification of MUST Category 2

MUST Category 1, e.g., direct superposition, is not able to ensure Gray mapping property in the composite constellation, thus requiring more advanced receivers such as code-word-level interference cancelation (CWIC) in order to achieve good performance. The implementation complexity of CWIC, as well as the processing delay and control signaling overhead, is too high for mobile terminals. Hence, MUST Category 1 was put in low priority.

Due to its ability to maintain Gray mapping property in the composite constellation, hence allowing the use of less advanced receivers, MUST Category 2 and MUST Category 3 became the mainstream in MUST study item. MUST Category 2 supports a more flexible power partition between the near and the far users, which is helpful in maximizing the system throughput. MUST Category 3 is a bit-partition-based solution and its supported power partition choices are not as many as MUST Category 2. In the first type of MUST Category 3, only one power partition is defined for each combination of modulation between the near and the far users, rendering the composite constellation to be one of the legacy constellations supported by LTE, e.g., 16QAM, 64QAM and 256QAM. This type can be considered as a special case for MUST Category 2 with more constrained choices of power allocation. The second type of MUST Category 3 emphasizes the bit mapping to modulation symbols. Although certain adjustment is involved in MUST Category 2, it does not fundamentally change the constellation layout of the near and the far users. Its standards impact is not huge. However, for the second type of MUST Category 3, a new composite constellation needs to be defined, as well as channel quality indicator reporting, etc. Therefore, it was not adopted in the Rel-14 MUST specification.

In Rel-14 MUST standardization, the work on the transmitter side focuses on how to achieve MUST Category 2 in Case 1 (including also the first type of MUST Category 3). The candidate solutions are mirror transformation, inclusive OR, table look-up, etc.

4.1.2 For Case 1 and Case 2, the Modulation Order of Far User Is Limited to QPSK

From numerous simulation studies, it is found that most of the time, the far user is scheduled with quadrature-phase shift keying (QPSK), as shown in Table 4.1 for three ranges of power allocations.

Successful decoding of the far user's signal would not only benefit the far user but also help the receiver of the near user to carry out the interference cancellation. When the far user is QPSK modulated, the near user can use a relatively simple receiver such as symbol-level interference cancelation to detect the far user's signal and thus increase the robustness of the NOMA system. Restricting the far user to QPSK only would also simplify the transmitter-side processing. As discussed in Chapter 2, no matter whether to use mirror transformation or bit inclusive OR methods, the representation of MUST Category 2 becomes much simplified. QPSK restriction for the far user can also simplify the downlink physical control signaling.

TABLE 4.1 Probabilities of QPSK for Far User

Power Ratio Ranges for Far User	Percentage of QPSK (%)	Percentage of Non-QPSK (%)
0.65–0.7	100.00	0
0.75–0.8	99.6	0.4
0.85–0.9	97.61	2.39

With such restrictions, the possible combinations of modulation are limited to three: QPSK+QPSK, QPSK+16QAM, and QPSK+64QAM (note that in LTE Rel-14, 256QAM is not generally supported for downlink).

For Case 3, due to the potentially different precoders for paired users which may not exhibit near-far effect, there is no restriction on the far user regarding the modulation.

4.1.3 Power Allocation for Case 1/Case 2, and Finalizing the Solution

Since the transmit power allocation for the near and far users would affect the composite constellation, the discussion of power allocation should balance both technical merits and business interests. From a purely technical point of view, more flexible power allocation is beneficial to the system performance, as illustrated in Figure 4.1 [3]. Here, the performance gain of MUST Category 2 over orthogonal multiple access (OMA) is shown, under a single power ratio (one for each modulation combination) and multiple power ratios, e.g., step size of 0.02, with a range of [0.7, 0.99]. When a single power ratio is used, the NOMA gain is about 12.5% for cell average and 18% for cell edge. When multiple power ratios are used, the NOMA gain is about 16% for the cell average and 28% for the cell edge.

While flexible power allocation helps to maximize the system throughput, the too-flexible allocation would not only bring diminishing performance benefits but also cause some issues:

- Increased overhead for control signaling. Normally the power allocation needs to be informed to the terminal if multiple allocations are possible for each modulation combination. Otherwise, the terminal

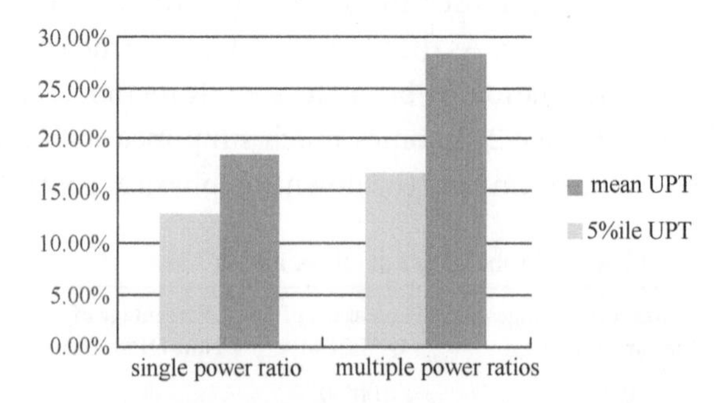

FIGURE 4.1 MUST Category 2 performance gain over the OMA system under single or multiple power allocations.

needs to perform blind decoding to figure out the actual power allocation, which increases the complexity of the receiver and is not engineering feasible.

- The complexity of the scheduler is related to the flexibility of power allocations. As seen in Section 2.2.4, the scheduler needs to search through all the combinations of power combinations. In many cases, MUST would operate jointly with MU-MIMO, thus further increasing the burden on the transmitter.
- As the flexibility of power allocation is beyond a certain level, the composite constellation may have many patterns and hardly be represented in a limited number of look-up tables. In another word, only mirror transformation or bit inclusive OR methods can efficiently represent the key processing for MUST Category 2 in the specification. This poses certain intellectual property right (IPR) risks to some companies, making them difficult to get around the IPRs of mirror transformation or bit inclusive OR methods.

After extensive discussions, a compromise was reached. That is, for each modulation pair, only 3 power ratios are to be defined, leading to two bits for power allocation indication. In Section 2.4, a wider range of power ratios are listed which can ensure no overlapping in the composite constellation and facilitate the use of less advanced receivers. During the work item stage, it is observed from the simulations that the range of power ratio can be further narrowed down to [0.7–0.95] and this range is applicable for all the modulation pairs.

The next step of the work is to decide the exact three power ratios for each modulation pair. There are two schemes. In the first scheme, several specific ratios are picked within the range of the power ratio so that the composite constellation would be a subset of a constellation of a uniform grid [4]. Its design principle is to select several constellation points from a super constellation consisting of even-spaced points, as listed in Table 4.2. The highest modulation order of near users is 64QAM and the modulation order of the far user is fixed to QPSK. The composite constellation has up to 256 points. Hence, the super constellation is a 16-by-16 grid.

In the second scheme, there is no restriction on the uniform grid for the composite constellation. Rather, the values of three power ratios can be equally spaced, to maximize the performance.

TABLE 4.2 Power Ratio Choices (the Power Ratio Values Are Adopted by 3GPP Standards, but the Way of Selection of Points of Super Constellation Is Not Adopted)

LTE Constellation	Composite MOD	Near UE MOD	Far UE MOD	Allocated Subset of Constellation Points Along One Dimension	Power Ratio Normalized	Composite Constellation Scaler [-]
16QAM	4	2	2	±1, ±3	0.8/0.2	$\sqrt{10}$
64QAM	4	2	2	±3, ±7	0.86207/0.13793	$\sqrt{58}$
64QAM	4	2	2	±3, ±5	0.94118/0.058824	$\sqrt{34}$
64QAM	4	2	2	±1, ±5	0.69231/0.30769	$\sqrt{26}$
64QAM	6	4	2	±1, ±3, ±5, ±7	0.7619/0.2381	$\sqrt{42}$
256QAM	6	4	2	±5, ±7, ±9, ±11	0.92754/0.072464	$\sqrt{138}$
256QAM	6	4	2	±3, ±5, ±7, ±9	0.87805/0.12195	$\sqrt{82}$
256QAM	6	4	2	±1, ±5, ±9, ±13	0.71014/0.28986	$\sqrt{138}$
256QAM	8	6	2	±1, ±3, ±5, ±7, ±9, ±11, ±13, ±15	0.75294/0.24706	$\sqrt{170}$

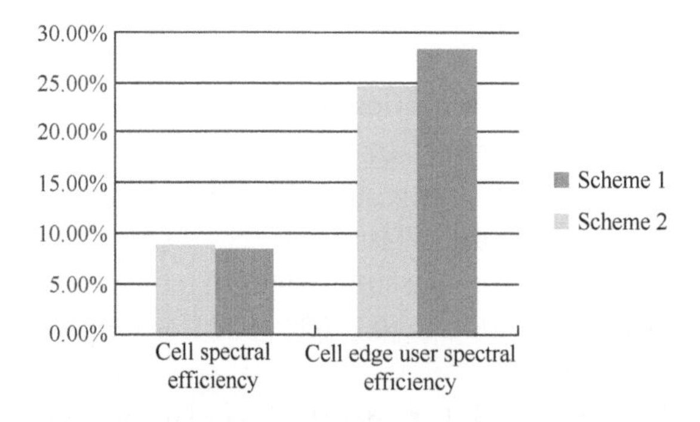

FIGURE 4.2 Performance gain of MUST Category 2 over OMA, Scheme 1 (equal distance of constellation points) vs. Scheme 2 (equal distance of power ratios).

In Figure 4.2, the system throughputs of MUST Category 2 using these two schemes are compared and the gains are respective to OMA [5]. Here the full buffer traffic is assumed, with wideband scheduling. It is observed that Scheme 1 (equal distance of composite constellation points) slightly outperforms Scheme 2 (equal spacing of power ratios) from the perspective of average cell throughput. Scheme 2 has significantly better performance than Scheme 1 in terms of cell edge throughput. If EVM is considered, the composite constellation of Scheme 1 is the legacy constellation of LTE, and thus can ensure the minimum distance between constellation points would not be less than that of 256QAM of LTE. However, in Scheme 2, its

TABLE 4.3 Power Ratios Adopted by Rel-14 MUST in 3GPP Specifications

Power Ratio Index	Modulation Order		
	QPSK	16QAM	64QAM
01	8/10	32/42	128/170
10	50/58	144.5/167	40.5/51
11	264.5/289	128/138	288/330

composite constellation is more flexible. In some cases, it has an impact on the legacy requirement for EVM. For instance, when a cluster of constellation points are more concentrated around the cluster center, the cell edge rate would be increased. If the legacy EVM, e.g., 4% is reused, constellation points within the cluster may deviate too much from their noise-free positions, thus degrading the performance of near users. With these considerations, the power ratios suggested by Scheme 1 were adopted by 3GPP, as listed in Table 4.3. Note that Scheme 1 itself was not adopted due to its potentially big impact on the specifications, for example, a new set of bit mapping tables would be introduced. Note that similar to the related chapters in TS 36.211 [6], the numbers in Table 4.3 are all exact values without rounding errors.

In LTE MUST standards, the way to achieve Gray mapping property in the composite constellation is specified as

$$x = e^{j\phi_0\pi} c(I-d) + e^{j(\phi_1+1/2)\pi} c(Q-d) \tag{4.1}$$

In the above equation, I and Q represent the in-phase and quadrature-phase components of the near user's constellation, respectively, which correspond to the legacy QPSK, 16-QAM and 64-QAM, respectively. $\phi_0, \phi_1 \in \{0,1\}$ represent the bit mapping of legacy QPSK. The term $e^{j(\phi_1+1/2)\pi}$ is essentially a flipping function along the axis of the coordination, e.g., mirror transformation. Parameters c and d depend on the power allocation. More specifically, parameter c represents the distance between constellation points within a cluster, e.g., the square root of the transmit power of the near user. Parameter d is related to the modulation pair and also depends on parameter c. The values of c and d are specified in TS 36.211 and listed in Table 4.4.

For example, when the power ratio index is 01 and the near user is QPSK modulated, if the far user has two bits "00" to transmit and the near user has two bits "01" to transmit, the values of parameters $\phi_0, c, I, d, \phi_1, Q$,

TABLE 4.4 Values of Parameters *c* and *d* in the Specification

Power Ratio Index	Modulation Order					
	QPSK		16QAM		64QAM	
	c	*d*	*c*	*d*	*c*	*d*
01	$\sqrt{1/5}$	$\sqrt{2}$	$\sqrt{5/21}$	$2\sqrt{2/5}$	$\sqrt{21/85}$	$4\sqrt{2/21}$
10	$2/\sqrt{29}$	$5/(2\sqrt{2})$	$3\sqrt{5/334}$	$17/(3\sqrt{10})$	$\sqrt{7/34}$	$3\sqrt{3/14}$
11	$7/\sqrt{1/578}$	$23/(7\sqrt{2})$	$\sqrt{5/69}$	$4\sqrt{2/5}$	$\sqrt{7/55}$	$2\sqrt{6/7}$

would be $1, \sqrt{1/5}, 1/\sqrt{2}, \sqrt{2}, 1, -1/\sqrt{2}$, respectively. Substituting them to Eq. (4.1), we can get $x=1/\sqrt{10}+j\cdot 3/\sqrt{10}$.

Hence, the transmitter-side processing for MUST is specified in the form of flipping along the axis of coordination or mirror transformation. This outcome merges the flexible power allocations of Category 2 and the engineering feasibility of the first type of MUST Category 3 and provides a good trade-off between the system performance, the overhead of control signaling, the scheduling complexity, the radio-frequency requirements and the conciseness of specifications.

In fact, from the power allocations listed in Table 4.3, the values of parameters c and d in Eq. (4.1) can be derived. For instance, the power ratio index is 01, the near user is QPSK modulated and the "00" bits of the far user correspond to "11" bits of the near user. The superposed signal corresponds to "0011" in 16QAM, i.e., $\sqrt{1-c^2}\cdot 1/\sqrt{2}+c\cdot 1/\sqrt{2}=3/\sqrt{10}$. Then, we get $c=\sqrt{1/5}$. Substituting ϕ_0, c, I, ϕ_1, Q, into Eq. (4.1), we can calculate $d=\sqrt{2}$. Note that Table 4.4 is illustrated in a chapter in TS 36.211 [6] that is different from the chapter for Table 4.3. Hence, to facilitate the reading of specifications, parameters c and d are explicitly listed.

4.2 BRIEF INTRODUCTION OF DOWNLINK PHYSICAL CONTROL SIGNALING FOR MUST

During the study item phase, many types of MUST categories were considered. There are no limitations on the receiver types, resource scheduling and modulation for far users. Because of this, downlink physical control signaling to support MUST can include much potential assistance information. There is a strong relationship between the MUST study item and the previous work item of network-assisted interference cancellation (NAIC). Many terminologies used in MUST are borrowed from NAIC, as well as some design principles for control signaling.

4.2.1 Identified Potential Assistance Information during the Study Item Phase

First of all, the basic information which is also needed even for the least advanced receiver such as MMSE-IRC includes

- The transmit power allocations of the physical downlink shared channel (PDSCH) for the far user and the near user

In addition, reduced maximum likelihood (ML) or symbol-level interference cancelation (SLIC) receiver can be used for either near or far users, and these types of receivers require the following assistance information:

- Presence or absence of the interference of the paired use, this may refer to each spatial layer
- Modulation order of the pair user
- Resource allocation of the paired user
- Demodulation reference signal (DMRS) information of the paired user
- TM of the paired user (if one user is with transmit diversity and the other is with closed-loop spatial multiplexing)
- Precoding matrix of the paired user

For the near user, a more advanced receiver such as CWIC can also be considered. For this type of receiver, in addition to the above-listed assistance information, more information is needed in order to decode and reconstruct the signal of the paired user so that it can be canceled:

- Transport block size of the paired user
- HARQ information of the paired user
- Assumption of limited buffer rate matching of the paired user
- Scrambling parameter of PDSCH of the paired user

For the second type of MUST Category 3, in addition to all the above assistance information, the following information is also needed:

- Modulation order of the composite constellation
- Bit mapping of the composite constellation

4.2.2 Criteria for Downlink Control Signaling Design

When MUST entered the work item stage, the consensus was built around the following design criteria for downlink control signaling design.

Criterion 1: MUST operation transparent to the far user

The power allocated to the near user is typically much less than that of the far user. It is very difficult for the far user to successfully decode the near user's signal and cancel it, even if advanced receivers are used. Hence, a less advanced receiver, such as MMSE-IRC, would be enough. During the early deployment of MUST, it is expected that there would be a large portion of legacy terminals which are not MUST-capable. For more users to participate in non-orthogonal transmission, it is desirable to pair a legacy user as the far user with a MUST-capable user as the near user. This would significantly expand the user pool for MUST and make it easier to pair users and improve the system throughput. Transparency to the user means that the far user would operate by following the legacy control signaling, without knowing that there is another paired user transmitting in the same time and frequency resources.

Criterion 2: to support dynamic switching between orthogonal transmission and MUST

The fast fading channel of each user is dynamically changing. The interference in each subframe would not be the same. The traffic arrival is also random and dynamic. Hence, whether the situation is suitable for MUST pairing, or which two users are to be paired, is dynamically changing. According to the description of the scheduler in Chapter 2, the capacity of non-orthogonal transmission is not always higher than orthogonal transmission at each time and for each sub-band. Dynamic pairing means that the power ratio is also changing. Therefore, user pairing information and power ratio information need to be carried in downlink control indication (DCI)

Criterion 3: The far user and the near user each should have their own DCI

Normally, the downlink wideband signal-to-noise ratios (SNRs) of the far user and the near user are quite different. If these two users share a common DCI, the information carried in this common DCI would be significantly more than that in the legacy DCI. For the far user, much assistance information in the common DCI is not necessary. Sending it to the far user would cause resource waste, especially considering that the SNR of the far user is low, and more physical resources (e.g., higher aggregation level) and power are required for PDCCH/EPDCCH transmission. Transmitting separate DCIs for different users can make the downlink control signaling design more flexible.

Criterion 4: not to introduce a new TM

Non-orthogonal transmission is a general technique that can be used in conjunction with different TMs (e.g., different antenna technologies). No matter whether Case 1 or Case 3, non-orthogonal transmission is not tied to a specific multi-antenna feature. In addition, there are ten TMs already specified in LTE, each having its own DCI format. It is infeasible to combine all these DCI formats into a new DCI format that is specifically used for MUST. Hence, there is no need to introduce a new TM for MUST or a new DCI format. Instead, it is more reasonable to add certain bit fields on top of each legacy DCI format. Higher-layer signaling can be used to activate the use of these "enhanced" DCI formats. Such practice is not unprecedented, e.g., the specification of NAIC followed a similar approach.

There are two options to add assistance information to legacy DCI formats, as illustrated in Figure 4.3. The first option is to take

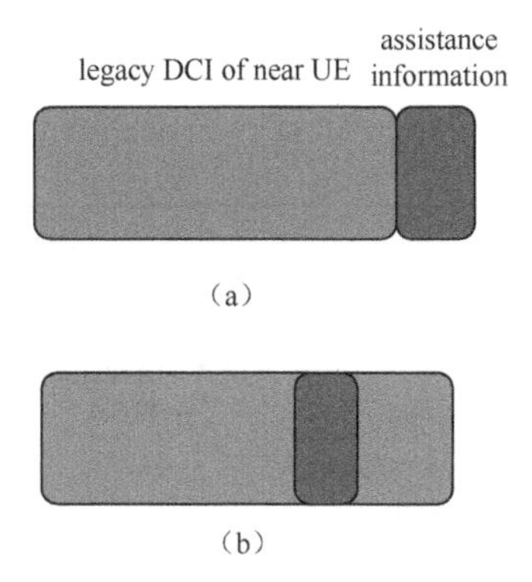

FIGURE 4.3 Two options to add assistance information to the legacy DCI formats. (a) Add the assistance information as a new bit field, without changing the legacy bit fields in DCI formats. (b) Assistance information embedded in DCI formats by re-defining certain bit fields.

the assistance information as a new bit field to be added to the legacy DCI. The new bit field would include the power ratio index and some information about the interfering user. The advantage of this option is the simpler design and the similar practice for all the DCI formats. Its disadvantage is the relatively large overhead. The second option is to re-define some bit fields in the legacy DCI formats. Its merit is less control signaling overhead. If some bit fields are reserved, the second option does not incur any overhead increase. However, its design is more intricate and requires customized redefinition for each DCI format. The specification work would be challenging, considering the vast difference between different DCI formats.

Regarding the option in Figure 4.3, here is an example of how it is actually designed. Assuming that the precoding matrix indicator (PMI) field in

TABLE 4.5 Bit Field Definition for Legacy Two Transmit Antennas

Single Code Word Code Word 0 Enabled, Code Word 1 Disabled		Double Code Word Code Word 0 Enabled, Code Word 1 Enabled	
Bit Field Mapped to Index	**Message**	**Bit Field Mapped to Index**	**Message**
0	2 layer: Transmit diversity	0–2	2 layer: precoding information
1–6	1 layer: precoding information	3–7	Reserved
7	reserved		

TABLE 4.6 Re-Defined Bit Field for PMI of Two Transmit Antennas (Candidate Solution, Not Specified)

Single Code Word Code Word 0 Enabled, Code Word 1 Disabled		Double Code Word Code Word 0 Enabled, Code Word 1 Enabled	
Bit Field Mapped to Index	**Message**	**Bit Field Mapped to Index**	**Message**
0	2 layer: Transmit diversity	0–2	2 layer: precoding information, *first layer*
1–6	1 layer: precoding information	3–5	2 layer: precoding information, *second layer*
7	reserved	6–7	2 layer: precoding information, *first and second layers*

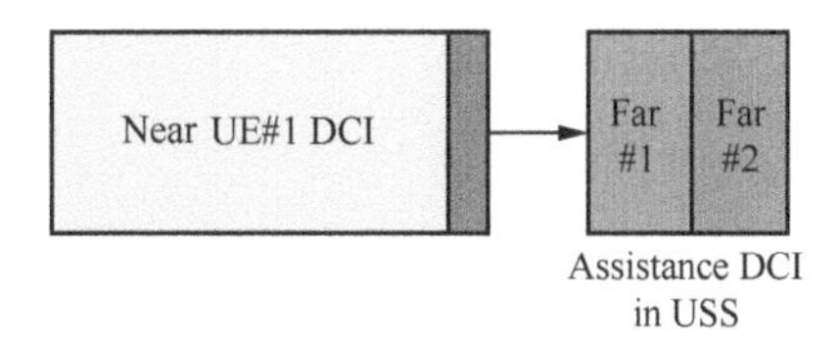

FIGURE 4.4 Two-step DCI for MUST signaling.

DCI 2 can be re-defined to indicate the presence of interference. Table 4.5 shows the definition of the legacy PMI bit field. Table 4.6 shows the re-defined PMI bit field.

In the case of a single code word, there is a single layer for transmission. If there is interference, the interference would be in that layer. In the case of a double code word, the range of values for "bit field mapped to index" is 0–2. This bit field not only can indicate the precoder information but also the interference in the first layer.

Similarly, the bit field of DCI format 2A under the 4-port PMI and "antenna ports, scrambling identity and number of layers indication" in DCI format 2C can also be re-defined in this manner to support additional indications for MUST.

The above four criteria are not independent and should be used jointly. For example, the far-user transparency in Criterion 1 facilitates the dynamic switching between OMA and NOMA in Criterion 2. Criterion 1 also implies Criterion 3, e.g., far user and near user would not share a common DCI. All these four criteria suggest no new signaling for far users, i.e., the assistance information would be defined only for the near user.

Apart from the above design criteria, a few detailed solutions were also discussed. For instance, two DCIs can be designed for the near user. In addition to blindly decoding the legacy DCI (the leftmost part in Figure 4.4), the near user also needs to decode the small DCI [7] (the second leftmost part in Figure 4.4). This small DCI can indicate the location and size of assistance information. It can be in the common search space or user-specific search space. The advantage of this solution is being able to indicate a large amount of assistance information by using less control overhead. However, in general, the essential assistance information for MUST operations is not big.

In addition, group companion DCI was proposed to include all the assistance information of paired users. The assistance information includes:

- RB allocation
- TM/RI/PMI index
- Power ratio index

This group companion DCI [8] can be broadcast. This solution assumes a rather complicated user pairing situation where for instance, a near user can be paired with multiple far users. The far users are allocated a subset of the shared resources with the near user.

4.2.3 Trimming of Potential Assistance Information

In the work item stage, the focus of transmitter-side processing is MUST Category 2 and the first type of MUST Category 3. Regarding the near user, a majority of terminal vendors believe that code word-level interference cancelation (SWIC) is too complicated. Hence, only the first six items of assistance information as well as the power ratio would be the focus for specification.

As the work item of MUST progressed, some of these seven items are also crossed out. For example, since the modulation order of the far user is fixed to QPSK in Case 1 and Case 2, there is no need to indicate the modulation order of the far user to the near user.

There are many ways of resource sharing for the far user (paired) and the near user (target). As illustrated in Figure 4.5, the physical resources of the far and the near user are complicated and overlapped in the left figure. Although it is not very flexible, such resource sharing can simplify the control signaling. That is, to the near user, there is only one paired user in this cell. Such sharing is also assumed in the scheduler in Chapter 2. In the middle figure, two far users can share the physical resources with one of the near users. This would provide more flexibility. However, for the near user, assistance information needs to include two paired (interfering) users, resulting in more signaling overhead and more complicated

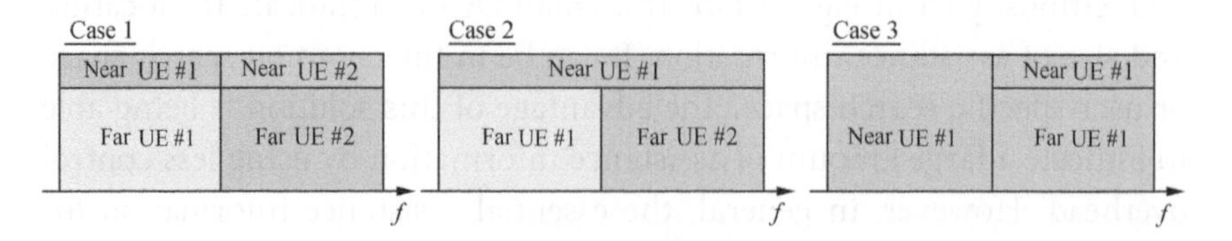

FIGURE 4.5 Several cases of resource sharing between the far user and the near user (Case 1 was adopted in the specification).

scheduling. The resource allocation in the right figure is very flexible and also requires a very big overhead for control signaling. Via comprehensive consideration of signaling overhead, the system performance and the scheduling complexity, 3GPP reached the consensus to support only the resource sharing in the left figure. That means there is no need to indicate the resource allocation of the paired user.

Regarding the resource allocation, there is another detailed aspect, e.g., the starting symbol of PDSCH. In normal operation, the far user and the near user are paired in the same subframe for scheduling. They would try to decode the same physical control format indicator channel (PCFICH) to figure out the starting symbol of PDSCH. There is no necessity to indicate this information to each other. However, the deployment of MUST may include carrier aggregation where cross-carrier scheduling would be supported. In this case, there would be a situation when the starting symbol of one user's PDSCH is indicated via PCFICH, while the starting symbol of the paired user is indicated by higher-layer signaling. These two indications may not carry the same value, causing mis-aligned resources in the time domain. To simplify the control signaling and reduce the complexity of the receiver, it is explicitly specified in TS 36.213 [9] that the starting symbol of PDSCH of the far user and the near user should be aligned.

There was a brief discussion on whether to pair a user operating in Case 1 and another user operating in Case 2. In practical deployment, such mixed use is not typical, because each has different use scenarios and with different numbers of transmit antennas. They seldom change in a dynamic manner. In addition, such paring would significantly increase the signaling overhead for the assistance information. Thirdly, it would make the scheduler more complicated. Therefore, a consensus was quickly reached that paired users should operate in the same MUST case.

The same precoder or transmit diversity should be used for the far user and the near user in Case 1 and Case 2. However, for MU-MIMO, the scheduling strategy is to pair users whose precoders have low cross-correlation, which is different from that of MUST. Hence, Case 1 and Case 2 would be mainly used for closed-loop single-user MIMO (SU-MIMO) such as common reference signal (CRS)-based TMs, e.g., TM2, TM3 and TM4, without the need for the information of DMRS of the paired user.

User pairs in Case 3 are often selected from those whose precoders are quite different, not necessarily satisfying the near-far criterion. The composite constellation would be quite arbitrary. No power allocation needs to be indicated. Nor the restriction on the modulation orders. Still, it is desirable to indicate the modulation order to reduce the complexity of the receiver.

For MUST Case 1/2 and Case 3, there was some discussion on whether to indicate the presence of interference in the assistance information or whether to solely rely on blind decoding without explicit signaling. After considering the receiver complexity and the performance of blind decoding, it was decided later to use signaling to indicate the presence of interference.

After the above trimming, the assistance information required for Case 1 and Case 2 is finally decided to include

- The transmit power allocation of PDSCH between the far user and the near user
- The presence of the interference of paired user, possibly for each spatial layer

For Case 3, the following assistance information should be indicated in the DCIs:

- The presence of the interference of paired user, possibly for each spatial layer
- Modulation order of the paired user
- DMRS information of the paired user
- PMI of the paired user

4.3 SIGNALING FOR MUST CASE 1/2

Control signaling for Case 1 and Case 2 is relatively simple, that is to add a small number of bits to the legacy DCI formats to indicate the assistance information for SLIC in the receiver of the near user. In the specification, there is no explicit mention of Case 1 and Case 2. Instead, these two deployment cases are implied in the newly introduced bit field in DCI format 1, format 2A and format 2, corresponding to TM2, TM3 and TM4, respectively. None of these three TMs supports MU-MIMO. If the feature of superposition transmission is turned on, paired users should use the same precoder.

From the system simulation results, as listed in Table 4.7, when the number of transmit antennas is 4 and TM4 is used, the gain of MUST is

TABLE 4.7 Performance Gains over OMA in TM4 with Four Transmit Antennas

Transmission Method	Resource Utilization	Mean UPT (Mbps)	5% UPT (Mbps)	50% UPT (Mbps)
SU-MIMO	0.8278	6.2457	0.7739	3.7425
MUST	0.8122	6.57985	0.82995	4.0695
Gain		5.35%	7.24%	8.74%

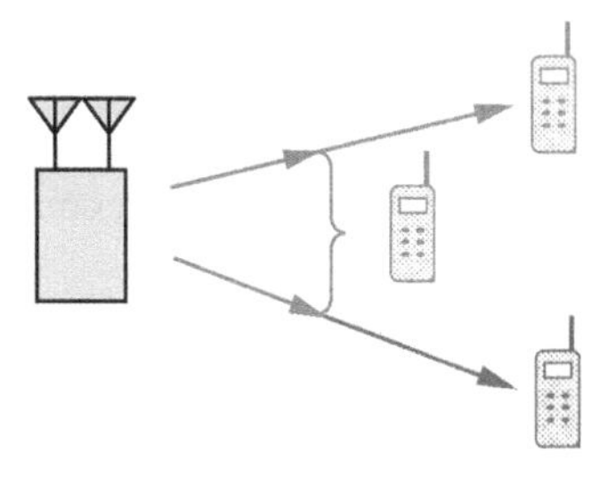

FIGURE 4.6 Near user of two spatial layers paired with two far users, each of single-spatial-layer transmissions.

not as significant as in Chapter 2 for two transmit antennas. Hence, in the specification, the maximum number of transmit antennas supported in MUST Case 1/2 is 2. This implies that the maximum number of spatial layers per user is 2.

It is observed that the power ratio can take three values. "Without the presence of interference" is also a state. There are a total of four states to be indicated, leading to exact two bits for joint indication. It is specified in Section 6.3.3 in TS 36.211 where "MUST interference presence and power ratio (MUSTIdx)"="00" denotes that no interference is present. The other three states are described in Table 4.3.

In order to be more compatible with the previous specifications, it is decided that in the MUST spec, if the near user and the far user are in two-spatial-layer transmission, the power should be equally split between the two spatial layers, which is reflected in Section 6.3.3 in TS 36.211, e.g., the scaling factor per-spatial layer "$\alpha^{(j)}$" does not depend on the spatial index "j".

This joint indication can be independently configured per each spatial layer. Figure 4.6 shows an example of an independent configuration where a near user of two-spatial-layer transmission is paired with two far users, each with the single-spatial-layer transmission. Since the SNRs of these two far users may not be the same, the power ratios may not be the same. It

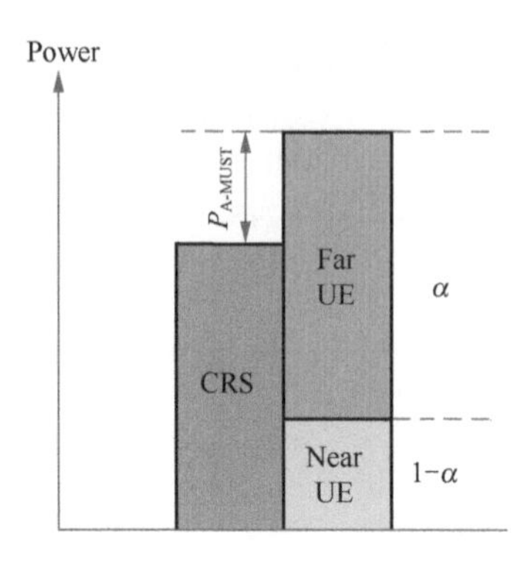

FIGURE 4.7 Newly introduced higher-layer power control parameter for MUST.

is also possible that sometimes, one of the far users (each with single-spatial-layer transmission) is not scheduled, leaving the near user interfered by only single-layer transmission. This situation is reflected in the new bit field in DCI format 2 of TS 36.211 where "MUST interference presence and power ratio (MUSTIdx)" is added to both "Transport Block 1" and "Transport Block 2".

In LTE specification, certain relationship should be maintained between the transmit power of PDSCH and the transmit power of CRS. This relationship is indicated by higher-layer configuration P_A. This parameter is used to adjust the other cell interference of PDSCH and the operating point of each user. It is a user-specific higher-layer signaling and can be different for different users. P_A can take values of {−6.00, −4.77, −3.00, −1.77, 0.00, 1.00, 2.00, 3.00} dB. In MUST, if the above values are reused, the SNR operating point for the far user would be too low. Hence, a new higher-layer parameter $P_{A\text{-MUST}}$ was introduced, as illustrated in Figure 4.7. In TS 36.213, the related parameter is "p-a-must-r14". In order to ensure that the actual transmit power of PDSCH for the far user and the near user is consistent with the power ratio indicated in DCI, the paired near and far users should have the same value of $P_{A\text{-MUST}}$.

4.4 SIGNALING FOR MUST CASE 3

The specification for MUST Case 3 is focused on the control signaling with the aim to reduce the receiver complexity in order to cancel the cross-layer interference in MU-MIMO. More specifically, via explicit signaling, the burden for blind detection can be reduced, for instance, whether there is interference, the antenna port from which the interference comes, and the modulation order of the interference so that the robustness and the performance of the systems can be improved. The progress of MUST Case 3 was relatively slow at the beginning of the work item when a series of system-level simulations were carried out to (1) verify the performance gain of MUST in Case 3, compared to OMA and (2) identify the suitable system configuration for Case 3.

Tables 4.8 and 4.9 show the system performance gains of MUST Case 3 and Case 1, for two Tx antennas and four Tx antennas, respectively. The baseline is OMA or SU-MIMO. Full buffer traffic is assumed, with wide-band scheduling. Note that MUST Case 1 can be considered as a special case of Case 3, with the constraint to pair the users whose precoder is the same. OMA or SU-MIMO here can be considered as a special case of MUST Case 1, e.g., there is no user pairing. In the system-level simulation of MUST Case 3, for the resource scheduling and user pairing, a decision should be made whether Case 3 is the best in terms of system capacity. If at a certain time instant, MUST Case 1 or OMA can offer better performance for some users, the transmission should fall back to MUST Case 1 or OMA in those situations.

According to Tables 4.8 and 4.9, MUST Case 3 can significantly improve the cell average throughput, but not the cell edge throughput. By contrast, the gain of MUST Case 1 is mainly reflected in the cell edge throughput, not in the cell average throughput.

TABLE 4.8 System Performance Gain of MUST Case 3 and Case 1, with two Tx Antennas, Full Buffer, Compared to OMA

Case	User Throughput (bps)	Baseline (SU-MIMO)	MUST Case 3	Gain (%)	MUST Case 1	Gain (%)
2Tx	Cell average	1.3240	1.6002	**20.85**	1.3886	**4.87**
	Cell edge	0.0241	0.0268	**11.31**	0.0287	**19.19**

TABLE 4.9 System Performance Gain of MUST Case 3 and Case 1, with four Tx Antennas, Full Buffer, Compared to OMA

Case	Spectral Efficiency (b/s/Hz/Sec)	Baseline (SU-MIMO)	MUST Case 3	Gain (%)	MUST Case 1	Gain (%)
4Tx	Cell average	1.5722	2.1235	**35.06**	1.6091	**2.35**
	Cell edge	0.0337	0.0358	**6.24**	0.0394	**17.07**

TABLE 4.10 Statistics in MUST Case 3 and Case 1 for two Tx Antennas

Case	Operation Mode	MUST Case 3 (%)	MUST Case 1 (%)
2Tx	SU-MIMO	5.88	40.00
	MUST with different precoding vectors	**64.05**	-
	MUST with the same precoding vector	30.07	60.00

TABLE 4.11 Statistics in MUST Case 3 and Case 1 for 4 Tx Antennas

Case	Operation Mode	MUST Case 3 (%)	MUST Case 1 (%)
4Tx	SU-MIMO	2.13	57.68
	MUST with different precoding vectors	**80.76**	-
	MUST with the same precoding vector	17.12	42.32

TABLE 4.11 Joint Coding of Interference Presence and Modulation Order

Bit field	Message
00	No interference presence
01	Interference is present with QPSK
10	Interference is present with 16QAM
11	Interference is present with 64QAM or 256QAM

Performances shown in Tables 4.8 and 4.9 can be explained by the statistics in Tables 4.10 and 4.11. In MUST Case 3, a majority of users have different precoders. As the number of transmit antennas increases, this ratio would increase to 80%, meaning that most of the users are in MU-MIMO. It is known that MU-MIMO would mainly benefit the performance of cell center users or users with moderate SNR. Hence, significant performance improvement is observed in cell average throughput. The above tables suggest that MUST Case 3 is suitable for MU-MIMO. In LTE, TM 8, TM9 and TM10 support DMRS-based PDSCH which typically operates in MU-MIMO. In contrast, a CRS-based TM cannot efficiently support MU-MIMO. Hence, in the spec., MUST Case 3 is not supported in TM2, TM3 and TM4.

In LTE, the TM is a higher layer configured in a semi-static fashion. The switching between TMs cannot be dynamic. MUST Case 1/2 is only supported in TM2, TM3 and TM3, whereas MUST Case 3 is only supported in TM8, TM9 and TM10. Hence, the dynamic switching between MUST Case3 and Case 1 assumed in the system simulation in Tables 4.8 and 4.9 is not supported by MUST specification. Those gains just reflect the upper bound of the system performance.

According to some performance analysis in 3GPP RAN4, if the DMRS antenna ports are not orthogonal, the performance of MUST Case 3 would be significantly degraded. Hence, in the MUST specification, the values of parameters $n_{\mathrm{SCID}}\, n_{ID}^{(n_{\mathrm{SCID}})}$ and orthogonal cover code (OCC) length of the paired users in MUST Case 3 would be the same. This is clearly stated in TS 36.213.

Due to various reasons, the detailed design of control signaling for MUST Case 3 had not been discussed till the end of the work item. Not many companies participated in the discussion, and many related conclusions were made in a rush. The required assistance information for Case 3, such as the presence of interference, the antenna port of the interference and the modulation order of the interference, would be tailored for specific DCI formats.

For DCI format 2B (corresponding to TM8, dual-layer beamforming), due to the small number of orthogonal DMRS antenna ports, the maximum number of bits to indicate the interference layer is 1. The signaling design for MUST Case 1/2 can be reused here, the presence of interference can be jointly coded with the modulation order. The signaling is specified in TS 36.212, as listed in Table 4.11, which is added to DCI format 2B. Such indication is per target layer. In DCI format 2B, a maximum of two antenna ports are supported, e.g., port 7 or port 8. The implicit indication is used here. When a user knows its own port, the other port would be the interference. It is noticed that the bit field "11" indicates two possible modulations. Considering the very low probability of 256QAM, such dual indication can significantly reduce the signaling overhead.

Regarding DCI format 2C and format 2D (corresponding to TM9 and TM10, respectively), the situation is a little more complicated. There are two situations: the maximum number of interfering layers is 1, and the maximum number of interfering layers is 3. The number of interfering layers depends on the user equipment capability (UE) capability and is semi-statically configured by radio resource control (RRC) signaling, e.g., higher-layer parameter k-max which is set to 1 or 3.

When the maximum number of interfering layers is 1, four new bits would be introduced, with joint coding. The two most significant bits are used to indicate the presence of interference and the antenna ports, as listed in Table 4.12, which is specified in TS 36.212. It should be pointed out that the possible antenna port set is {7, 8, 11, 13}. If the target user's own

TABLE 4.12 Joint Coding of Interference Presence and Interfering Antenna Ports

Bit field	Message
00	No interference presence
01	First antenna port
10	Second antenna port
11	Third antenna port

TABLE 4.13 Indication of Interference Presence and Interfering Antenna Ports

Antenna Port of Target UE	Port with Assistance Information			
	$B = 00$	$B = 01$	$B = 10$	$B = 11$
7	No assistance information	8	11	13
8		7	11	13
11		7	8	13
13		7	8	11

Note: When $B = 00$, UE ignores the assistance information of MOD.

TABLE 4.14 Indication of Modulation Order of Interference

Bit Field	Message
00	QPSK
01	16QAM
10	64QAM
11	256QAM

TABLE 4.15 Indication of Interfering Antenna Ports

Number of Layers of Desired Signal = 1 and OCC Length = 2		Number of Layers of Desired Signal = 1 and OCC Length = 4		Number of Layers of Desired Signal = 2 and OCC Length = 4	
Antenna port of target UE	Ordered ports with assistance information	Antenna port of target UE	Ordered ports with assistance information	Antenna ports of target UE	Ordered ports with assistance information
7	8	7	8, 11, 13	7, 8	11, 13
8	7	8	7, 11, 13	11, 13	7, 8
		11	7, 8, 13		
		13	7, 8, 11		

antenna port is 7, two bits can be used to indicate one of the four states, interference is absent, port 8, port 11 and port 13, as listed in Table 4.13.

The other two least significant bits (LSBs) are used to indicate the modulation order of the interfering signal, as listed in Table 4.14, which is also specified in TS 36.212.

When the maximum number of interfering layers is 3, six bits are introduced for the indication. Among them, two bits are used to indicate the modulation order of one interfering antenna port. If the target user's own antenna port is 7, two bits are used to indicate the modulation order of antenna port 9, two bits for the modulation order of antenna port 11, and two bits for the modulation order of antenna port 13, as listed in Table 4.15. Note that 6 bits are just an upper limit, the two or four LSB may be reserved in the case of two or one interfering layer, respectively. In some cases, not all of them would be used. For instance, as shown in Table 4.15, when there is only single-layer transmission, and the OCC length is four, the number of interfering layers is 3 and each layer can have two bits for modulation order indication. The total number of states is $2^2 \times 2^2 \times 2^2 = 64$. If the target user has two-layer transmission, and the OCC length is four with only two layers of interference, the number of states is $2^2 \times 2^2 = 16$. If the target user has only single-layer transmission, and the OCC length is 2, this would be similar to Format 2B, e.g., there are only four states, corresponding to the four modulation orders.

REFERENCES

1. 3GPP, TR 36.859, Study on Downlink Multiuser Superposition Transmission (MUST) for LTE (Release 13).
2. 3GPP, RP-160680, New Work Item proposal: downlink multiuser superposition transmission (MUST) for LTE, MediaTek, RAN#71, March 2016, Gothenburg, Sweden.
3. 3GPP, R1-164281, Multiuser superposition transmission scheme for LTE, ZTE. RAN1#85, May 2016, Nanjing, China.
4. 3GPP, R1-1609616, On standardization of the composite constellations for MUST Case 1 and 2, Nokia. RAN1#86bis, October 2016, Lisbon, Portugal.
5. 3GPP, R1-1608678, MUST system performance with different power ratio choices, ZTE. RAN1#86bis, October 2016, Lisbon, Portugal.
6. 3GPP, TS 36.211, E-UTRA Physical channels and modulation.
7. 3GPP, R1-167688, On network assistance and operation for MUST, Nokia, RAN1#86, August 2016, Gothenburg, Sweden.
8. 3GPP, R1-166278, Mechanisms for efficient operation for MUST, Qualcomm, RAN1#86, August 2016, Gothenburg, Sweden.
9. 3GPP, TS 36.213, E-UTRA Physical layer procedures.

CHAPTER 5

General Discussion of Uplink Non-Orthogonal Multiple Access

Yifei Yuan, Zhifeng Yuan, Nan Zhang, Weimin Li, Ziyang Li, Qiujin Guo and Jian Li

5.1 GRANT-FREE ACCESS

5.1.1 Scenario Analysis

The first two generations of cellular systems are mainly for voice services of low and constant rates. There, the circuit switch is used and the radio resource scheduling is limited to the radio resource control (RRC) level, e.g., semi-statically configured. Hence, the control overhead of the physical layer is very small. The drawback of semi-static configuration is the low resource utilization rate and its inability in fast link adaptation. In 3G and 4G networks, data services become dominant, especially in orthogonal frequency division multiple access (OFDMA)-based 4G, link adaptation and resource scheduling have become more mature, to sufficiently handle the channel variation due to small fading, random bursty traffic, cross-user interference, etc. so that the system performance can be further improved.

One important prerequisite to achieve high spectral efficiency via dynamic scheduling is that the packet size should be large enough, e.g., much larger than that of physical control signaling. For big data services such as file transfer and high-resolution video/audio, this condition can usually be met. However, for some data services, their packet

DOI: 10.1201/9781003336167-5

size is limited. Hence, the physical control overhead would have a noticeable impact on the overall system performance. Figures 5.1 and 5.2 are from a simulation related to uplink data transmission and the associated downlink control [1]. More specifically, Figure 5.1 shows the cumulative density function (CDF) of the resources used for the downlink physical control channel (PDCCH). Figure 5.2 is the resource utilization of uplink resources. The traffic model is FTP Model 3 of Poisson arrival, and the packet size is fixed to 600 Bytes. There are a total of 36 physical resource blocks (PRBs) in the simulation bandwidth. There are 20 users in a cell. Proportional fair scheduling is assumed, together with MU-MIMO and sub-scheduling (at the granularity of 6 PRBs).

It is observed from Figure 5.1 that when the system loading is increased from 10 to 40 packets per second, about 12% of the time, the entire 36 PRBs have to be used in order to carry the downlink physical control channels. At the same time, as seen in Figure 5.2, the uplink resource utilization rate is only about 50%. From these results, we can see that when the packet size is not very big, the increased system loading would significantly increase the control signaling overhead, thus negatively affecting the system throughput.

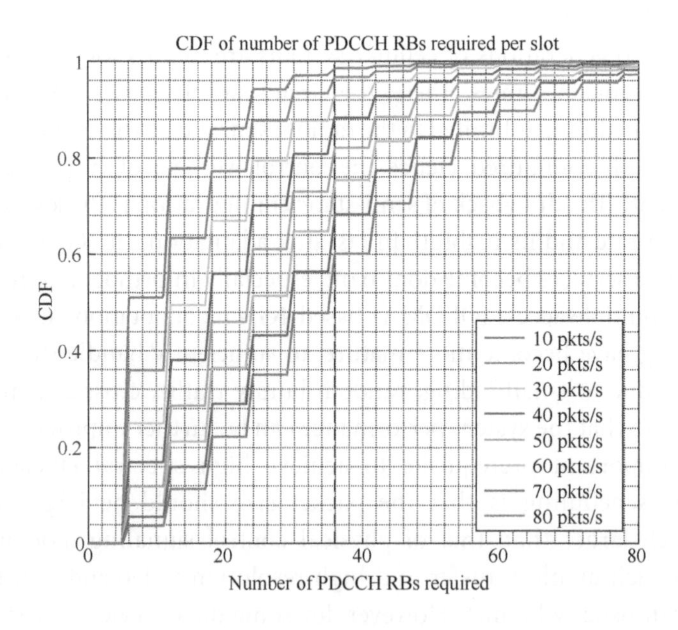

FIGURE 5.1 CDFs of occupied resources for the downlink physical control channel.

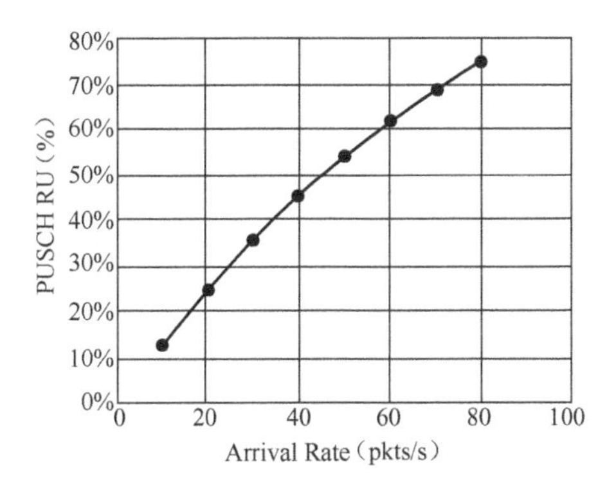

FIGURE 5.2 RU of the uplink traffic channel (PUSCH).

In 4G long-term evolution (LTE), there is a scheduling mechanism called semi-persistent scheduling (SPS). Its aim is to reduce the physical control overhead for small data packets, especially for periodic traffics such as voice-over-IP (VoIP). During a talk spurt, the data rate of VoIP remains constant, e.g., a voice packet generated every 20 ms. The average duration of a talk spurt is about 1–2 seconds, each consisting of about 50–100 voice packets. The small-scale fading can be compensated by closed-loop power control so that the signal-to-noise ratio (SNR) at the receiver can be kept more or less constant. This allows the modulation coding scheme (MCS) to be unchanged over some time, as well as the resource allocation. All these mechanisms can save the dynamic signaling. SPS can be considered as an enhanced semi-static configuration, mainly suitable for periodic and small packets with fixed packet sizes.

SPS normally operates in RRC Connected, e.g., after the terminal completes the initial access procedure. Even though the scheduling frequency is far lower than the packet arrival rate, in most cases, the transmissions are non-contention-based, e.g., no resource collisions between different users. In 5G systems, SPS can be used for ultra-reliable low-latency communication (uRLLC) scenarios, to ensure high reliability and reduce the latency in the user plane. In such a case, grant-free can also be called configured grant. It is very similar to SPS in LTE, with some new features added, for instance, no need for physical layer control signaling to activate or de-activate SPS and its configuration. Configured grants can be considered as a special case of grant-free, in the sense that the dynamic scheduling request can be saved. Essentially, the configured grant is dynamic-scheduling-free. It should be pointed out that in this SPS-based grant-free transmission, the resources of different users are pre-configured by the base station, instead of being randomly selected by each user. Furthermore, for this type of contention-free grant-free, while the resources of the traffic channels of different users can overlap (collide), the demodulation reference signal (DMRS) should be pre-configured by the base station, e.g., to ensure orthogonal allocation of DMRS, as well as the distinct ID for each user.

While SPS or Configured Grant can reduce the overhead of physical control signaling, there are some issues:

1. SPS or Configured Grant requires pre-configuration of physical resources and can only work when the user is in RRC Connected. For non-periodic or infrequent packets, the resource utilization of SPS or Configured Grant would be quite low. In a scenario of massive machine-type communication (mMTC), it is not realistic to keep each terminal device in RRC Connected.

2. In SPS or Configured Grant, if the initial transmission is not successfully decoded, it is not easy to retransmit the packet. For instance, when the initial transmission of a packet by a user experiences deep fade and its reference signal is not detected, the base station would not know whether the user has tried to access it. Then, there is no way for the base station to indicate negative acknowledgement (NACK) and the resources for retransmission to the terminal.

3. For those services whose packet sizes are not fixed, SPS has some problems. As long as the size of a packet exceeds the transport block size per SPS transmission, block segmentation would be needed. When a packet is divided into multiple smaller packets, the latency of the transmission would be increased.

4. Lastly, when a user switches between cells, the user should be connected to the target cell and request the physical resources of SPS. For the massive and infrequent small packets, in order to improve the spectral efficiency, a longer period of SPS may be considered. The longer period of SPS means reduced system throughput, and complicates the complexity of the system.

To reduce the burden on the system and the energy consumption at the devices, a terminal would stay in RRC Idle most of the time. Only when it has data to transmit, would the terminal enter RRC Connected state and start the transmission, usually based on the scheduling grant by the serving base station. The problem with this mechanism is that normally four steps are needed in a legacy random access process: the transmission of preamble, random access response (RAR), L2/L3 control signaling and Message 4 transmission, as illustrated in Figure 5.3. The entire process is not only time-consuming but also costs a lot of energy. It is apparently not efficient for infrequent small packet transmissions.

If a terminal can keep staying in RRC Idle, once the traffic data arrives from the upper layers to the physical layer, this terminal would automatically start the transmission without any coordination or interaction with the base station. After the data transmission is completed, this

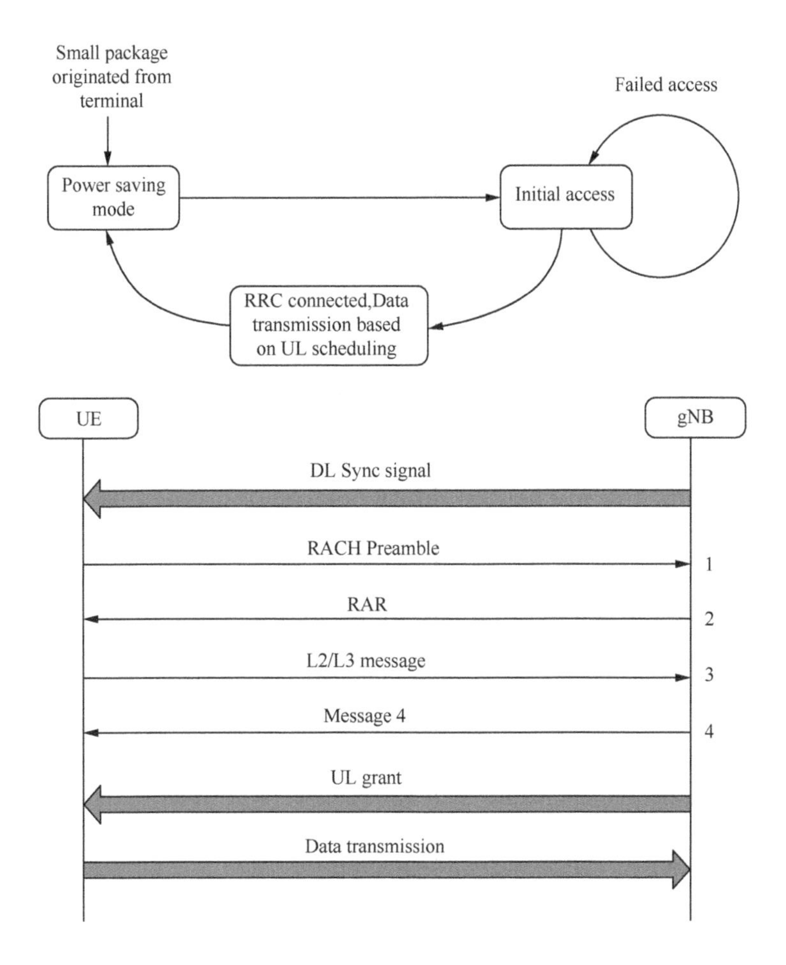

FIGURE 5.3 Legacy 4-step random access procedure to support the transition from RRC Idle to RRC Connected, as well as the data transmission in RRC Connected.

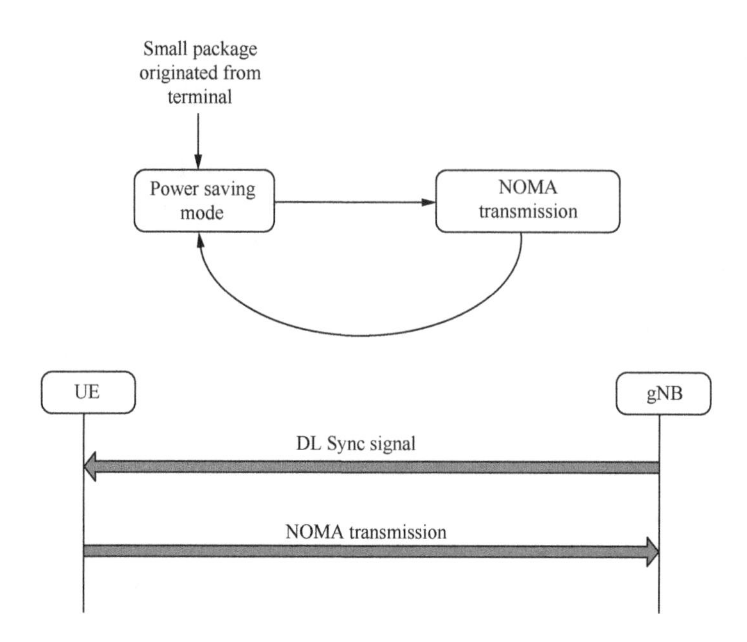

FIGURE 5.4 Grant-free transmission in RRC Idle.

terminal would still be in RRC Idle, as Figure 5.4 shows. Because there is no need to perform the burdensome four-step random access in order to switch to RRC Connected and process the scheduling grant to get the indication of resource allocation, the control signaling overhead can be significantly reduced, as well as the power consumption and control plane latency. This mechanism is very suitable for massive machine-type communication (mMTC) and enhanced mobile broadband (eMBB) of small packets because their traffic characteristics are infrequent and of small payloads. This grant-free transmission in RRC Idle can achieve super-simple uplink transmission. As there is no coordination by the base station, all the resources related to the transmission, such as reference signal, spreading sequence, or various types of signatures, would be autonomously selected by the terminals. The access would be contention-based, which would lead to resource collisions. This poses big challenges in multi-user detection. Autonomous or contention-based grant-free will be elaborated on in Chapter 8.

As discussed above, uplink grant-free has two working modes, as shown in Figure 5.5. The first mode operates in RRC Connected, with some similarity to grant-based transmission, for instance, to maintain the precise synchronization in the uplink by adjusting the timing advance, to maintain a relatively constant instantaneous SNR, to ensure no collision of

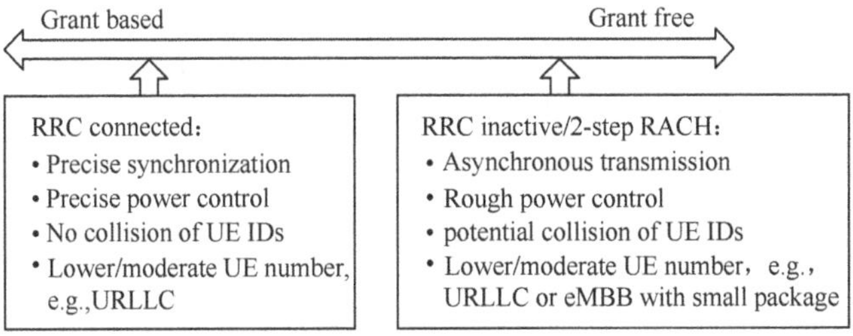

Grant based
Grant free

RRC connected:
• Precise synchronization
• Precise power control
• No collision of UE IDs
• Lower/moderate UE number, e.g.,URLLC

RRC inactive/2-step RACH:
• Asynchronous transmission
• Rough power control
• potential collision of UE IDs
• Lower/moderate UE number, e.g., URLLC or eMBB with small package

FIGURE 5.5 Two working modes of uplink grant-free transmission.

preamble/reference signal/signature via pre-configuration. This content-free mode is more suitable for uRLLC.

The second mode of grant-free is quite different from grant-based transmission in the following aspects:

1. Since the terminal remains in RRC Idle or RRC Inactive, it is difficult to adjust the timing advance. Hence, at the base station receiver, signals from different users may not be synchronized. The timing difference increases with the cell size and can exceed the cyclic prefix, causing cross-user interference.
2. When a terminal is in RRC Idle, only open loop power control can be carried which is based on the measurement of downlink long-term wideband SNR and the target received power. Without the closed-loop power control, the SNR at the receiver of a user would vary with the fast fading and would no longer be constant. Also considering the limitation of maximum transmit power of the terminal, even the average SNRs at the receiver among different users may not be the same. In all, no matter whether it is large scale or small scale, the near-far effect is expected at any time instant.
3. When a terminal is in RRC Idle or RRC Inactive, a network normally would not pre-allocate the physical resources or signatures, otherwise, it would cause resource waste. If a terminal has some data to send, it has to randomly pick a physical resource or signature. When multiple terminals want to transmit at the same time, resource/signature collisions would be inevitable, which can significantly impact the performance.

The second working mode of grant-free is contention-based, quite suitable for massive machine-type communication (mMTC) and small-data enhanced mobile broadband (eMBB). This type of service does not have a stringent requirement for latency, but has very stringent requirements for power consumption, especially for mMTC. With such a massive number of terminals in a network, the system cannot pre-configure each user with unique time and frequency resources and physical signatures.

In the grant-free transmission, no matter whether it is content-free or contention-based, non-orthogonal multiple access (NOMA) can be used to improve the spectral efficiency or number of accessed users of a system. Let us take a look at a system-level simulation [2] where the performances of legacy grant-based transmission (e.g., random access + dynamic scheduling) vs. grant-free-based (e.g., contention-based grant-free with NOMA) are compared. Background traffic of eMBB considered here is featured small packets and infrequent arrival. The traffic model is described in Ref. [3]. The model is based on the extensive measurement of real networks and data and reflects practical situations. There are two types: light background traffic and heavy background traffic. Figure 5.6 shows the CDFs of the packet arrival interval of these two types of background traffic. It can be seen that the medium value of packet arrival interval is about 2 seconds for light background traffic and about 0.1 seconds for heavy background traffic.

Grant-free-based and grant-based scheduling are compared in Table 5.1. In the grant-free solution, when the uplink buffer size is below a certain threshold, the user would not trigger the random access procedure; instead, it remains in RRC Idle or RRC Inactive and starts data transmission immediately once it has the data in the transmit buffer. Note that it is possible that multiple users transmit at the same time and frequency, thus causing collision. When the uplink buffer size exceeds that threshold, the user would kick off the two-step random access channel (RACH) where the transmission of the first message uses the content-based NOMA. In grant-free, the data is transmitted only in RRC Idle or RRC Inactive. By contrast, for the grant-based, regardless of the uplink buffer size, grant-based transmission relies on the legacy four-step random access where the data transmission can be carried in Msg 3 and Msg 5. After the connection is set up, the user enters into RRC Connected. If there is still data to be transmitted in the buffer, dynamic scheduling can be used, until the buffer is flushed out. Then, the user can return to RRC Idle, waiting for the arrival of the next data packet.

Simulation results are shown in Table 5.2. It is observed that for light background service, compared to the grant-based scheme, grant-free has the performance advantage in the average delay, the signaling overhead

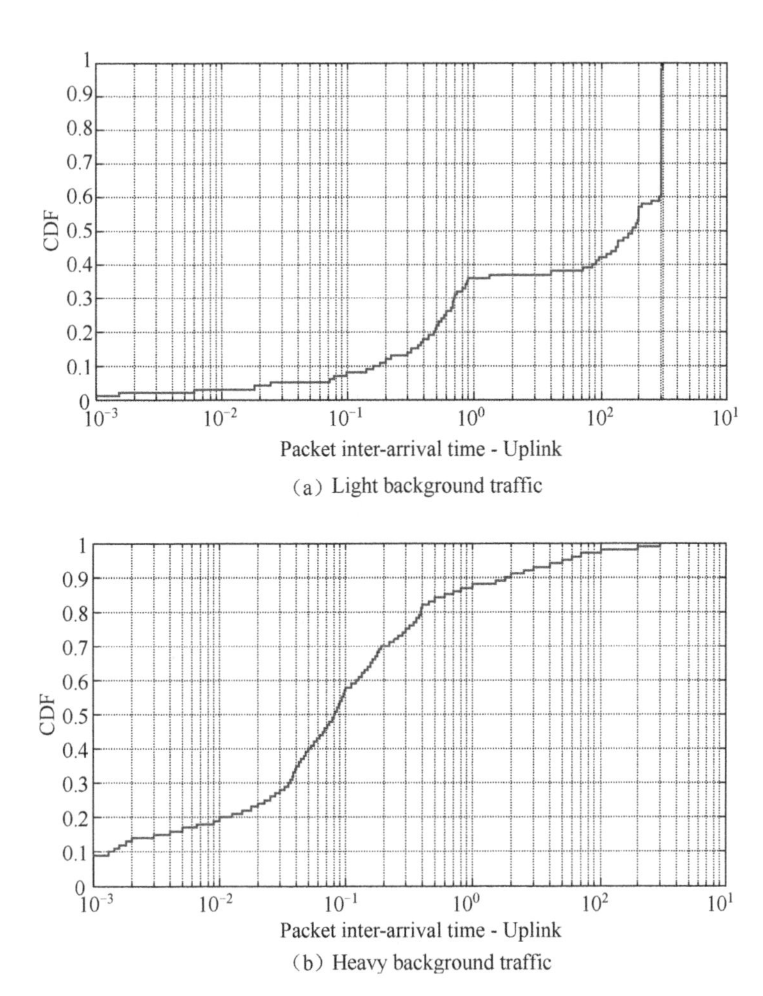

FIGURE 5.6 Distributions of packet arrival interval of light background traffic and heavy background traffic.

TABLE 5.1 Grant-Free-Based vs. Grant-Based Transmissions

Solution/Condition	Grant-Free Transmission	Grant-Based Transmission
Uplink buffer size is below a certain threshold	No random access is performed. Data transmission in RRC Idle directly	No threshold is set. To perform 4-step RACH. Data can be sent in Msg. 3 and Msg. 5
Uplink buffer size exceeds a certain threshold	Two-step random access, contention-based transmission for Msg 1 using NOMA	
Data transmission in RRC Connected	No	Dynamic uplink scheduling

TABLE 5.2 Simulation Results of Grant-Free and Grant-Based Schemes

	Light Background Service		Heavy Background Service	
Scheme/Results	**Grant-Free**	**Grant-Based**	**Grant-Free**	**Grant-Based**
Average delay (ms)	4.4	9.1	5.2	5.9
Signaling overhead (%)	15	53	11	7
Terminal power consumption (J/s)	47	677	321	2142

and the power consumption. As the traffic load increases, the benefit of grant-based becomes more pronounced. For heavy background service, the signaling overhead is even lower than that of grant-free. But still, in terms of the average delay and power consumption, grant-free outperforms grant-based.

Apart from the above-mentioned mMTC, small-data eMBB (e.g., background services) and uRLLC scenarios, NOMA can also be used for V2X. It is known that currently the vehicle-to-vehicle communication is mostly broadcast/multicast where each vehicle would automatically select the radio resources for broadcast or multicast. There is no master node to coordinate and schedule each user's transmission. Hence, there is a chance of collision which can affect the data rate and coverage range. By using NOMA, even when collisions occur, the receivers still have a high chance to correctly decode the data, thus being able to improve the reliability of V2V communications.

5.1.2 Basic Procedure

In the current new-radio access technology (NR) systems, the key difference between grant-free transmission and legacy grant-based transmission is that the configuration for grant-free uplink transmission is indicated via higher-layer RRC signaling and the subsequent transmission would follow the pre-configured setting. Its advantage is the saving of dynamic uplink measurement and scheduling compared to legacy grant-based one. Such type of grant-free transmission guarantees the reliability of uplink transmission, suitable for the scenario where the number of simultaneously transmitting users is not large, for example, uRLLC.

However, such a grant-free mode has a very low resource utilization rate and requires a large amount of signaling at the beginning, especially for mMTC scenarios, as well as for those traffics whose arrival time is not predictable, such as small-data eMBB. If each user needs to be allocated a unique uplink resource, it would incur too much waste. Furthermore, with

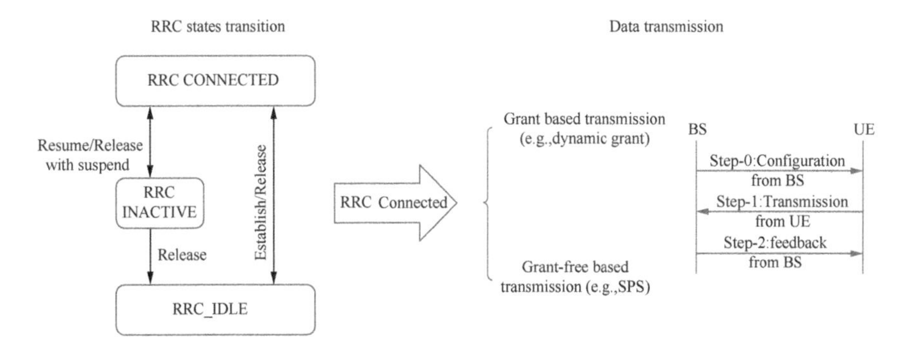

FIGURE 5.7 Switching of RRC states and block diagram of data transmission.

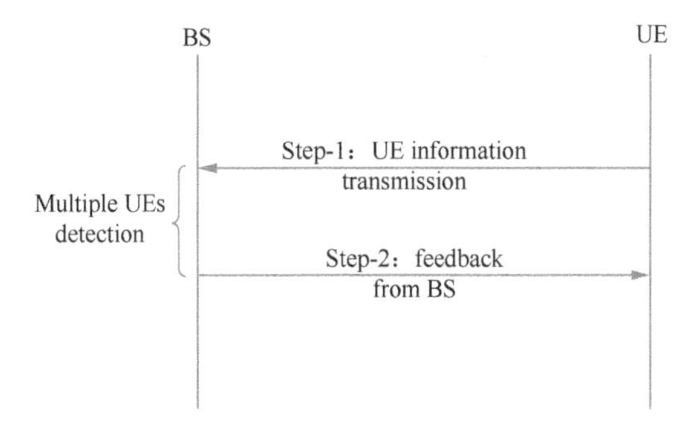

FIGURE 5.8 Autonomous uplink transmission.

the ever-changing traffic, this grant-free mode leads to frequent re-configuration at higher layers. As shown in Figure 5.7, this content-free grant-free is similar to the grant-based transmission in the sense that a user needs to enter RRC Connected before sending the data. This is often achieved by random access. Therefore, for a user who has just a small amount of data to send, the signaling overhead, the power consumption and the associated delay would be significant and cannot be ignored.

Regarding the above issues, with the non-orthogonal transmission, we can consider the procedure outlined in Figure 5.8.

Furthermore, considering the signaling overhead to switch between different RRC states, autonomous transmission can have two flavors: data transmission in RRC Inactive and two-step random access channel (2-step RACH).

5.1.2.1 Transmission in RRC Inactive

Compared to the legacy grant-based or content-free grant-free transmission, transmission in RRC Inactive can reduce the signaling overhead and transmission latency. In this mode, the user's data can be sent with the legacy physical channel, together with the NOMA. The E-ID information may be embedded in the payload data at the physical layer as shown in Figure 5.9.

According to the current specification of 5G, before a user enters RRC Inactive, it has already gone through the random access procedure to enter RRC Connected. The uplink timing advance (TA) adjustment is already completed. In the subsequent transmission in RRC Inactive, the change in uplink TA is mostly due to the terminal mobility. In scenarios such as machine-type communication, or small-data eMBB, uplink TA may not change significantly and most of the processing at the receiver would be to detect the users and decode the data. Here, the user detection includes detecting whether a user is in transmission, based on the reference signal, as well as interpreting a user ID from the physical layer payload. There are two cases here:

1. When the reference signal or the resource allocation is unique across a network, the mapping from user ID to the reference signal or the resource allocation is one-to-one.

2. When the reference signal or the resource allocation is not one-to-one mapped, after the user detection, the user ID may be determined by uplink control indication (UCI), media access control (MAC) CE or scrambled data.

   After these, the base station would send the indication to the user, based on whether it has detected any user or successfully decoded the data.

5.1.2.2 Two-Step Random Access (2-step RACH)

Compared to the aforementioned transmission in RRC Inactive, 2-step RACH-based grant-free and data transmission have broader applications.

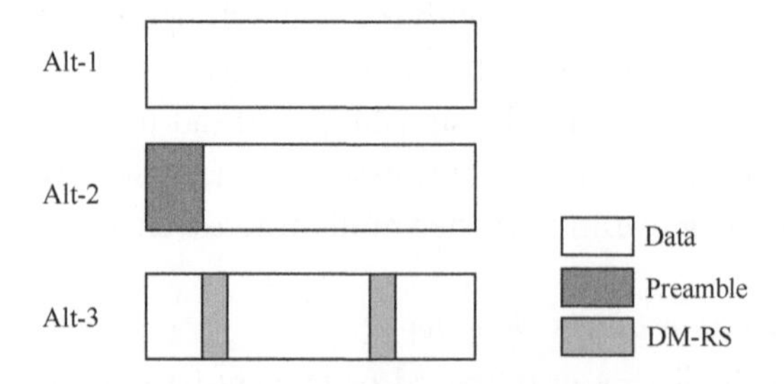

FIGURE 5.9 Data to be sent via transmission in RRC Inactive.

As an enhancement of the legacy 4-step RACH, only two steps are required in 2-step RACH as shown in Figure 5.8. More specifically,

1. Considering the general deployment scenarios of RACH, users may not have valid TA information. Such a case would occur during the initial access or the RACH procedure triggered by the loss of synchronization. Hence, in the first step, traffic data bearing physical packets should have corresponding preambles so that the base station can perform user detection/identification and TA estimation. When the terminals are assigned user-specific preambles, user detection and identification can be achieved by preamble detection. Otherwise, the base station needs to decode the payload in order to identify the users. Currently, the major designs of physical channel structures of 2-step RACH are shown in Figure 5.10.

   Regarding 2-step RACH, for terrestrial networks, the preamble design can follow that for 4-step RACH. For other scenarios such as non-terrestrial networks or unlicensed spectrum services, due to the impact of excessive frequency offset or listen-before-talk mechanism, enhancement of preamble can be considered.

2. In order to fulfill the functionality of random access and data transmission, as illustrated in Figure 5.10, the channel structure should include the resources of the data transmission and reference signals. To be more compatible with the legacy 4-step RACH, the preamble for 2-step RACH can reuse the resource configuration for the preamble and the data of the legacy 4-step RACH. The resource configuration for data part may be enhanced to support more users.

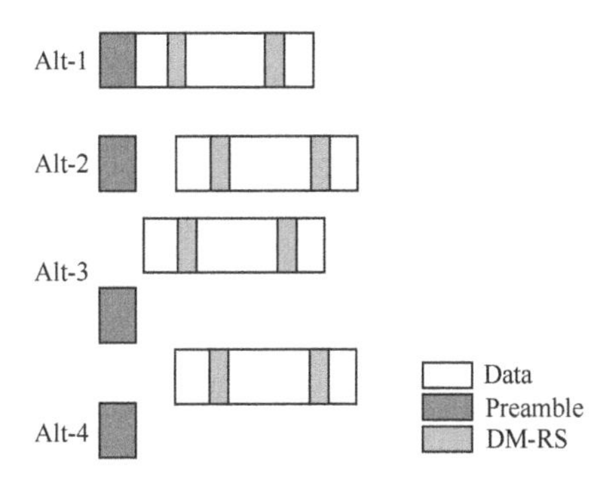

FIGURE 5.10 Physical channel structures in 2-step RACH.

Different deployment scenarios should be considered, for instance, to support different sizes of the payload. The following information may be included:

- Common control information: can be further divided into RRC setup signaling, RRC reconstruct message, RRC system information request, RRC response, etc. The payload size would normally be between 50 and 70 bits.
- Dedicated control information: including the specific confirmation message for RRC state update, UE capability, etc.
- MAC CE-related service request: to support the subsequent faster scheduling, can report the terminal buffer information during the data transmission.
- User data payload: to support high-speed transfer. The actual information rate depends on the potential resources to be used and the channel conditions. The legacy UCI can also be used to support fast ACK/NACK and channel-state information (CSI) reporting.

In the practical resource configurations, multiple sets of resources can be defined in a system to better match the requirements. There are a couple of solutions for the resource selection and configuration of the preamble and data part:

- Solution 1: resources of the preamble and data part are defined in pairs. A user can directly choose one of the pairs.
- Solution 2: the mapping between the resources of the preamble and the resources of the data part is defined by the base station. Once a user selects the preamble, it would figure out the resources for data transmission based on the mapping.

The resource mapping between the preamble and the data part can be one-to-one, many-to-one, or one-to-many, depending on the requirements. When the mapping from data to preamble is one-to-many, it means that multiple users would share the same resource block. In this case, apart from the allocation based on the orthogonal reference signals, non-orthogonal transmission can also be used to improve the demodulation performance under multi-user scenarios.

If the channel does not change significantly between the preamble transmission and the subsequent data transmission, the demodulation reference signal in the data part can be saved in order to reduce the overhead.

3. After the base station completes the uplink data decoding, it would send the feedback information to the terminals, based on the results of user detection, identification and demodulation. The following situation may occur:

   a. Failure of preamble detection: in this situation, the base station does not know whether the user has transmitted the data. Hence, it would not provide any feedback. Similar to 4-step RACH, when a terminal does not receive any feedback, it would try to send it again.

   b. Data detection failure: in this situation, although the base station cannot identify which user has sent the data, it can indicate a NACK that corresponds to the preamble detected to all users via broadcasting. Once a user detects this message, if the preamble has been used before, the user can start the next transmission or fall back to 4-step RACH.

   c. When both the preamble and the data are successfully received, the base station can have the following two ways for the feedback:

      i. Broadcasting the feedback: the feedback can be scrambled by RA-RNTI and carried in downlink control information (DCI). The message in the feedback can include the ID information which can be used to resolve the collision (e.g., the ID of the successfully decoded user). This is mainly for the case when multiple users share a preamble. It would facilitate that the unsuccessful user can start another attempt as soon as possible.

      ii. Unicast the feedback: the feedback is scrambled by C-RNTI and carried in DCI. For those users whose data are successfully received, this feedback can include ACK/NACK, as well as the downlink and uplink scheduling grant. In addition, to support the subsequent uplink transmissions (UL TA) can also be carried. Based on this, if a user has not received any corresponding feedback within a configured time duration, the user would assume that the previous transmission fails, due to either data decoding error and/or preamble detection failure.

In wireless communications, when a base station schedules an uplink or a downlink transmission, to improve the system performance, the scheduler would group users based on their geographic locations, channel characteristics, service types, etc., for instance, the user pairing for MU-MIMO. For non-orthogonal transmission, user grouping can be used to improve the efficiency of resource sharing, reduce cross-user interference and increase the detection probability in interference cancelation-based receivers.

5.1.2.2.1 Criteria for User Grouping

1. Grouping based on channel characteristics: because the channels experienced by different users are diverse, the received power and the fading characteristics can vary drastically between users. The near-far effect illustrated in Figure 5.11 can be exploited to allow fast convergence of multi-user detection. Hence, the grouping of these two users can ensure that in successive interference cancelation receivers, Terminal 1 can be detected first and its interference can be canceled.

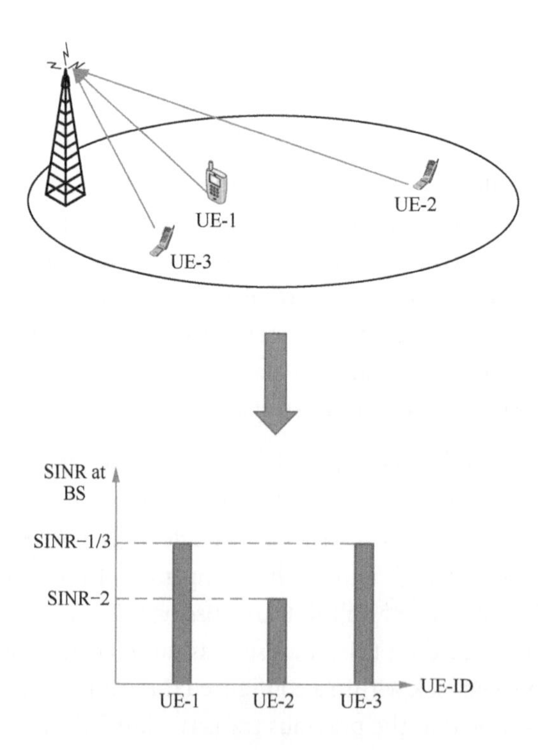

FIGURE 5.11 User grouping based on received power levels.

In addition, received power-based user grouping can also be facilitated by scheduling. For example, as shown in Figure 5.12, even if the received signal noise power (SNRs) of Terminals 1 and 3 are the same; however, to ease the demodulation, the base station can configure an extra power offset. In this case, non-orthogonal codes may be applied for intra-group users, while orthogonal codes would be applied for inter-group users. Alternatively, users whose SNRs are not very different would be grouped together and allocated with orthogonal codes, whereas users between different groups can use non-orthogonal codes.

2. Grouping based on service types: different service types require a different number of bits for the same resource allocation and MCS level, which leads to different code rates. Hence, the required SNR would increase as the code rate is increased. If paring two users with very different code rates together, the base station can first decode the user with the lower code rate, cancel it and then try to decode the user with the higher code rate.

3. Grouping based on different transmission assumptions: in case of NOMA, when a certain number of users fail the decoding, those users would retransmit using the same resources and assumption (configuration). This would lead to very poor performance. In order to improve the success rate of the retransmission, the network may adjust the configuration, for instance, using the lower MCS, or power ramping. Hence, users of initial transmission and users of retransmission can be put into different groups, in order to make retransmissions more robust.

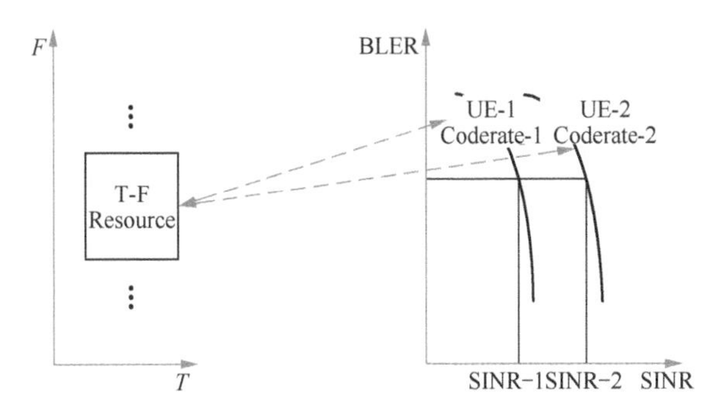

FIGURE 5.12 User grouping based on traffic types.

5.1.2.2.2 Implementation of User Grouping

User grouping can be implemented in the following two ways:

1. Via the configuration signaling and explicit implementation. This method is suitable for RRC Connected users. In this case, the base station can perform the user grouping based on real-time measurement. When the user receives the configuration signaling, it would start the data transmission based on the configuration.
2. Via user self-organized manner. This method is suitable for a grant-free scenario and can be carried out regardless of the RRC state of the user. As illustrated in Figure 5.13, the base station can signal multiple sets of potential configurations.

When the user receives these configuration sets, it would choose the best configuration set, based on its own channel condition, e.g., reference signal receiving power (RSRP). In this case, a user would not know which configuration set other users have chosen. Next, the base station groups the users based on the RSRP report.

5.2 BRIEF DISCUSSION ON EVALUATION METHODOLOGY

5.2.1 Overall Configuration of Link-Level Simulations and Evaluation Metrics

There are two types of metrics in link-level evaluation: (1) performance metrics and (2) metrics related to implementation.

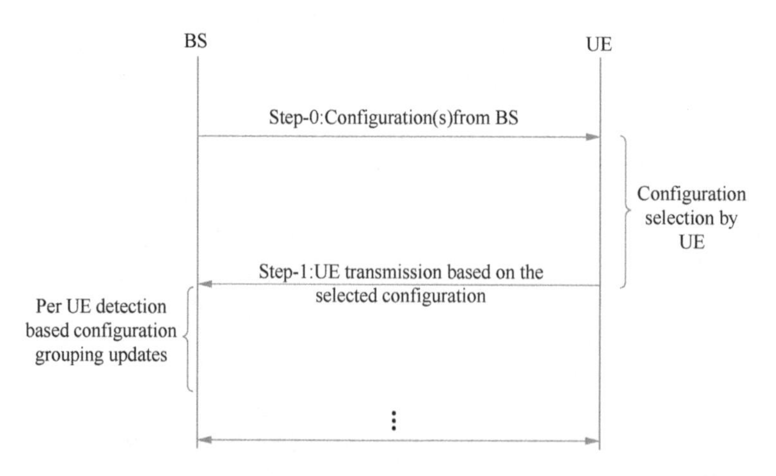

FIGURE 5.13 Self-organized user grouping.

Performance metrics should include at least:

1. Given the per-user spectral efficiency and the number of users, the block error rate (BLER) vs. per-user SNR. This is the most widely used metric. Quite often, BLER vs. SNR performance is primarily determined by the design of the transmitter and the receiver algorithm. In addition, appropriate modulation and code rate would also play an important role.
2. Given the per-user spectral efficiency, the number of users and the target BLER, the sum rate vs. total SNR. This metric is often used to quantify the throughput of a cell and can be translated to BLER vs. SNR.
3. Maximum coupling loss: to measure the coverage of a base station.

Metrics related to implementation:

1. Peak to average power ratio (PAPR)/cubic metric: In general, the dynamic range of power amplifiers is limited. If PAPR/CM is large, the implementation cost for terminals would be high
2. Receiver complexity and processing delay: Receiver complexity directly impacts the cost. The higher the complexity, the higher the implementation cost, which also affects the processing delay. For delay-sensitive services such as automatic driving, there is a very stringent requirement for latency.

5.2.2 General Simulation Setting for System-Level Simulations and Evaluation Metrics

For uplink NOMA, three scenarios, mMTC, eMBB small data and uRLLC are simulated at the system level. The corresponding traffic models, evaluation methodology and performance metrics are listed below.

5.2.2.1 mMTC Scenario

- Traffic model
  The traffic model for mMTC is based on the traffic model for the performance evaluation of NB-interference over thermal (IoT) [1]. The packet size of each user follows the Pareto distribution of range 20–200 bytes, with Pareto index $\alpha = 2.5$. An additional 29 bytes

protocol overhead needs to be considered for the higher layer. Poisson arrival is assumed for each user's traffic.

When a user is in data transmission, packet segmentation can be carried out at the radio link control (RLC) layer where a relatively large packet can be segmented into multiple transport blocks (TBs). The signaling overhead at the RLC layer and MAC layer is assumed to be 5 bytes in total.

- Evaluation methodology
  For contention-free grant-free transmission, the baseline is to use configured grant type 1 or configured grant type 2. The time-frequency resources and DMRS are semi-statically configured for each user. There is no spreading carried out at the transmitter, e.g., following the current signal processing chain in legacy 5G. In the case of NOMA, each user's time-frequency resources, DMRS and NOMA signature is semi-statically configured. In this way, DMRS collision and/or NOMA signature collision can be avoided, as long as the DMRS pool and NOMA signature pool are large enough.

  For contention-based grant-free, in the baseline system, each user can randomly select its time-frequency resources and DMRS. For the uplink NOMA case, the time-frequency resources, DMRS and NOMA signatures are also randomly selected by users. DMRS collision and NOMA signature collision are inevitable.

- Performance metrics
  In the mMTC scenario, an important performance metric is the packet drop rate (PDR) at a certain packet arrival rate (PAR). In another word, at a certain PDR (e.g., 1%), how high the PAR can be supported. There are also some other metrics such as transmission latency, IoT, resource utilization (RU), etc.

  Here, the PDR is defined as the number of dropped packets as opposed to the number of generated traffic packets. Dropped packets refer to either when the dropping timer is reached, the traffic packet is still not successfully decoded or when the maximum number of hybrid automatic retransmission request (HARQ) transmissions is reached, the packet is still not successfully decoded.

5.2.2.2 eMBB Small-Data Scenario

- Traffic model
  The traffic model for eMBB small data is based on the outcome of the 3GPP LTE eDDA (enhancements for diverse data application) study. In that study, typical small-data services are analyzed, including the background services and real-time message services. From the analysis, it is found that the packet size of this type of service follows the Pareto distribution. After further fitting and modification, the packet size for eMBB is assumed to be 50–600 bytes (including higher-layer overhead) of Pareto distributed with a Pareto index $\alpha = 1.5$. Each user's traffic follows Poisson arrival.

  Similar to the mMTC scenario, when a user is in data transmission, a relatively large packet can be segmented into multiple TBs. The overhead of the RLC and MAC header is assumed 5 bytes in total.

- Evaluation methodology
  Similar to the mMTC scenario, for grant-free transmission, the baseline is configured grant type 1 or configured grant type 2 in 5G NR where the time-frequency resources and DMRS of each user are semi-statically configured. For the uplink NOMA, the time-frequency resources, DMRS and NOMA signatures are also semi-statically configured. In this way, DMRS collisions and NOMA signature collisions can be avoided as long as the DMRS pool or the NOMA signature pool are large enough.

  Dynamic scheduling can also be simulated in eMBB. In this case, the control signaling should be considered; hence, the simulation platform should be able to simulate downlink and uplink simultaneously.

- Performance metric
  For eMBB, in the case of grant-free transmission, the key performance indicator is similar to that of mMTC, that is, the PDR at a certain PAR or the supportable packet arrival rate when a certain packet error rate (e.g., 1%) can be maintained. The definition of packet dropping is similar to that of mMTC. Other metrics include transmission latency, IoT, and RU.

When dynamic scheduling is used, the relevant performance metric is the user experienced rate at a certain PAR or the experienced rate when a certain PDR (e.g., 1%) can be maintained.

5.2.2.3 uRLLC Scenario

- Traffic model
  The typical packet size for uRLLC is relatively small. In uRLLC, the size of the packet is fixed, e.g., 60 bytes or 200 bytes, if excluding higher-layer signaling. Both Poisson arrival and periodic arrival can be assumed, although the former is more widely used. In the simulations of Chapters 7 and 8, the Poisson arrival model is assumed.
  The key requirement for uRLLC services is ultra-high reliability and very low latency. In the simulation, the reliability target is 99.999%. For the packet size of 60 bytes, the latency requirement is 1 ms. For the packet size of 200 bytes, the latency requirement is 4 ms.
- Evaluation methodology
  In contention-free grant-free transmission, configured grant type 1 or configured type 2 is used as the baseline where the time-frequency resources and DMRS of each user are semi-statically configured. For uplink NOMA, the time-frequency resources, DMRS and NOMA signature are semi-statically configured for each user. In this way, the collision of DMRS and the collisions of NOMA signatures can be avoided as long as the pool size of DMRS and the pool size of NOMA signatures are large enough.
  Because of the reliability requirement of 99.999%, the simulation complexity can be very high if literally following the general methodology. Instead, a simplified method is used where the averaged BLER is considered as the reliability: to use the average BLER as the reliability.
- Performance indicator
  As mentioned above, the key requirements for uRLLC include reliability and latency. Considering the simplified method, the main performance metric can be the percentage of users satisfying reliability and latency requirements at a certain PAR or when the reliability and latency requirements are met by a certain percentage of users (e.g. 95%) and how high the PAR can be supported. Other metrics include the reliability distribution among users and IoT and RU.

5.3 BRIEF INTRODUCTION OF THE NOMA TRANSMITTER AND RECEIVER

There are three major categories of NOMA, featured in the specific transmitter-side processing, as illustrated in Figure 5.14. In this figure, the blocks without color filling are already supported in the 5G NR specification, which is basically for orthogonal multiple access. Those processing blocks include a channel encoder, legacy bit interleaver, legacy bit scrambler, legacy modulation, DFT precoding (e.g., DFT-s-OFDM) and resource mapping. In order to support NOMA, new blocks can be added, filled with gray color [4]. These new blocks include

- Symbol-level linear spreading: symbol-level spreading is performed after the legacy modulation, in order to separate the modulation symbols between different users. The data transmission rate is related to the spreading length. In general, the longer the spreading length, the lower the data rate is, and the lower the cross-correlation between spreading sequences.
- Bit-level processing: after channel coding, a user-specific bit interleaver is applied to each user. Note, here, that the bit interleaver is different from the legacy bit interleaver in the sense that the former is user-specific, while the latter is common to all the users. When different users have the same code rate and information block size, exactly the same bit interleaver would be applied for all the users in the legacy system.

  Multi-dimension modulation: signal modulation and symbol-level spreading are jointly designed, using one set of codebooks, so that the code bits would be mapped directly to the spread modulation symbols.

In Figure 5.14, there is also a block called resource sparse mapping which is different from the legacy system. Sparse mapping can be achieved by symbol-level spreading or a multi-dimensional codebook. Hence, it is not shown to be specifically for one type of solution. In all the above-mentioned three categories of NOMA, a user can have one branch (layer of data), or multiple branches which are to be elaborated on in Chapter 6.

Legacy mobile communications systems are orthogonal multiple access (OMA)-based. Hence, the channel codes adopted in the standards so far are optimized for a single user. For multiple-user NOMA transmission,

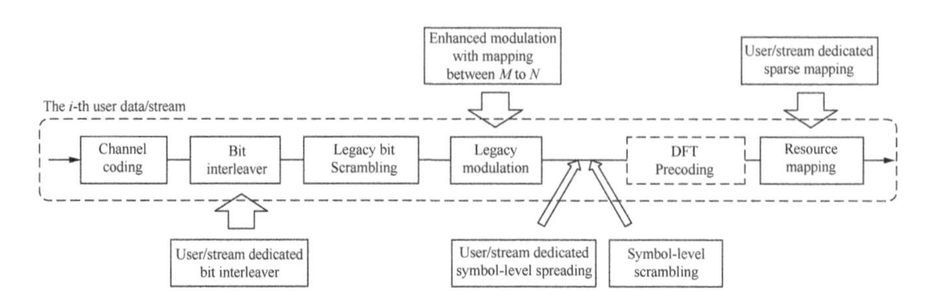

FIGURE 5.14 Block diagram for the NOMA transmitter.

there are some related studies [5]. At the beginning of the NOMA study item, some companies proposed to optimize the low-density parity check (LDPC) to improve the NOMA system capacity [6]. However, considering that the scope of the NOMA study item is mainly to study signal processing such as spreading, modulation, interleaving and scrambling, there is no enhancement in channel coding.

While the transmitter-side processing sets the foundation of NOMA, the receiver implementation is also very crucial in order to achieve enough good performance. For NOMA, an advanced receiver is expected. The general structure of the NOMA receiver is illustrated in Figure 5.15, made up of a detector, channel decoder and interference cancelation. The primary function of the detector is to demodulate the signal and output the log-likelihood ratio (LLR) and soft bits. The channel decoder is responsible for channel coding and can output the hard decision (whether the bit is one or zero) or soft bits (to be further refined). Interference cancelation is self-explainable. The reason for using dashed block is that interference cancelation would often reside in the detector module for iterative receivers.

Usually, the legacy receivers for single-user transmission only have a detector and a decoder where interference cancelation is rarely seen and neither are the iterations between detectors and decoders. In 3GPP, the legacy receiver is often called MMSE-IRC (minimum mean squared error – interference rejection combining). Note here that MMSE is in the sense

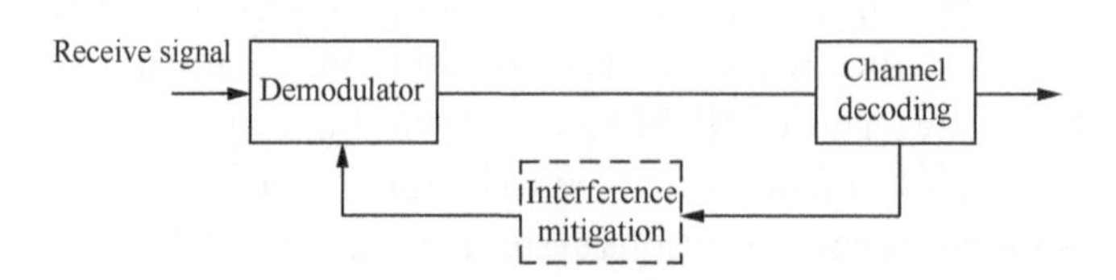

FIGURE 5.15 General block diagram of the advanced receiver for NOMA.

of spatial MMSE for multiple receive antennas. IRC refers to the rejection capability to inter-cell interference via spatial MMSE in the linear domain, rather than using nonlinear methods to directly cancel the interference.

There are three types of receivers for NOMA:

- MMSE-hard interference cancellation (IC): in the detector module, MMSE is carried out jointly in the spreading code domain and spatial domain, which can significantly suppress the cross-user interference and then output the LLR. In the decoder module, the output is the hard bits. In the interference cancelation module, the output bits from the decoder are used to reconstruct the interference signal to be canceled from the received signal. Multi-user interference cancelation can be done in a serial fashion, or parallel, or hybrid. This type of receiver is typically used for a symbol-level linear spreading category of NOMA. The entire architecture of the MMSE-hard IC receiver is similar to that of the legacy receiver, e.g., MMSE-IRC, and its computation complexity and implementation cost are lower than in two other types of advanced receivers.
- ESE + SISO: the detector module is comprised of spatial MMSE and an elementary signal estimator (ESE) at the bit level. The decoder should be capable of outputting soft bits to be fed back to the detector. Via ESE and the iteration between the detector and the decoder, the "belief" of each symbol is increased. This type of receiver is suitable for the bit-level processing category of NOMA. Its computation complexity is higher than MMSE-hard IC but lower than extended pedestrian A model (EPA)+SISO.
- EPA + SISO: the detector is comprised of spatial MMSE and EPA requiring multiple inner iterations. The decoder should be capable of outputting soft bits to be fed back to the detector. Via the inner iteration inside EPA and the outer iteration between the detector and the decoder, the "belief" of each symbol is increased. This type of receiver is suitable for the multi-dimensional modulation category of NOMA. However, its computation complexity is the highest among the three receiver types.

NOMA receiver type and typical corresponding transmitter solutions are listed in Table 5.3. More details on transmitter-side solutions will be elaborated on in Chapter 6.

TABLE 5.3 NOMA Receiver Type and Typically Corresponding Transmitter Solutions

NOMA Category at Transmitter	Specific Solution	Typical Receiver Type
Symbol-level linear spreading	Multi-user shared access (MUSA) Non-orthogonal code access (NOCA) Welch-bound spreading multiple access (WSMA), Resource-shared multiple access (RSMA) Non-orthogonal coded multiple access (NCMA) User-grouped multiple access (UGMA) Pattern-defined multiple access (PDMA)	MMSE-hard IC (general use) EPA + SISO (for high spectral efficiency case)
Bit-level processing	Low code rate spreading (LCRS) Interleaver-division multiple access (IDMA) Asynchronous spreading multiple access (ASMA) Interleaved grid multiple access (IGMA)	ESE + SISO (general use) LCRS can also use MMSE-hard IC
Multi-dimensional modulation	Sparsely coded multiple access (SCMA)	EPA + SISO

Symbol-level linear spreading can trace back to code division multiple access (CDMA) in the 3G uplink. However, the spreading length in 3G is typically long, e.g., 16, 32, 64, suitable for low-rate voice type of services. For the long-spreading code, the requirement for structured design is quite low. A typical random sequence would be enough due to high processing gain. The receiver can be rather a simple Rake type to achieve maximum ratio combining. In this case, the receiver power of different users should be maintained to similar levels where closed-loop power control is needed and the systems would operate in RRC Connected. All of these are quite different from the situation of 5G grant-free NOMA.

REFERENCES

1. 3GPP, R1-1809437, System level performance evaluation for NOMA, Qualcomm, RAN1#94, August 2018, Gothenburg, Sweden.
2. 3GPP, R2-1701932, Quantitative analysis on UL data transmission in inactive state, ZTE, RAN2#97, February 2017, Athens, Greece.
3. 3GPP, TR 36.822, LTE radio access network (RAN) enhancements for diverse data applications.
4. 3GPP, TR 38.812, Study on non-orthogonal multiple access (NOMA) for NR.
5. Y. Zhang, K. Peng, and J. Song, "Enhanced IDMA with rate-compatible raptor-like quasi-cyclic LDPC code for 5G," in *Proceedings of* IEEE Globecom Workshops, Singapore, December 2017, pp. 1–6.
6. 3GPP, R1-1801888, Spectral efficiency of NOMA-optimized LDPC code vs. LTE turbo code for UL NOMA, Hughes, RAN1#92, February 2018, Athens, Greece.

REFERENCES

CHAPTER 6

Uplink Transmitter-Side Solutions and Receiver Algorithms

Yifei Yuan, Zhifeng Yuan, Li Tian, Chen Huang, Yuzhou Hu, Chunlin Yan and Ziyang Li

6.1 SHORT SEQUENCE-BASED LINEAR SPREADING AND TYPICAL RECEIVER ALGORITHMS

Figure 6.1 shows the transmitter-side processing for NOMA that is based on symbol-level spreading. Compared to legacy orthogonal transmission, spreading-based NOMA schemes exploit the spreading code domain and power domain dimensions and rely on the interference cancelation capability of advanced receivers. In these types of NOMA schemes, after the coded bits are modulated using legacy modulation, different users would use non-orthogonal spreading sequence to stretch either in time or in frequency domain, and then transmit in the same time and frequency resources. At the receiver, interference cancelation would normally be carried out in a serial or parallel fashion. The basic principle is to suppress the cross-user interference by the process of dispreading so that the stronger user whose signal-to-noise ratio (SNR) is higher can be decoded first. Its signal can be reconstructed based on the correctly decoded bits and then canceled from the receiver signal. Then, the signal to interference and noise ratio (SINR) for the rest of the users (with weaker SNR) would be increased.

DOI: 10.1201/9781003336167-6

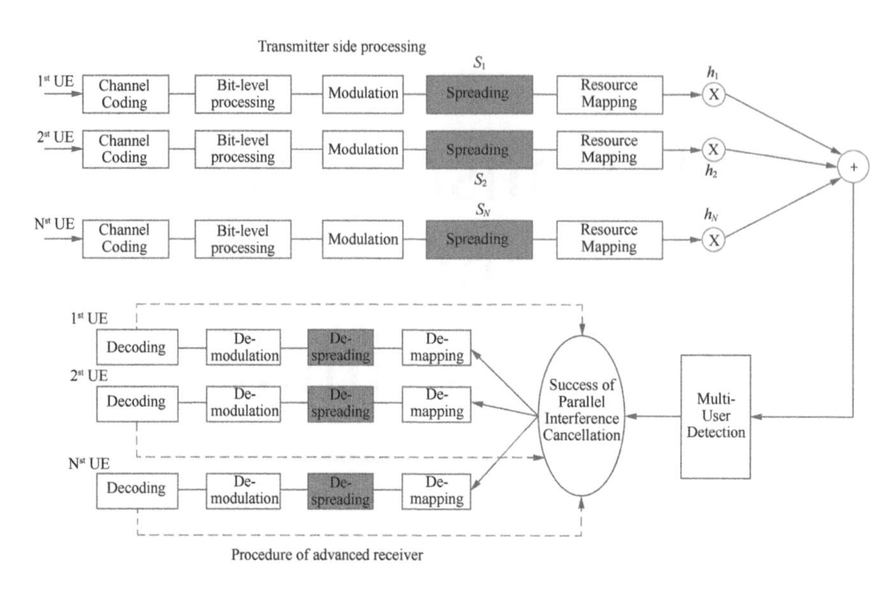

FIGURE 6.1 NOMA transmitter-side processing for symbol-level spreading.

6.1.1 Design Principles

The primary goal for short linear spreading codes is to improve the overloading capability and operational flexibility. This would pose challenging requirements for cross-correlation, codebook size, nested property, sparsity, etc. In addition, peak to average ratio (PAPR) and complexity of receiver should also be considered.

A codebook here refers to a set of spreading codes (or spreading sequences) with normalized power. A codebook consisting of K spreading codes. Each of length L code (or sequence) can be represented as $\{s_k\}$, $s_k \in \mathbb{C}^L$, $\left\| s_k^2 = 1 \right\|$ $k = 1,\ldots, K$, or simply $C(L, K)$. Here the elements of the codebooks are assumed complex-valued.

For a codebook itself, its most important property is cross-correlation. There is a constraint between the size of a codebook and the cross-correlation. That is, the larger the size of a codebook, the higher the correlation is between the codewords. When linear spreading is used for NOMA, codebooks with larger size is often quite desirable in order to support more users transmitting simultaneously. However, this brings more cross-user interference.

Welch bound is the most fundamental bound for the cross-correlation of codebooks such as $C(L, K)$. Detailed derivation can be referred in Ref. [1]. According to Ref. [1], two types of codebooks can be defined and both of them are Welch-bound-related.

6.1.1.1 TSC-Bound Equality (TBE) Codebooks

The design criterion for this type of codebook is to minimize the total squared correlation (TSC) $T_c \triangleq \sum_{i,j} |s_i^H s_j|^2$. Note that the definition of TSC includes auto-correlation. Specifically, for a codebook $C(L, K)$, there is a lower bound of TSC, that is, $T_c \geq K^2 / L$. Hence, codebooks that satisfy TSC bound $T_c = K^2 / L$ are called TBE codebooks.

TBE codebooks can only achieve the minimization of the sum of squared correlation. It does not impose any constraint on the cross-correlation between any two codewords. Because of this, in some TBE codebooks, the cross-correlation can vary significantly, e.g., some have very low cross-correlation and some others have very high cross-correlation, even close to 1. A very simple but very extreme example is to repeat $L*L$ unitary matrix by R times to get a codebook $C(L, L*R)$ that consists of $L*R$ codewords, each of length L. Apparently, this codebook is not useful, since the range of cross-correlation represents the extreme case: cross-correlation between some codewords is 0 and between other codewords is 1. The TSC of this codebook is $T_c = L * R^2 = \frac{(L * R)^2}{L}$, satisfying TBE.

From this extreme case, it is seen that for practical TBE codebooks, the maximum cross-correlation should be kept as small as possible, e.g., the range of cross-correlation should not be too wide. Several practical codebooks are described below.

- Equal cross-correlation multiple unitary matrix (ECMU) codebooks
- Low cross-correlation multiple unitary matrix (LCMU) codebooks
- Computer-generated (CG)-TBE codebooks

6.1.1.1.1 ECMU Codebook

While it is not meaningful to construct codebooks by repeating unitary matrices, it is indeed meaningful to construct codebooks by mixing different unitary matrices, which can be called multi-unitary-matrix codebooks. It is not difficult to verify that multi-unitary-matrix codebooks are TBE codebooks.

In multi-unitary-matrix codebooks, codewords derived from different $L*L$ unitary matrices have the same cross-correlation which is $1/\sqrt{L}$ and can have a maximum of $L + 1$ number of $L*L$ unitary matrices. The widely used Zafoff-Chu sequences can form an ECMU codebook after the power normalization. The L sequences of the same root of Zadoff-Chu (ZC) are orthogonal to each other and form an $L*L$ unitary matrix. There are L different roots and the cross-correlation between their sequences is $1/\sqrt{L}$. Hence, after the power normalization, an ECMU codebook can be obtained which contains $L - 1$ unitary matrices, each of $L*L$.

ECMU codebook has a fairly good cross-correlation property, since

- For an ECMU codebook composed of $L + 1$ different $L*L$ unitary matrices, the cross-correlation of $1/\sqrt{L}$ with different unitary matrices is the smallest in theory.
- When ECMU codebooks are used for multiple access, user grouping can be readily supported. That is, each $L*L$ unitary matrix can be L users as a group. Hence, intra-group users are orthogonal, making sure that intra-group interference is minimum. User grouping can be used in the following scenarios:
  a. Users with similar receiver SNRs can be in a group, while users with very disparate SNRs are placed in different groups. By doing this, code domain and power domain dimensions can be fully utilized to improve the performance of multiple access. Certainly, in order to get this benefit, base station scheduling is required to get precise and in-time information about the power. This may not be feasible for some systems, e.g., grant-free ones or those with very fast mobility.
  b. To ease the successive interference cancelation (SIC) in multi-user detection because in each round of SIC, we can start from the strongest set of orthogonal users, therefore reducing the number of iterations in SIC and simplifying the implementation.
  c. If the users are gradually added in, the first few users can be allocated with orthogonal sequences in ECMU codebooks so that they can be quickly decoded. Non-orthogonal sequences (from different roots) would be used for later joined-in users.

6.1.1.1.2 LCMU Codebooks

Although ECMU codebooks have very good cross-correlation property, there is a size limit of the codebook, that is, $K \leq L^2 + L$. For short-length codewords (e.g., small L), K is not large. This would limit the use of ECMU codebooks for high overloading scenarios. In addition, for some L, no complete set of unitary matrices (e.g., $L + 1$) have been found so far with the cross-correlation of $1/\sqrt{L}$. For instance, when $L = 6$, only three matrices, each being a 6*6 unitary matrix, have been found, with cross-correlation of $1/\sqrt{6}$.

If the cross-correlation requirement between different unitary matrices can be relaxed, for instance, cross-correlation larger than $1/\sqrt{L}$ is allowed, more than $L + 1$ unitary matrices, each of L*L, can be found. Certainly, multi-unitary matrices should be of practical cross-correlation value, rather than based on pure unitary characteristics. The general approach to achieve minimization of maximum cross-correlation of LCMU codebooks is still an open problem yet to be solved. However, some practical methods can be considered.

Take the MUSA codebook with 64 entries of length-4 codewords as an example, it is essentially an LCMU(4,64) codebook consisting of 16 unitary matrices and each unitary matrix is 4*4. The maximum cross-correlation between different unitary matrices is 0.7906. If the identity matrix is added, we can get an LCMU(4,68) codebook consisting of 16 unitary matrices and each unitary matrix is 4*4. Its maximum cross-correlation between different unitary matrices is still 0.7906. In contrast, for ECMU codebooks, there are at most five unitary matrices, each of size 4*4, which means 20 codewords, or ECMU(4,20). Apparently, LCMU codebooks composed of MUSA sequences are much larger than that of ECMU(4,20). However, the maximum cross-correlation between 16-unitary matrices for MUSA is 0.7906, which is larger than 1/2 for ECMU(4,20). However, still, the larger size of LCMU codebooks is very important for scenarios such as high overloading or contention-based grant-free transmission.

If we group 16 unitary matrices of the MUSA codebook into two sub-codebooks, with the maximum cross-correlation between unitary matrices in each group to be $1/\sqrt{2}$, together with an identity matrix, an LCMU(4, 36) codebook can be constructed. Apparently, LCMU(4, 36) is much larger than ECMU(4, 20). While its maximum cross-correlation is $1/\sqrt{2}$, a little larger than 1/2, such LCMU(4, 36) codebook has practical usage.

Furthermore, if we group 16 unitary matrices of the MUSA codebook into four sub-codebooks, with the maximum cross-correlation between unitary matrices in each group to be 1/2, together with an identity matrix, they essentially form ECMU (4, 20).

It should be pointed out that LCMU codebooks such as MUSA, not only can have a much larger size than ECMU codebooks but also inherit the benefit of user grouping: (1) to exploit the code domain and power domain dimension, (2) to ease SIC implementation and (3) to better support the gradual increase of access users.

6.1.1.1.3 CG-TBE Codebooks

The above-mentioned ECMU codebooks and LCMU codebooks are all made up of a set of unitary matrices, with low cross-correlation between different unitary matrices. They have certain structuredness and can be designed via analytical means. However, for some L and K, computer generation seems to be a viable y way to find codebooks to meet the total square correlation bound equity (TBE) criterion. There are two ways to generate TBE via computers:

1. If the codeword is relatively short and elements of the codeword can take only limited values, computers can be used to search through all possible permutations. Certain criteria can be imposed, e.g., to eliminate those codewords with high cross-correlations.
2. Iterative construction. There are multiple ways of iteration for TBE codebook generation. For instance, interference avoidance in [2] and the minimum mean squared error (MMSE) filter method in [3].

6.1.1.2 Welch Bound Codebooks and Equiangular Tight Frame (ETF) Codebooks

Welch bound defines the lower bound of the maximum cross-correlation of a codebook C(L, K) which is $\max\{|s_i^H s_j|\} \geq \sqrt{\frac{K-L}{L(K-1)}}$. Hence, any codebook that satisfies Welch bound is called a WBE codebook. In some literature, such type of codebook is also called Maximum Welch-Bound-Equality codebook.

WBE codebooks have a unique characteristic, that is, $|s_i^H s_j| = \sqrt{\frac{K-L}{L(K-1)}}$, meaning that the cross-correlation between any two of the codewords is the same. Hence, WBE codebooks are also called ETF codebooks. A WBE codebook is indeed a TBE codebook. However, the opposite is not true, e.g., many TBE codebooks do not satisfy WBE, which can also be seen from the above discussion of TBE codebooks. In summary, WBE is a tight bound for cross-correlation between codewords, whereas TBE can be considered as a loose bound for cross-correlation.

For short codeword spreading, WBE/ETF has two additional key properties:

1. The size of a WBE/ETF codebook has an upper limit, that is, $K \le L^2$, which is similar to that of ECMU codebooks. However, the size of a WBE/ETF codebook is often smaller than that of ECMU codebooks, meaning the overloading capability of WBE/ETF sequences has an upper limit, $\frac{K}{L} \le L$, which can cap the performance. When $K = L^2$, the cross-correlation between any two codewords of ETF is $1/\sqrt{L+1}$. For instance, for $L = 4$ ETF codebooks, at most there are 16 codewords. The cross-correlation is $1/\sqrt{5}$.
2. WBE/ETF codebooks are sparse [4], meaning that for many pairings of (L, K), ETF $C(L, K)$ does not exist, as shown in Table 6.1 marked with "-". For entries marked with "R", it means real number codebooks exist. "C" means complex number codebooks. For instance, there are no ETF codebooks for $L = 4$ with a total of 12 codewords. However, there exist ETF codebooks for $L = 4$ with total 13 codewords. These characteristics of ETF impose a certain constraint on the systems.

It should be emphasized that for some reason, in some literature and standard contributions, codebooks satisfying the TSC criterion $T_c = K^2 / L$ are called WBE codebooks. Strictly speaking, it is not consistent with the common definition of Welch bound which aims to study lower bounds on the maximum cross-correlation signals, instead of total-squared correlation. The common definition of Welch bound is to satisfy

$$\max\left\{\left| \mathbf{s}_i^H \mathbf{s}_j \right|\right\} = \sqrt{\frac{K-L}{L(K-1)}}$$

6.1.1.3 Specific Design Criteria Considering Deployment Scenarios

The main goal of non-orthogonal linear spreading is high overloading for multiple access. Such overloading can be embodied in three scenarios:

1. Grant-based scheduling with fixed overloading: the number of simultaneously transmitting users is arranged by the base station either dynamically or semi-statically. Each user is allocated with a sequence of length L, where $K > L$. A codebook for spreading is needed which contains at least K codewords. If semi-statically configured or pre-configured, the activation probability is 1.

2. Based on semi-statically configured, with variable loading: there are K potential users to be configured by the base station. The base station would allocate a sequence of length L to each user. K should be larger than L. Not all the users would access the system each time, e.g., the activation probability λ can be <1. A codebook for spreading is needed which contains at least K codewords. "Overloading" can have two meanings, either when $K > L$ or when $\lambda K > L$.
3. Grant-free overloading: each access is automatically triggered by the users. Spreading sequences are selected by users from a codebook that contains K sequences each of length L. The overload, in this case, means that the average number of active users is larger than L. Due to the autonomous selection of spreading sequences, multiple users may choose the same sequence, causing collision. To reduce the collision probability, K is often set several times bigger than L.

Obviously, the spreading sequences in the first two scenarios are allocated by the base station, or in another word, without contention. By contrast, in the third scenario, spreading sequences are automatically selected by each user, and the process is content-based.

While all the above three scenarios involve non-orthogonal access with overloading, each scenario has its own characteristics and therefore has different requirements for the codebook design. In the following, specific design considerations are elaborated for these three scenarios. To ease the representation, K, as the number of users, will be denoted as K_1, K_2 and K_3, for Scenario 1, Scenario 2 and Scenario 3, respectively.

6.1.1.3.1 Grant-Based Scheduling with Fixed Overloading

The main characteristics of this scenario are

All K_1 users that have been allocated with spreading codes would access the system. The codebook for the access can be pre-configured.

If the number of users K_1 and the spreading length L have a certain relationship, that is, when the pair (L, K_1) has to meet a certain criterion, the choice of the codebook can be based primarily on the performance, for instance, WBE/ETF codebooks. If the base station has only one receive antenna, and the received powers for all the users are the same, choosing the optimal codebook in the sense of cross-correlation, e.g., WBE/ETF, would lead to the best performance. In this case, there is no multi-user discrimination in the spatial domain or power domain. WBE/ETF codebooks ensure equal and minimum cross-correlation, thus minimizing the cross-user interference and providing the best discrimination in the spreading code domain.

However, WBE/ETF codebooks do not exist for an arbitrary number of users, as seen in Table 6.1. For example, if the system requires length $L = 4$ spreading, and also 12 users to access the system, there is no WBE/ETF (4, 12). Hence, even with the fixed number of accessing users, WBE/ETF still has flexibility issues.

If a base station has multiple receive antennas, or there is a near-far effect (e.g., different users have different receive powers), certain multi-user discrimination capability in spatial and power domains at the receivers can reduce the need for codebooks having the best characteristics in terms of cross-correlation. Either ECMU or LCMU codebook can be used to take advantage of intra-group orthogonality and inter-group low correlation.

6.1.1.3.2 Based on Semi-Static Configuration, with Variable Load

This scenario has two characteristics:

a. The number of potential users K_2 is quite large.

b. K_2 users that have been allocated the spreading codes may not try to access the system each time.

TABLE 6.1 Existence of C(*L*, *K*) Codebooks

K	L					K	L				
	2	3	4	5	6		2	3	4	5	6
3	R	R	-	-	-	20	-	-	-	-	-
4	C	R	R	-	-	21	-	-	-	C	-
5	-	-	R	R	-	22	-	-	-	-	-
6	-	R	-	R	R	23	-	-	-	-	-
7	-	C	C	-	R	24	-	-	-	-	-
8	-	-	C	-	-	25	-	-	-	C	-
9	-	C	-	-	C	26	-	-	-	-	-
10	-	-	-	R	-	27	-	-	-	-	-
11	-	-	-	C	C	28	-	-	-	-	-
12	-	-	-	-	C	29	-	-	-	-	-
13	-	-	C	-	-	30	-	-	-	-	-
14	-	-	-	-	-	31	-	-	-	-	C
15	-	-	-	-	-	32	-	-	-	-	-
16	-	-	C	-	R	33	-	-	-	-	-
17	-	-	-	-	-	34	-	-	-	-	-
18	-	-	-	-	-	35	-	-	-	-	-
19	-	-	-	-	-	36	-	-	-	-	C

For the first characteristic, the size of the codebook (L, K_2) should be large. This poses a significant challenge for the codebook optimization. If $K_2 > L^2$, WBE/ETF codebooks would not exist. If $K_2 > L^2 + L$, ECMU codebooks would no longer exist. When K_2 is very large, even if some (L, K_2) codebooks can satisfy the TBE criterion, the cross-correlation between codewords can be quite high.

The second characteristic means that during each time of access, only a subset of (L, K_2) will be used. This poses a dilemma. Even if a codebook (L, K_2) has been optimized, it would be difficult to guarantee that any subset of (L, K_2) is also optimal.

Hence, in this scenario, the benefit of the minimum cross-correlation property of WBE/ETF codebooks is significantly discounted. Instead, the small size of WBE/ETF codebooks shows its disadvantages. The size of ECMU codebooks is slightly larger than that of WBE/ETF codebooks but still may not support a large number of potential users. Therefore, LCMU codebooks, whose size is much larger than that of ECMU codebooks, can be considered. If large codebooks can be found, CG-TBE can also be considered.

6.1.1.3.3 Overloading Capability in Grant-Free

The unique characteristic of this scenario is that a large number of users can automatically select their spreading codes.

Codebook size and collision probability are directly related. Collision can significantly degrade the performance of multi-user detection. Hence, if WBE/ETF codebooks or ECMU codebooks are used, the collision probability can be excessively high, causing severe performance loss. The collision issue completely outweighs the merit of low cross-correlation of WBE/ETF or ECMU codebooks. In this scenario, codebooks of large sizes, e.g., LCMU codebooks or CG-TBE codebooks, should be considered while trying to make the cross-correlation as small as possible.

6.1.1.4 Other Design Criteria

6.1.1.4.1 Peak to Average Power Ratio

For uplink transmission, there are some requirements for PAPR. In this sense, SC-FDMA is a good choice due to its low PAPR. When symbol-level spreading is introduced, it is desirable not to increase the PAPR of SC-FDMA. Regarding the design of the codebook and the entire spreading process, the following two aspects can be considered:

- Firstly, the elements of the spreading codes should be a constant module.
- Secondly, the spreading should be spreading over the entire SC-FDMA symbol. Take $L=4$ as an example, assuming that the spreading sequence is $[1, -1, j, -j]$ which is a constant module. In addition, assuming the SC-FDMA symbol is $\boldsymbol{x}$. The meaning of "spreading over the entire SC-FDMA symbol" is repeating the SC-FDMA symbol by 4 times and multiplying by the four elements in the corresponding codeword, and getting $\boldsymbol{x}, -\boldsymbol{x}, j^*\boldsymbol{x}, -j^*\boldsymbol{x}$. It is still a constant module, and its PAPR would not change. There are multiple ways of implementation for "spreading over the entire SC-FDMA symbol".

6.1.1.4.2 Complexity of Receiver Implementation

With the introduction of spreading, related processing should be added, for instance, symbol-level spreading at the transmitter, de-spreading at the receiver, and memory management for codebooks. The burden rests more on the receiver, whereas for the multi-user detector, many more correlations should be carried out. For the data-only blind receiver [5,6], the number of correlations is even more.

Both spreading and correlation involve multiplication for each element of the codeword. If the elements are complex numbers, complex multiplication should be carried out which is complicated. Hence, if a certain cross-correlation criterion can be met, it is better to use the spreading sequence whose elements are simple, e.g., with finite values. Multi-user shared access (MUSA) and non-orthogonal code access (NOCA) are two examples which are the simplest complex codebooks that can satisfy cross-correlation requirements. A common character of these two codebooks is that the real and imaginary components are all selected from {1, 0, −1} so that no multiplication is actually needed in the processing. Certainly, the binary pseudo noise (PN) sequence (elements are chosen from {1, −1}) is also very simple. However, when the spreading length is small, the number of such PN sequences is small, or with poor cross-correlation, and cannot be used for the heavily overloaded case.

6.1.2 Description of Specific Codebooks

For symbol-level spreading, the design is focused on the codebook construction. To reduce the cross-user interference, normally sequences with low cross-correlation or with sparsity would be chosen. Due to the requirement for high spectral efficiency, 5G systems would operate with high overloading. Hence, a short spreading length (e.g., <12) is more preferable. This also helps to reduce the complexity of MMSE in the receiver. In this subsection, several schemes based on short-length spreading are discussed.

6.1.2.1 Codebooks with Highly Quantized Elements (MUSA and NOCA)

Compared to the widely used binary PN sequence (elements can take {1, –1}), complex-valued sequences can provide more codewords, under the same length of the sequence. As illustrated in Figure 6.2, the real and imaginary components of the elements can take {–1, 0, 1}, corresponding to the 9-QAM constellation, or just take {–1, 1}, corresponding to the quadrature phase-shift keying (QPSK) constellation.

In theory, there are total 9^L sequences with length L whose elements can be chosen from 9-QAM. If the constellation is QPSK, there are 4^L sequences. In practical engineering, detection complexity and cross-correlation property should be considered jointly. As listed in Table 6.2, the more stringent requirement for cross-correlation, the less number of sequences would be that can satisfy the requirement. The looser the requirement, the more sequences, but with more severe cross-user interference.

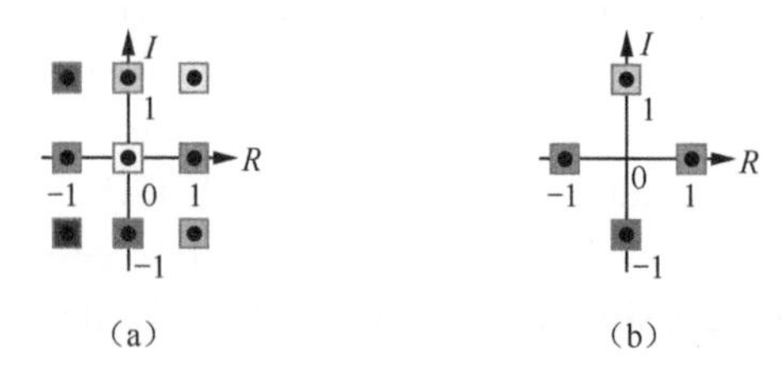

FIGURE 6.2 Complex-valued elements of codewords shown in constellations. (a) 9-QAM and (b) QPSK.

TABLE 6.2 Pool Size Comparison between Sequences with QPSK Element and PN Sequences ($L = 4$)

Maximum Cross-Correlation within the Pool $\max\{\lvert s_i^H s_j \rvert\}$	Pool Size of Sequences Satisfying the Cross-Correlation Criterion: QPSK Complex Sequence	PN Sequence
0.5	20	8
0.7071	64	8
0.7906	156	8

In general, the number of complex-valued sequences is much larger than that of binary PN sequences.

Typical quantization-based complex spreading sequences include MUSA sequences and NOCA sequences. The detailed construction methods for those sequences are described below.

6.1.2.1.1 MUSA Sequences

MUSA sequences [7] can be obtained by extending either orthogonal or non-orthogonal base sequences. For instance, the set of sequences with length L should include at least L base sequences, e.g., Hadamard sequences, Walsh sequences, identity matrix sequences, discrete Fourier Transform sequences, and sequences with a certain number of zero elements. On top of these base sequences, the codebook size can be expanded by rotating the elements of each sequence by a certain phase angle, e.g., 1, j, −1, $-j$, and $e^{j\alpha\pi}$, or setting to zero.

For instance, sequences in Tables 6.3, 6.4b and 6.5 are all extended from orthogonal sequences. These three groups of MUSA sequences are LCMU codebooks. Among them, the codebooks in Tables 6.3 and 6.4b are also ECMU codebooks.

Another example is shown in Table 6.4a where the base sequences are [1, 1, 0], [1, 0, 1],[0, 1, 1]. By multiplying the two non-zero elements by ω and ω^2, or ω^2 and ω, respectively. Totally 9 non-orthogonal sequences can be obtained which have very good cross-correlation property, e.g., in fact, they are WBE/ETF codebooks (Table 6.6).

TABLE 6.3 Examples of MUSA Sequences, $L = 2$, $K = 6$

Sequence Index	Element 1	Element 2
1	1	1
2	1	−1
3	1	j
4	1	$-j$
5	1	0
6	0	1

TABLE 6.4a Examples of MUSA Sequences, $L = 3$, $K = 9$

$$\frac{1}{\sqrt{2}}\begin{bmatrix} 1 & \omega^2 & \omega & 1 & \omega^2 & \omega & 0 & 0 & 0 \\ 1 & \omega & \omega^2 & 0 & 0 & 0 & 1 & \omega^2 & \omega \\ 0 & 0 & 0 & 1 & \omega & \omega^2 & 1 & \omega & \omega^2 \end{bmatrix}, \omega = e^{j\frac{2\pi}{3}}$$

TABLE 6.4b Examples of MUSA Sequences, $L = 3$, $K = 12$

$$\frac{1}{\sqrt{3}}\begin{bmatrix} 1 & 1 & 1 & 1 & 1 & 1 & 1 & 1 & 1 & 1 & 0 & 0 \\ 1 & \omega & \omega^2 & \omega^2 & 1 & \omega & \omega & \omega^2 & 1 & 0 & 1 & 0 \\ 1 & \omega^2 & \omega & \omega^2 & \omega & 1 & \omega & 1 & \omega^2 & 0 & 0 & 1 \end{bmatrix}, \omega = e^{j\frac{2\pi}{3}}$$

TABLE 6.5 Examples of MUSA Sequences, $L = 4$, $K = 64$

Sequence Index	Element 1	Element 2	Element 3	Element 4	Sequence index	Element 1	Element 2	Element 3	Element 4
1	1	1	1	1	**33**	1	1	1	–*j*
2	1	1	–1	–1	**34**	1	1	–1	*j*
3	1	–1	1	–1	**35**	1	–1	1	*j*
4	1	–1	–1	1	**36**	1	–1	–1	–*j*
5	1	1	–*j*	*j*	**37**	1	1	–*j*	1
6	1	1	*j*	–*j*	**38**	1	1	*j*	–1
7	1	–1	–*j*	–*j*	**39**	1	–1	–*j*	–1
8	1	–1	*j*	*j*	**40**	1	–1	*j*	1
9	1	–*j*	1	*j*	**41**	1	–*j*	1	1
10	1	–*j*	–1	–*j*	**42**	1	–*j*	–1	–1
11	1	*j*	1	–*j*	**43**	1	*j*	1	–1
12	1	*j*	–1	*j*	**44**	1	*j*	–1	1
13	1	–*j*	–*j*	–1	**45**	1	–*j*	–*j*	*j*
14	1	–*j*	*j*	1	**46**	1	–*j*	*j*	–*j*
15	1	*j*	–*j*	1	**47**	1	*j*	–*j*	–*j*
16	1	*j*	*j*	–1	**48**	1	*j*	*j*	*j*
17	1	1	1	–1	**49**	1	1	1	*j*
18	1	1	–1	1	**50**	1	1	–1	–*j*
19	1	–1	1	1	**51**	1	–1	1	–*j*
20	1	–1	–1	–1	**52**	1	–1	–1	*j*
21	1	1	–*j*	-j	**53**	1	1	–*j*	–1
22	1	1	*j*	j	**54**	1	1	*j*	1
23	1	–1	–*j*	j	**55**	1	–1	–*j*	1
24	1	–1	*j*	-j	**56**	1	–1	*J*	–1
25	1	–*j*	1	-j	**57**	1	–*j*	1	–1
26	1	–*j*	–1	j	**58**	1	–*j*	–1	1
27	1	*j*	1	j	**59**	1	*j*	1	1
28	1	*j*	–1	-j	**60**	1	*j*	–1	–1
29	1	–*j*	–*j*	1	**61**	1	–*j*	–*j*	–*j*
30	1	–*j*	*j*	–1	**62**	1	–*j*	*j*	*j*
31	1	*j*	–*j*	–1	**63**	1	*j*	–*j*	*j*
32	1	*j*	*j*	1	**64**	1	*j*	*j*	–*j*

TABLE 6.6 Examples of MUSA Sequences, $L = 6$, $K = 16$

Sequence Index	Element 1	Element 2	Element 3	Element 4	Element 5	Element 6
1	1	1	1	1	1	1
2	1	1	1	1	−1	−1
3	1	1	1	−1	1	−1
4	1	1	1	−1	−1	1
5	1	1	−1	1	1	−1
6	1	1	−1	1	−1	1
7	1	1	−1	−1	1	1
8	1	1	−1	−1	−1	−1
9	1	−1	1	1	1	−1
10	1	−1	1	1	−1	1
11	1	−1	1	−1	1	1
12	1	−1	1	−1	−1	−1
13	1	−1	−1	1	1	1
14	1	−1	−1	1	−1	−1
15	1	−1	−1	−1	1	−1
16	1	−1	−1	−1	−1	1

6.1.2.1.2 NOCA Sequence

The way of generating NOCA sequences [8] is very similar to that of the demodulation reference signal (DMRS) sequences in long-term evolution (LTE). For sequences of length L, their generation formula is $r_{u,v}(n) = \exp\left(\frac{j\phi(n)\pi}{4}\right)$, $0 \le n \le L-1$, where the value of the root and the corresponding phase of each QPSK element are shown in Tables 6.7–6.9. These values are obtained by computer search to find a good set of sequences with relatively low cross-correlation and low PAPR. For each root sequence, there are L cyclic shifts. The total number of sequences in the codebook is the number of root sequences K and the number of cyclic shifts for each root sequence.

6.1.2.2 Sequences Satisfying Total-Squared-Correlation Bound (TBE)

For some TBE codebooks such as Welch-bound equality spread multiple access (WSMA), RSMA and MUSA, their design criterion is to minimize the sum of cross-correlation $T_c \triangleq \sum_{i,j} |\boldsymbol{s}_i^H \boldsymbol{s}_j|^2$ so that the cross-user interference can be reduced. For a given spreading length of L and codebook size K, according to Cauchy–Schwarz Inequality, the constraint on sum cross-correlation can be represented as $K^2 / L \le T_c$. Welch bound defines the

TABLE 6.7 Examples of NOCA Root Sequences, $L = 4$, $K = 40$

u	$\varphi(0),\ldots,\varphi(3)$			
0	3	3	1	3
1	−3	−3	−3	1
2	−3	−1	−1	−1
3	−1	−1	−3	−3
4	1	3	−1	−1
5	1	−1	−1	−3
6	−3	1	−1	−3
7	1	1	3	−3
8	1	3	1	−3
9	−1	3	1	−3

minimum sum of cross-correlation, which is $B_{\text{Welch}} \triangleq K^2 / L$. It should be pointed out that WSMA [9], although being the acronym of "Welch bound equality spreading multiple access", is in fact based on total-squared cross-correlation equality, as stated in [9].

> The WBE sequences are designed to meet the bound on the total-squared cross-correlations of the vector set with equality $B_{\text{Welch}} \triangleq K^2 / N$. We call such sequences Welch bound equality spread multiple access (WSMA).

Hence, WSMA codebooks belong to TBE sequences. The detailed design of WSMA sequences be found in [9]. Tables 6.10 and 6.11 provide some examples.

RSMA is based on the Chirp sequence (often used in Radar and underwater communications, it has been recently adopted by low-power systems such as IEEE 802.15.4a). RSMA sequences satisfy the TBE criterion and RSMA codebook of size K and spreading length L can be represented in a closed form

$$s_k(l) \triangleq \frac{1}{\sqrt{L}} \exp\left(j\pi \left(\frac{(k+l)^2}{K} \right) \right); \; 1 \le k \le K, 1 \le l \le L \qquad (6.1)$$

According to Eq. (6.1), the cross-correlation can be calculated [10] as shown in Figure 6.3. It can be observed that RSMA sequences can satisfy the minimum sum of cross-correlations.

TABLE 6.8 Examples NOCA Root Sequences, $L = 6$, $K = 180$

u	$\varphi(0),\ldots,\varphi(5)$					
0	−1	−3	3	−3	3	−3
1	−1	3	−1	1	1	1
2	3	−1	−3	−3	1	3
3	3	−1	−1	1	−1	−1
4	−1	−1	−3	1	−3	−1
5	1	3	−3	−1	−3	3
6	−3	3	−1	−1	1	−3
7	−1	−3	−3	1	3	3
8	3	−1	−1	3	1	3
9	3	−3	3	1	−1	1
10	−3	1	−3	−3	−3	−3
11	−3	−3	−3	1	−3	−3
12	3	−3	1	−1	−3	−3
13	3	−3	3	−1	−1	−3
14	3	−1	1	3	3	1
15	−1	1	−1	−3	1	1
16	−3	−1	−3	−1	3	3
17	1	−1	3	−3	3	3
18	1	3	1	1	−3	3
19	−1	−3	−1	−1	3	−3
20	3	−1	−3	−1	−1	−3
21	3	1	3	−3	−3	1
22	1	3	−1	−1	1	−1
23	−3	1	−3	3	3	3
24	1	3	−3	3	−3	3
25	−1	−1	1	−3	1	−1
26	1	−3	−1	−1	3	1
27	−3	−1	−1	3	1	1
28	−1	3	−3	−3	−3	3
29	3	1	−1	1	3	1

6.1.2.3 Cyclic Difference Set ETF and Grassmannian Sequence (NCMA)

For a given L and K, ETF sequences can be constructed by searching cyclic differences. The difference set is defined as: taking a positive integer K as the module, the L non-congruent integers form the set $D \equiv \{d_1, d_2, \ldots, d_L\} \pmod K$. If for each $a \neq 0 \pmod K$, there are γ_{-} ordered pairs (d_i, d_j) in D, such that $a \equiv d_i - d_j \pmod K$. Then D is called a

TABLE 6.9 Examples of NOCA Root Sequences, $L=12$, $K=360$

u	$\varphi(0),\ldots,\varphi(11)$											
0	−1	1	3	−3	3	3	1	1	3	1	−3	3
1	1	1	3	3	3	−1	1	−3	−3	1	−3	3
2	1	1	−3	−3	−3	−1	−3	−3	1	−3	1	−1
3	−1	1	1	1	1	−1	−3	−3	1	−3	3	−1
4	−1	3	1	−1	1	−1	−3	−1	1	−1	1	3
5	1	−3	3	−1	−1	1	1	−1	−1	3	−3	1
6	−1	3	−3	−3	−3	3	1	−1	3	3	−3	1
7	−3	−1	−1	−1	1	−3	3	−1	1	−3	3	1
8	1	−3	3	1	−1	−1	−1	1	1	3	−1	1
9	1	−3	−1	3	3	−1	−3	1	1	1	1	1
10	−1	3	−1	1	1	−3	−3	−1	−3	−3	3	−1
11	3	1	−1	−1	3	3	−3	1	3	1	3	3
12	1	−3	1	1	−3	1	1	1	−3	−3	−3	1
13	3	3	−3	3	−3	1	1	3	−1	−3	3	3
14	−3	1	−1	−3	−1	3	1	3	3	3	−1	1
15	3	−1	1	−3	−1	−1	1	1	3	1	−1	−3
16	1	3	1	−1	1	3	3	3	−1	−1	3	−1
17	−3	1	1	3	−3	3	−3	−3	3	1	3	−1
18	−3	3	1	1	−3	1	−3	−3	−1	−1	1	−3
19	−1	3	1	3	1	−1	−1	3	−3	−1	−3	−1
20	−1	−3	1	1	1	1	3	1	−1	1	−3	−1
21	−1	3	−1	1	−3	−3	−3	−3	−3	1	−1	−3
22	1	1	−3	−3	−3	−3	−1	3	−3	1	−3	3
23	1	1	−1	−3	−1	−3	1	−1	1	3	−1	1
24	1	1	3	1	3	3	−1	1	−1	−3	−3	1
25	1	−3	3	3	1	3	3	1	−3	−1	−1	3
26	1	3	−3	−3	3	−3	1	−1	−1	3	−1	−3
27	−3	−1	−3	−1	−3	3	1	−1	1	3	−3	−3
28	−1	3	−3	3	−1	3	3	−3	3	3	−1	−1
29	3	−3	−3	−1	−1	−3	−1	3	−3	3	1	−1

difference set (K,L,λ). Assuming $K=13$, then $D=(1,2,4,10)$ is a difference set of $(13,4,1)$, which can be verified as follows:

$$1\equiv 2-1,\ 2\equiv 4-2,\ 3\equiv 4-1,\ 4\equiv 1-10,\ 5\equiv 2-10,\ 6\equiv 10-4,$$

$$7\equiv 4-10,\ 8\equiv 10-2,\ 9\equiv 10-1,\ 10\equiv 1-4,\ 11\equiv 2-4,\ 12\equiv 1-2$$

Assuming $K=11$, then $D=\{2,6,7,8,10,11\}$ is a difference set of $(11,6,3)$, which can be verified as follows:

TABLE 6.10 Examples of WSMA Sequences, $L = 4$, $K = 8$

	Sequence Index	1	2	3	4
Element Values	1	−0.6617 + 0.1004*i*	−0.0912 + 0.4191*i*	0.4151 − 0.3329*i*	0.2736 − 0.4366*i*
	2	0.0953 + 0.4784*i*	−0.4246 − 0.0859*i*	0.2554 − 0.3140*i*	0.5452 + 0.2068*i*
	3	−0.4233 − 0.1399*i*	−0.4782 + 0.3752*i*	−0.3808 − 0.1569*i*	−0.4690 − 0.2225*i*
	4	−0.1265 + 0.3153*i*	0.4936 + 0.1233*i*	0.6130 − 0.0873*i*	−0.3399 + 0.0974*i*
	Sequence Index	**5**	**6**	**7**	**8**
Element Values	1	−0.4727 − 0.1234*i*	−0.3413 + 0.1257*i*	0.4216 + 0.1187*i*	0.4603 + 0.2142*i*
	2	0.0592 − 0.6432*i*	0.3671 − 0.1430*i*	−0.0241 − 0.5620*i*	0.0048 − 0.4244*i*
	3	0.3493 − 0.1988*i*	0.6514 − 0.0660*i*	−0.4507 + 0.0958*i*	0.4047 + 0.1601*i*
	4	−0.0975 − 0.4161*i*	0.2174 + 0.4864*i*	−0.5167 + 0.1116*i*	−0.4908 + 0.3629*i*

TABLE 6.11 Examples of WSMA Sequences, $L = 4$, $K = 12$

	Sequence Index	1	2	3	4
Element Values	1	−0.2221 + 0.3220*i*	−0.0690 − 0.5020*i*	−0.4866 + 0.3090*i*	0.4007 − 0.3034*i*
	2	0.1709 − 0.3679*i*	−0.2222 − 0.2729*i*	−0.4148 − 0.2589*i*	−0.3206 − 0.0231*i*
	3	0.4335 − 0.4253*i*	0.0875 − 0.3912*i*	0.5181 + 0.0067*i*	−0.6714 − 0.0514*i*
	4	−0.2877 + 0.4804*i*	0.6669 − 0.1183*i*	−0.3439 − 0.2048*i*	−0.2117 − 0.3819*i*
	Sequence index	**5**	**6**	**7**	**8**
Element Values	1	0.0525 − 0.6492*i*	−0.3121 + 0.4136i	0.1887 − 0.5138*i*	0.3628 − 0.5556*i*
	2	0.2786 + 0.2173*i*	−0.5533 + 0.2843*i*	−0.5603 + 0.0403*i*	−0.2496 − 0.3482*i*
	3	0.4058 − 0.3688*i*	−0.3497 + 0.2042*i*	0.3714 − 0.0660*i*	0.4539 − 0.0605*i*
	4	−0.0586 − 0.3831*i*	0.4123 + 0.1027*i*	0.3124 + 0.3807*i*	−0.2014 − 0.3549*i*
	Sequence Index	**9**	**10**	**11**	**12**
Element Values	1	−0.4067 − 0.0166*i*	−0.2969 − 0.2084*i*	0.3160 + 0.0753*i*	0.3612 − 0.2061*i*
	2	0.5821 − 0.2559*i*	−0.5414 − 0.1665*i*	−0.7029 − 0.1267*i*	0.3525 − 0.0158*i*
	3	0.1316 − 0.2310*i*	−0.1075 + 0.6412*i*	0.3540 − 0.2274*i*	−0.4880 − 0.1396*i*
	4	0.5222 − 0.2944*i*	0.2613 − 0.2380*i*	−0.3490 − 0.2925*i*	−0.5884 − 0.3142*i*

$$1 \equiv 7-6 \equiv 8-7 \equiv 11-10;\ 2 \equiv 8-6 \equiv 10-8 \equiv 2-11;$$

$$3 \equiv 10-7 \equiv 11-8 \equiv 2-10;\ 4 \equiv 6-2 \equiv 10-6 \equiv 11-7;$$

$$5 \equiv 7-2 \equiv 11-6 \equiv 2-8;\ 6 \equiv 8-2 \equiv 2-7 \equiv 6-11;$$

$$7 \equiv 2-6 \equiv 6-10 \equiv 7-11;\ 8 \equiv 10-2 \equiv 7-10 \equiv 8-11;$$

$$9 \equiv 11-2 \equiv 6-8 \equiv 8-10;\ 10 \equiv 6-7 \equiv 7-8 \equiv 10-11;$$

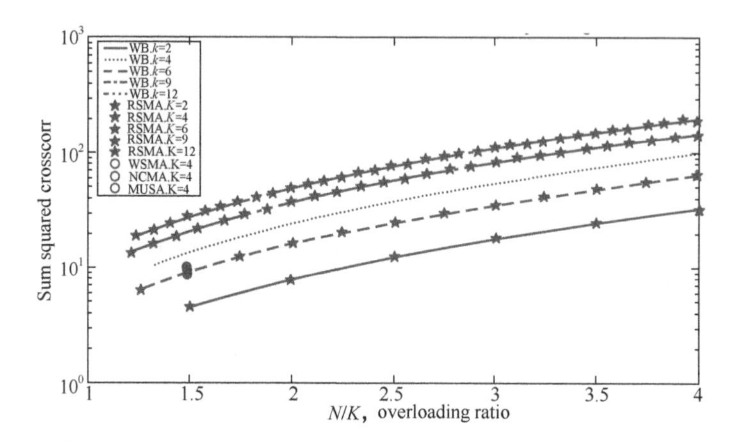

FIGURE 6.3 Cross-correlation of Chirp sequences for RSMA.

Once the difference set D is obtained, the ETF sequence set can be calculated as

$$s_k(l) \triangleq \frac{1}{\sqrt{L}} \exp\left(j2\pi \left(\frac{d_l * (k-1)}{K} \right) \right); \quad 1 \le k \le K, 1 \le l \le L, \quad d_l \in D. \quad (6.2)$$

It should be noted that the necessary condition of the existence of a different set is $\gamma(K-1) = L(L-1)$. For some values of L and K, such conditions cannot be met.

The Grassmannian sequences of NCMA in Ref. [11] also satisfy the ETF criterion. A few typical NCMA sequences are shown in Tables 6.12 and 6.13.

6.1.2.4 General Total Squared Correlation Bound Equality (GTBE) Sequences, e.g., UGMA

In realistic NOMA systems, because of the near-far effect and the fast fading, the minimum sum-squared may not be equivalent to the minimum of cross-user interference. Hence in theory, it makes more sense to introduce a general meaning of the TSC-bound criterion. UGMA are the sequences that satisfy this generalized criterion [12], that is, $\min_{s_k^H s_k = 1 \ \forall k} R_x = \boldsymbol{S}^H \boldsymbol{P} \boldsymbol{S}_F^2 = \sum_{i=1}^{K} \sum_{j=1}^{K} P_i P_j \left| \boldsymbol{s}_i^H \boldsymbol{s}_j \right|^2$ where P_i and P_j denote the signal power of user I and user j, respectively. Considering the weighted Cauchy–Schwarz inequality $\sum_{i=1}^{K} \sum_{j=1}^{K} P_i P_j \left| \boldsymbol{s}_i^H \boldsymbol{s}_j \right|^2 \ge \frac{\left(\sum_{k=1}^{K} P_k \right)^2}{L}$, only when

TABLE 6.12 Examples of NCMA Sequences, $L = 4$

Size K	Sequence Sets (Each Column Corresponds to a Sequence)					
8	−0.3769 −0.1993i	−0.4946 + 0.0729i	−0.0349 − 0.1744i	−0.4983 − 0.2361i		
	0.0071 − 0.4246i	0.0484 + 0.2172i	−0.4864 + 0.5118i	0.3678 − 0.0002i		
	−0.7438 − 0.2074i	0.1526 − 0.5642i	−0.1478 + 0.1545i	0.6445 + 0.1123i		
	0.0662 − 0.1932i	0.1281 − 0.5852i	−0.3512 + 0.5484i	−0.1883 + 0.3118i		
	−0.0589 −0.2775i	−0.3141 − 0.2162i	−0.3118 − 0.2513i	−0.6128 + 0.4861i		
	0.6654 − 0.2483i	0.2752 + 0.0869i	−0.0147 + 0.3864i	−0.3671 + 0.3724i		
	−0.4067 + 0.4932i	−0.2122 − 0.4038i	−0.3986 + 0.2848i	−0.1428 − 0.0632i		
	0.072 + 0.0362i	−0.5858 + 0.4691i	0.5659 − 0.3604i	0.282 + 0.104i		
12	−0.1211 + 0.1742i	−0.1864 + 0.1486i	−0.4450 − 0.2565i	−0.1650 + 0.3506i	−0.4503 + 0.2070i	−0.3310 − 0.2575i
	0.5284 − 0.0028i	0.5630 − 0.0523i	−0.5537 + 0.0264i	0.2754 + 0.1722i	0.0650 − 0.1528i	−0.5335 + 0.6004i
	0.1518 − 0.5314i	0.2665 − 0.4503i	0.3965 + 0.2446i	−0.2259 + 0.3311i	−0.1173 + 0.3294i	−0.1120 − 0.2999i
	0.3043 + 0.5270i	0.2024 − 0.5556i	0.3116 − 0.3387i	−0.3280 − 0.6900i	0.6983 + 0.3420i	−0.1290 + 0.2449i
	−0.2344 − 0.1865i	−0.4251 + 0.0869i	−0.2091 − 0.5656i	−0.8263 − 0.3684i	−0.5363 − 0.1981i	−0.6964 − 0.1831i
	0.1663 − 0.2439i	0.6626 − 0.4120i	−0.1403 − 0.1177i	0.1024 + 0.0356i	0.3090 − 0.5397i	0.1029 + 0.2755i
	0.7183 − 0.0739i	−0.0365 − 0.0355i	−0.0380 − 0.3106i	0.2040 − 0.3275i	−0.1106 + 0.2210i	−0.0585 + 0.6228i
	−0.4388 − 0.3303i	−0.3826 + 0.2322i	−0.1052 − 0.7027i	0.1073 + 0.0961i	0.2328 − 0.4136i	−0.0382 − 0.0488i

TABLE 6.13 Examples of Quantized NCMA Sequences, L = 4

Size K	Sequence Sets (Each Column Corresponds to a Sequence)											
8	−5 − 3i	−7 + 1i	−1 − 3i	−7 − 3i	−1 − 3i	−5 − 3i	−5 − 3i	−7 +7i				
	1 − 5i	1 + 3i	−7 + 7i	5 − 1i	7 − 3i	3 + 1i	−1 + 5i	−5 + 5i				
	−7 − 3i	1 − 7i	−1 + 3i	7 + 1i	−5 + 7i	−3 − 5i	−5 + 3i	−1 − 1i				
	1 − 3i	1 − 7i	−5 + 7i	−3 + 5i	1 + 1i	−7 + 7i	7 − 5i	3 + 1i				
12	−1 + 3i	−3 + 1i	−5 − 3i	−3 + 5i	−5 + 3i	−5 − 3i	−3 − 3i	−5 + 1i	−3 − 7i	−7 − 5i	−7 − 3i	−7 − 3i
	7 − 1i	7 − 1i	−7 + 1i	3 + 3i	1 − 1i	−7 + 7i	3 − 3i	7 − 5i	−1 − 1i	1 + 1i	5 − 7i	1 + 3i
	1 − 7i	3 − 5i	5 + 3i	−3 + 5i	−1 + 5i	−1 − 3i	7 − 1i	−1 − 1i	−1 − 5i	3 − 5i	−1 + 3i	−1 + 7i
	3 + 7i	3 − 7i	5 − 5i	−5 − 7i	7 + 5i	−1 + 3i	−5 − 5i	−5 + 3i	−1 − 7i	1 + 1i	3 − 5i	−1 − 1i

in equality can the sequences be claimed to meet the general sense of the TSC bound. More specifically, given the spreading length of L and the size of the codebook to be K, and the weights of each user $\{P_1, P_2, \ldots, P_K\}$, the sequence design of UGMA is as follows:

1. To find the set $\mathcal{K}$, so for each $k \in \mathcal{K}$, its power satisfies $P_k > \dfrac{\sum_{i=1}^{K} P_i \cdot \text{sign}(P_i > P_k)}{L - \sum_{i=1}^{K} \text{sign}(P_i \geq P_k)}$;
2. Based on the generalized Chan-Li Algorithm or Bendel-Mickey Algorithm, construct the matrix $Q \in C^{(K-|K|)\times(K-|K|)}$ whose diagonal elements are $\{P_i \mid i \notin K\}$, and eigenvalues are $\left[\dfrac{\sum_{i \notin K} P_i}{L - |\mathcal{K}|} 1_{L-|\mathcal{K}|}^T, 0_{(K-L)\times 1}^T\right]^T$;
3. To carry out the singular value decomposition of the matrix Q, e.g., $Q = \text{UVU}^H$;
4. To choose non-zero elements from V to get $\breve{\Lambda} = \dfrac{\sum_{i \notin K} P_i}{L - |K|} \boldsymbol{I}_{L-|K|}$, as well as the corresponding eigenvectors $\breve{U} \in C^{(K-|K|)\times(K-|K|)}$;
5. To construct the vector $\breve{S} = \breve{\Lambda}\breve{U}\breve{P}^{-\frac{1}{2}}$, where $\breve{P} = \text{diag}\{P_i \mid i \notin K\}$;
6. To construct the set of sequences $\mathbf{S} = C_{\text{orth}} \begin{bmatrix} I_{|K|} & 0_{|K|\times(L-|K|)} \\ 0_{(L-|K|)\times|K|} & \breve{S} \end{bmatrix}$, where $C_{\text{orth}} \mathbb{C}^{L\times L}$ is any unitary matrix that satisfies $C_{\text{orth}} C_{\text{orth}}^H = I_L$

Tables 6.14–6.17 list several examples of GTBE sequences.

6.1.2.5 Sparse Spreading Sequences, e.g., PDMA

Sparse spreading sequences can help to reduce the cross-user interference by setting some elements in the sequences to zero. There are multiple ways to achieve sparsity. One straightforward way is to let each sequence have only two-valued elements: 1 or 0, and the number of zero-valued elements per sequence is the same. Such design ensures equal interference between any two users. However, it significantly limits the number of sequences. For instance, there are only six sequences of a length of 4 and 50% sparsity.

TABLE 6.14 Examples of UGMA Sequences, $L = 4$, $K = 8$, Two Groups with Power Difference = 6 dB

Sequence Index	High-Power Group		Sequence Index	Low-Power Group	
1	$-0.3068 - 0.4002i$	$-0.1823 - 0.2575i$	5	-0.5	-0.5
	$0.2787 + 0.4238i$	$0.5287 - 0.3308i$		-0.5	-0.5
2	$-0.0229 + 0.3563i$	$0.869 - 0.2734i$	6	$-0.8869 - 0.2366i$	$0.1684 - 0.0164i$
	$0.0574 + 0.021i$	$-0.0142 - 0.1965i$		$0.1991 + 0.2087i$	$-0.1898 + 0.0979i$
3	$-0.1936 - 0.4658i$	$0.2822 - 0.0885i$	7	$-0.118 - 0.1499i$	$-0.0994 - 0.2647i$
	$-0.0736 + 0.3716i$	$-0.7151 + 0.0569i$		$0.377 - 0.381i$	$-0.6336 - 0.4415i$
4	$-0.3066 + 0.3717i$	$-0.2369 - 0.2155i$	8	$0.5835 - 0.2329i$	$-0.407 + 0.0047i$
	$-0.7377 + 0.2445i$	$0.0236 - 0.2465i$		$0.3753 - 0.453i$	$-0.2874 + 0.1048i$

TABLE 6.15 Examples of Quantized UGMA Sequences, $L = 4$, $K = 8$, Two Groups with Power Difference = 6 dB

Sequence Index	High-Power Group				Sequence Index	Low-Power Group			
1	$-1-i$	$-i$	$1+i$	$2-i$	5	-1	-1	-1	-1
2	i	$2-i$	0	$-i$	6	$-2-i$	1	$1+i$	-1
3	$-1-i$	1	I	-2	7	0	$-i$	$1-i$	$-2-i$
4	$-1+i$	$-1-i$	$-2+i$	$-i$	8	$2-i$	-1	$1-i$	-1

$$\begin{bmatrix}1\\1\\0\\0\end{bmatrix},\begin{bmatrix}0\\0\\1\\1\end{bmatrix},\begin{bmatrix}1\\0\\1\\0\end{bmatrix},\begin{bmatrix}0\\1\\0\\1\end{bmatrix},\begin{bmatrix}1\\0\\0\\1\end{bmatrix},\begin{bmatrix}0\\1\\1\\0\end{bmatrix}.$$

A more flexible design of sparse spreading sequences would allow a varying number of zeros in each sequence, for instance in PDMA sequences [13]. Its design principle is to allow different weights in different columns so that the cross-user interference has certain randomness, therefore benefiting the SIC (Table 6.18).

In addition, sparse spreading can also have complex numbers as the non-zero elements, which can help to enlarge the codebook size, as Table 6.19 shows.

6.1.2.6 Summary

In the above discussion, spreading sequences are roughly categorized based on their main design criterion. For realistic sequence designs, usually

TABLE 6.16 Examples of UGMA Sequences, $L = 4$, $K = 12$, Two Groups with Power Difference = 6 dB

Sequence Index	High-Power Group		Sequence Index	Low-Power Group	
1	$0.1904 + 0.0145i$	$0.7272 - 0.514i$	7	-0.5	0.5
	$-0.2103 - 0.2594i$	$0.2418 - 0.024i$		0.5	-0.5
2	$0.3224 - 0.3367i$	$-0.0836 + 0.6834i$	8	$-0.4686 - 0.1212i$	$-0.3517 - 0.4534i$
	$-0.1767 + 0.375i$	$0.3294 - 0.1684i$		$-0.059 + 0.1422i$	$0.5211 - 0.3757i$
3	$0.1415 + 0.2701i$	$0.3498 + 0.0601i$	9	$0.1485 + 0.1008i$	$-0.0978 - 0.1169i$
	$-0.5172 + 0.3126i$	$0.1143 - 0.6347i$		$-0.5555 + 0.3244i$	$-0.5181 + 0.5121i$
4	$0.0739 + 0.3791i$	$0.4316 + 0.0877i$	10	$-0.7443 + 0.2562i$	$-0.0277 - 0.3199i$
	$-0.4466 + 0.191i$	$-0.1158 + 0.6383i$		$-0.017 + 0.0495i$	$0.4254 - 0.3058i$
5	$-0.4567 + 0.3433i$	$0.0599 + 0.0521i$	11	$-0.3393 - 0.0337i$	$-0.0657 - 0.1985i$
	$0.5939 + 0.4209i$	$0.2542 + 0.2697i$		$0.1794 - 0.4201i$	$-0.4464 - 0.6574i$
6	$0.0308 - 0.8351i$	$0.0303 - 0.0103i$	12	$0.3945 - 0.0972i$	$0.3061 + 0.5307i$
	$0.026 - 0.4612i$	$0.0799 + 0.2843i$		$-0.1173 - 0.2804i$	$0.0612 - 0.6028i$

TABLE 6.17 Examples of Quantized UGMA Sequences, $L = 4$, $K = 12$, Two Groups with Power Difference = 6 dB

Sequence Index	High-Power Group				Sequence Index	Low-Power Group			
1	1	$2-i$	$-1-i$	1	7	-1	2	2	-1
2	$1-i$	$2i$	i	1	8	-1	$-1-i$	I	$2-i$
3	$1+i$	1	$-1+i$	$1-2i$	9	1	0	$-2+i$	$-1+2i$
4	i	2	$-1+i$	$2i$	10	$-2+i$	$-i$	0	$2-i$
5	$-1+i$	0	$2+2i$	$1+i$	11	-1	0	$1-i$	$-1-2i$
6	$-2i$	0	$-i$	i	12	1	$1+2i$	$-i$	$-2i$

TABLE 6.18 Examples of PDMA Sequences, $L = 4$, Element Values Can Be {0, 1}

Codebook Size K	Sparse Spreading Sequences (Not Normalized)
6	$G_{\text{PDMA,Type1}}^{[4,6]} = \begin{bmatrix} 1 & 1 & 0 & 1 & 1 & 0 \\ 1 & 0 & 1 & 1 & 0 & 1 \\ 0 & 1 & 1 & 0 & 1 & 1 \\ 1 & 1 & 1 & 0 & 0 & 0 \end{bmatrix}$
8	$G_{\text{PDMA,Type1}}^{[4,8]} = \begin{bmatrix} 1 & 0 & 0 & 1 & 1 & 0 & 0 & 0 \\ 0 & 1 & 1 & 0 & 0 & 1 & 0 & 0 \\ 0 & 1 & 0 & 1 & 0 & 0 & 1 & 0 \\ 1 & 0 & 1 & 0 & 0 & 0 & 0 & 1 \end{bmatrix}$
12	$G_{\text{PDMA,Type1}}^{[4,12]} = \begin{bmatrix} 1 & 0 & 1 & 1 & 1 & 0 & 0 & 0 & 1 & 0 & 0 & 0 \\ 0 & 1 & 1 & 0 & 0 & 1 & 1 & 0 & 0 & 1 & 0 & 0 \\ 1 & 1 & 0 & 1 & 0 & 1 & 0 & 1 & 0 & 0 & 1 & 0 \\ 1 & 1 & 0 & 0 & 1 & 0 & 1 & 1 & 0 & 0 & 0 & 1 \end{bmatrix}$

multiple criteria are considered, in order to get the best trade-off between the performance and the implementation complexity. For instance, in MUSA, the TBE criterion is used to screen the complex-valued sequence set. However, for NCMA or UGMA sequences, quantized elements of sequences would facilitate the system implementation, even though their cross-correlation does not meet WBE/ETF criterion or even the TBE criterion. Also, the combination of sparse sequences and complex-valued elements are helpful to expand the sequence pool.

Cross-correlation characteristics of several types of spreading sequences are shown in Table 6.20. The spreading length is 4 and the pool size is 8 and 12 as examples. The cumulative density functions (CDF) of the absolute value of cross-correlation are shown in Figure 6.4.

TABLE 6.19 Examples of PDMA Sequences, $L = 4$, Element Values Can Be $\{0, 1, -1, j, -j\}$

k	g_k				k	g_k				k	g_k			
0	1	j	−1	−j	**32**	1	1	j	−1	**64**	1	−1	−j	0
1	1	−j	−1	j	**33**	1	j	−1	−1	**65**	1	j	−j	1
2	1	−1	1	−1	**34**	1	0	0	−1	**66**	1	−1	j	0
3	1	−1	−j	j	**35**	1	1	−j	−1	**67**	0	1	−j	j
4	1	−1	j	−j	**36**	1	−j	−1	−1	**68**	0	1	−1	j
5	1	−1	−1	1	**37**	0	1	0	j	**69**	1	−j	J	0
6	1	j	−j	−1	**38**	1	0	j	0	**70**	1	−1	1	1
7	1	−j	j	−1	**39**	1	1	0	−1	**71**	1	1	−1	1
8	1	j	−j	j	**40**	1	0	−1	−1	**72**	1	−j	j	1
9	1	−1	1	j	**41**	1	j	j	−1	**73**	1	−1	j	−1
10	1	j	0	−1	**42**	1	j	−1	j	**74**	1	j	1	−1
11	1	0	j	−1	**43**	1	−j	1	j	**75**	1	−j	1	-1
12	1	−j	j	−j	**44**	1	−1	0	1	**76**	1	−1	−j	−1
13	1	−1	1	−j	**45**	1	0	−1	1	**77**	1	1	−1	j
14	1	0	−1	j	**46**	1	j	1	−j	**78**	1	−j	j	j
15	1	−j	0	j	**47**	1	−j	−1	−j	**79**	1	−1	0	−j
16	1	j	−1	0	**48**	0	1	0	−j	**80**	1	0	j	−j
17	0	1	j	−1	**49**	1	0	−j	0	**81**	1	1	−j	0
18	1	j	−1	1	**50**	1	−j	−j	−1	**82**	0	1	1	−j
19	1	−1	−j	1	**51**	1	−j	0	0	**83**	1	−j	−j	0
20	0	1	−1	1	**52**	0	0	1	−j	**84**	0	1	−j	−j
21	1	−1	1	0	**53**	0	1	−j	0	**85**	1	−1	−1	−j
22	1	0	−j	−1	**54**	0	1	−j	1	**86**	1	j	j	−j
23	1	−j	0	−1	**55**	0	1	j	1	**87**	0	1	j	j
24	1	j	0	−j	**56**	1	j	1	0	**88**	1	j	j	0
25	0	1	−j	−1	**57**	1	−j	1	0	**89**	1	1	j	0
26	1	0	−1	−j	**58**	0	0	1	j	**90**	1	1	j	−j
27	1	−j	−1	0	**59**	0	1	j	0	**91**	0	1	1	j
28	1	0	−1	0	**60**	1	j	0	0	**92**	1	−1	−j	−j
29	0	1	0	−1	**61**	0	1	j	-j	**93**	0	1	−1	0
30	1	−j	−1	1	62	0	1	−1	−j	94	1	−1	0	0
31	1	−1	j	1	63	1	j	−j	0	95	0	0	1	−1

In Figure 6.5, link-level performances under different spectral efficiencies and the number of users for several spreading sequences are compared. Since these results are from the same simulation platform, it is easier to see the impact of a specific design on the performance. The parameter setting is shown in Table 6.21. Here, the average receiver power of different users

TABLE 6.20 Cross-Correlation Property of Several Typical Spreading Sequences

Type of Spreading Sequences	[min, max] $L = 4, K = 12$ (WBE/ETF = 0.4264)	Total Squared $L = 4, K = 12$ (TBE = 24)	[min, max] $L = 4, K = 8$ (WBE/ETF = 0.378)	Total Squared $L = 4, K = 8$ (TBE= 8)
MUSA (first K sequences)	[0, 0.5]	24	[0, 0.5]	8
RSMA	[0, 0.8365]	24	[0, 0.6533]	8
WSMA	[0.035 0.7038]	24.003	[0.083, 0.6166]	8.003
NOCA (optimal roots)	[0, 0.5]	24	[0, 0.5]	8
NCMA (64QAM quantized)	[0.1819, 0.5838]	26.2344	[0.2199, 0.5123]	8.3166
NCMA (un-quantized)	[0.4233, 0.4491]	25.9995	[0.3801, 0.3917]	8.4754
UGMA	[0.0481, 0.9050]	26.2849	[0.036, 0.7362]	9.3788
PDMA (complex elements, first K sequences)	[0, 0.8165]	32	[0, 0.7071]	10

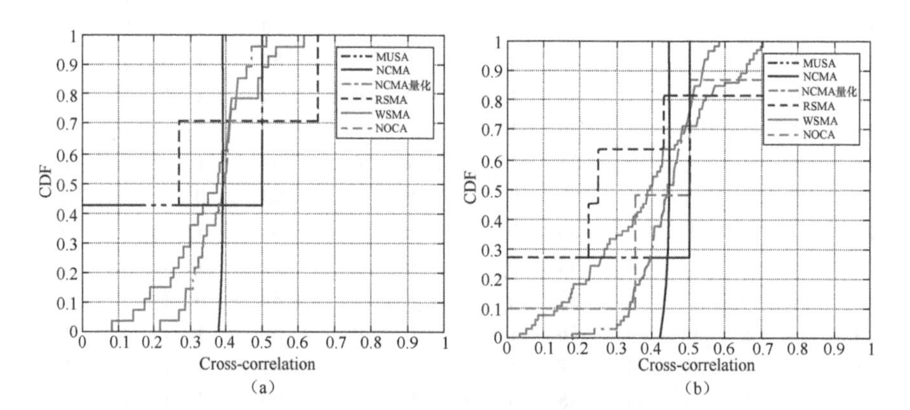

FIGURE 6.4 CDFs of cross-correlations of several typical spreading sequences. (a) $L = 4, K = 8$ and (b) $L = 4, K = 12$.

is assumed the same. Realistic channel estimation is simulated, and the receiver type is MMSE SIC.

In general, it is observed that at low spectral efficiency, the performances of these short spreading sequences are quite close. However, for higher spectral efficiency, because of the increased cross-user interference, for those sequences that cannot satisfy the theoretical optimization criteria, their performance degradation in terms of SNR can be 1 dB at BLER = 0.1. More results can be found in Chapter 7.

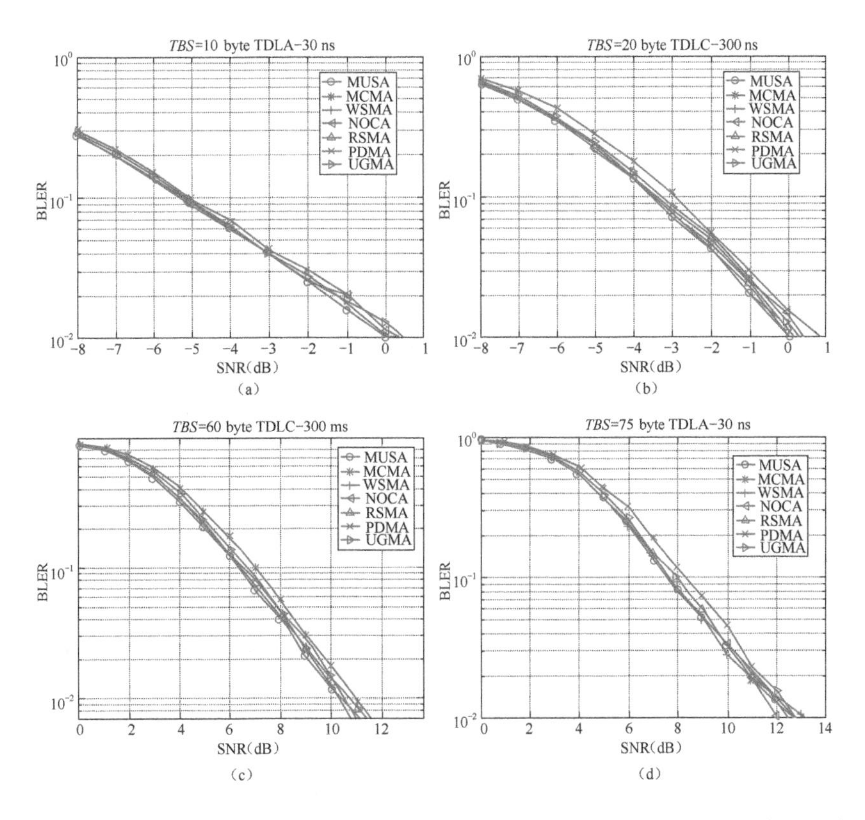

FIGURE 6.5 Link-level performance comparison between several typical spreading sequences. (a) TBS = 10 bytes, 12 UEs, QPSK, (b) TBS = 20 bytes, 12 UEs, QPSK, (c) TBS = 60 bytes, 8 UEs, 16QAM and (d) TBS = 75 bytes, 8 UEs, 16QAM.

TABLE 6.21 Several Simulation Cases of Link-Level Performance Evaluation

Simulation Case	Packet Size (Bytes)	Number of Users	Modulation Order	Channel Model	Total Spectral Efficiency (bps/Hz)
(a)	10	12	QPSK	TDL-A 30 ns	1.33
(b)	20	12	QPSK	TDL-C 300 ns	2.67
(c)	60	8	16QAM	TDL-C 300 ns	5.33
(d)	75	8	16QAM	TDL-A 30 ns	6.67

6.1.3 Symbol-Level Scrambling

For NOMA schemes that are based on short-length spreading, there are many ways to carry out the spreading, for instance, frequency-domain or time-domain spreading. For frequency-domain spreading, it can be at the resource element (RE) level or resource block (RB) level. For time-domain spreading, it can be at the OFDM symbol level or subframe level.

Time-domain OFDM symbol-level and frequency-domain RE-level spreading is shown in Figure 6.6.

Frequency-domain RE-level spreading is more widely used and has the advantage of in-time and de-spreading process at the receiver. Compared to time-domain spreading, frequency-domain spreading can reduce the waiting time. However, frequency-domain RE-level spreading would increase PAPR compared to the regular OFDM waveform because the spread symbols are mapped to the adjacent frequency locations, leading to a high correlation between the neighboring subcarriers.

One effective way to solve the issue of higher PAPR with frequency-domain spreading is to perform symbol-level scrambling after the spreading, as shown in Figure 6.7. Its basic principle is to use a pseudo-randomly generated scrambling sequence to randomize the phase change between the adjacent symbols. Alternatively, phase randomization can also be achieved by using different spreading sequences per modulation symbol. A merit of this method is that the cross-correlation between spread sequences can be maintained after the scrambling. Hence, even when different scrambling sequences are applied in neighboring cells, the cross-user interference between neighboring cells can still exhibit a certain structure in the spreading domain so that the other-cell interference can be effectively suppressed by the MMSE type of receivers to improve the system performance. Details can be found in Chapter 7.

From Figure 6.8 it is seen that PAPR distributions, i.e., complement cumulative distribution function (CCDF), of time-domain spreading, are

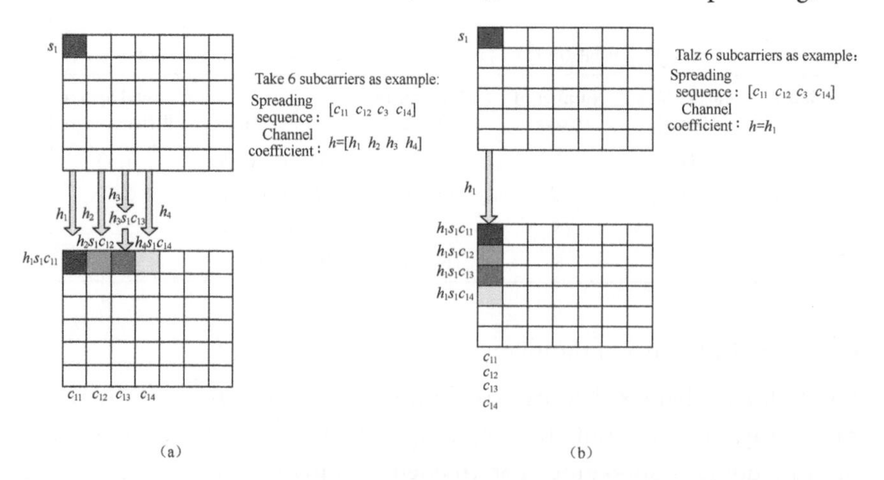

FIGURE 6.6 Illustration of different ways of spreading. (a) Time-domain spreading and (b) frequency-domain spreading.

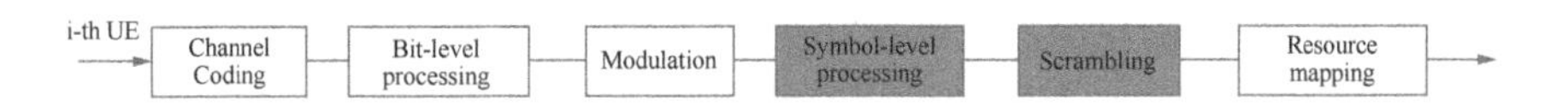

FIGURE 6.7 Transmitter-side processing for symbol-level scrambling.

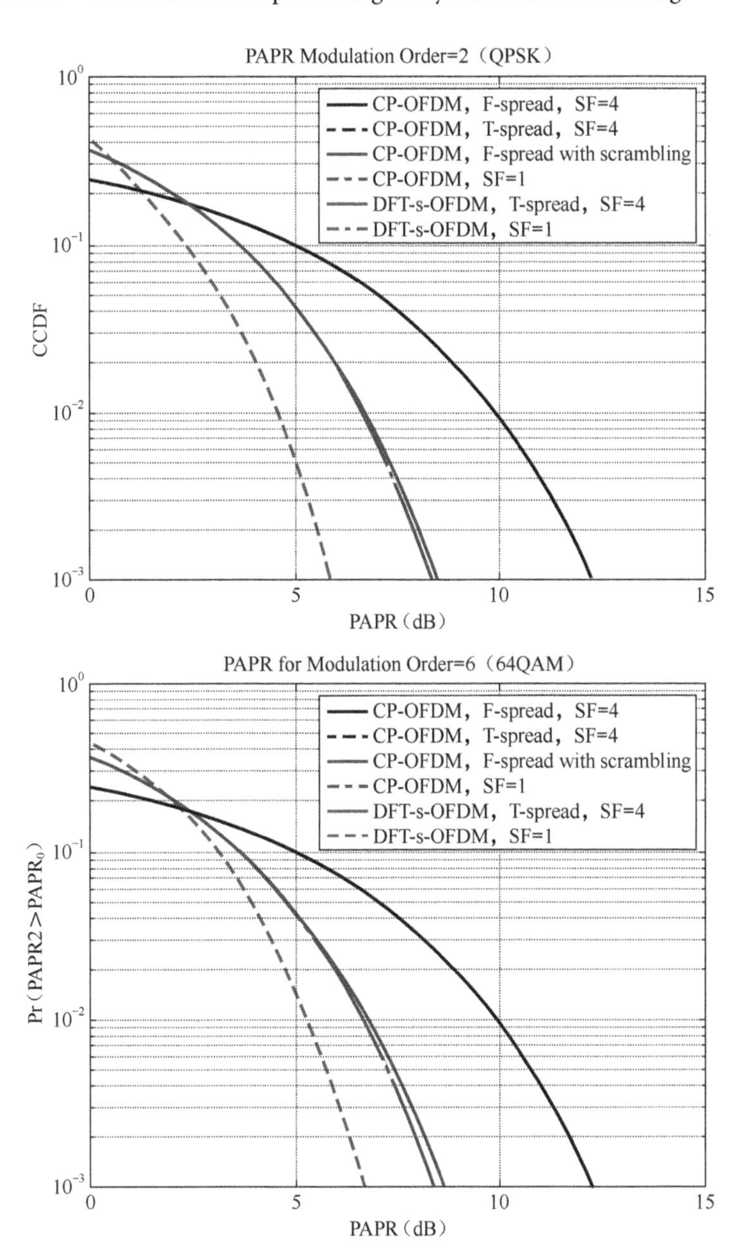

FIGURE 6.8 PAPR distribution for different time-domain and frequency-domain spreading, with and without symbol-level spreading.

almost the same as those without spreading, for both CP-OFDM and DFT-s-OFDM waveforms. However, PAPR distributions of frequency-domain spreading are higher than those without spreading for the CP-OFDM waveform, regardless of whether QPSK or 64-QAM modulations. At CCDF = 10^{-3}, the increase is about 5 dB, as seen in the CCDF curves. Such PAPR can be brought back to that of the original CP-OFDM, by carrying out symbol-level scrambling.

6.1.4 MMSE Hard IC Receiver Algorithms and Complexity Analysis

6.1.4.1 MMSE Hard Interference Cancelation Receiver

In this type of receivers, the interference is canceled via MMSE detection. Hard interference cancelation refers to that at the receiver, the interference is reconstructed based on the "hard" decoded bits. The cancelation can be carried out in successive, parallel or mixed, e.g., to select a few users whose SINR/SNR are the highest.

- In SIC, user ordering would be carried out for all those users that have not been decoded successfully. Signals (or interference) of users that have been decoded successfully would be reconstructed and canceled out.
- In parallel interference cancelation (PIC), multiple rounds of detection and decoding would be carried out for all those users that have not been decoded successively. Signals (or interference) of users that have been decoded successfully in each round would be reconstructed and canceled.
- In mixed interference cancelation, multiple rounds of detection and decoding would be carried out. In each round, user ordering will be performed for all those users that have not been decoded successfully. Signals (or interference) of users that have been decoded successfully would be reconstructed and canceled.

It should be pointed out that for SIC, one enhancement is that when a user is not decoded successfully in the first attempt, the entire decoding process would not terminate. Rather, the users whose SNR is lower would be tried. This mechanism allows SIC to use a smaller number of decoding to approach the performance of PIC that requires much more decoding. The flow charts of classic SIC and the enhanced SIC are shown in Figures 6.9 and 6.10, respectively.

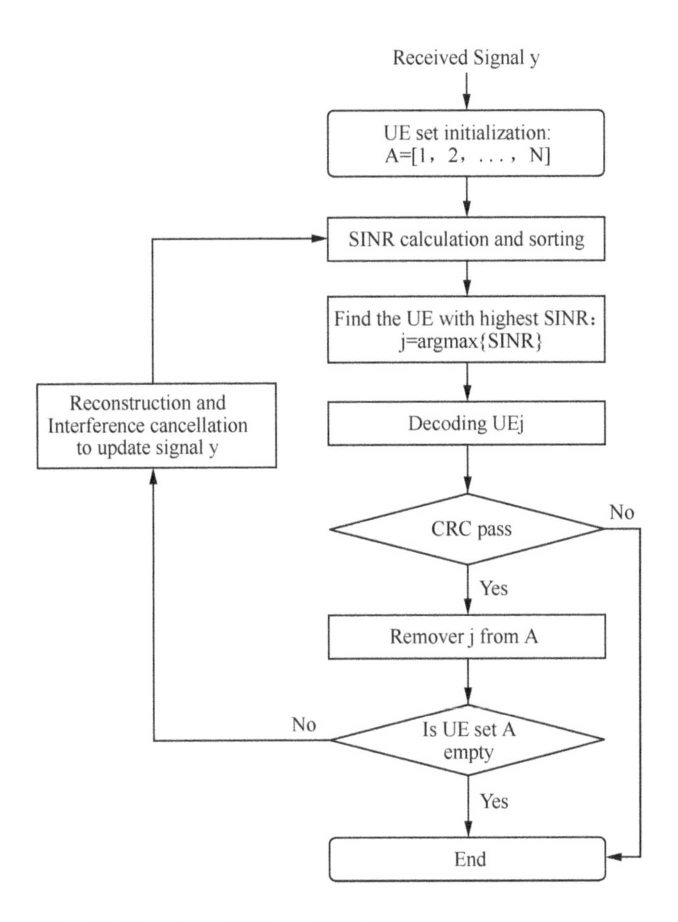

FIGURE 6.9 Classic MMSE hard SIC receiver.

Several other enhancements of this type of receivers include:

1. From the perspective of reducing the complexity, the user ordering can be based on SNR, rather than SINR [14]. Even for scenarios with a strong near-far effect, as mentioned in [14], the performance difference between SNR-based ordering and SINR-based ordering is negligible for the UGMA scheme. The calculation of SNR is simpler than that of SINR since in the former only the channel gain needs to be considered.

2. From the perspective of performance improvement, channel estimation can be done twice, as more users are decoded successfully. The quality of channel estimation can be improved by utilizing the reconstructed signals [15,16], which helps to reduce residual errors.

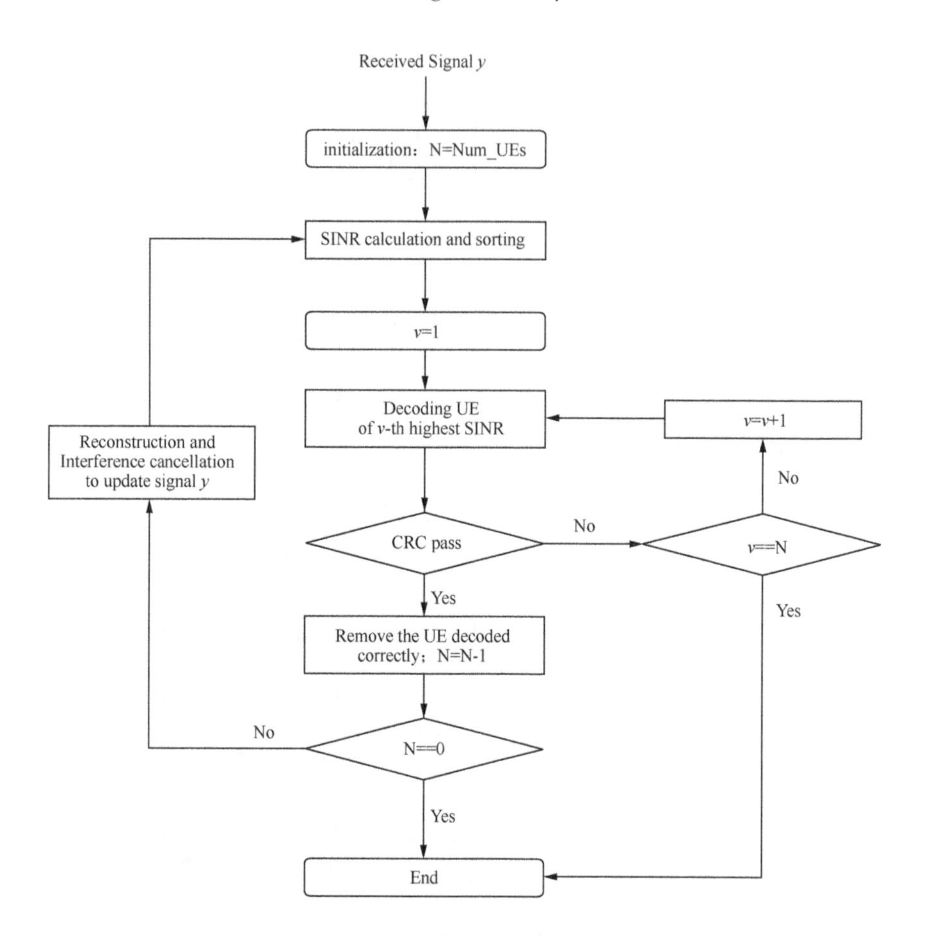

FIGURE 6.10 Enhanced MMSE hard SIC receiver.

6.1.4.2 Analysis of Computation Complexity

The complexity of MMSE hard IC receiver comes mainly from three modules: detector, decoder and interference cancelation. Normally, the complexity of the decoder is higher than that of the detector and interference cancelation. If measured quantitatively, the complexity of detector and interference cancelation can be well defined in the number of required summations and multiplications, whereas the complexity of the decoder is related to the number of binary "addition" and "comparison" in each computation element. Hence, a natural way of estimating the complexity is to add the complexity of the detector with the complexity of the interference cancelation, while treating the decoder separately as seen in Table 6.24.

Detailed analysis of detector, decoder and interference cancelation can be found in Ref. [17]. In the detector module, there are a number of key processing steps: the computation of covariance matrix, the computation of MMSE weights, and demodulation.

- The covariance matrix can be calculated as:

$$\mathbf{R}_{yy} = \mathbf{H}\mathbf{H}^* + \sigma^2\mathbf{I}$$

where the number of receiver antennas is denoted as N_{rx}. The spreading length is L. The number of users is N_{ue}. The number of Res is N_{RE}^{data}. The dimension of the covariance matrix is $N_{rx} * L$. If only complex multiplication is counted, the complexity of the covariance matrix computation is $(N_{rx} * L)^2/2$. The division by 2 is because the covariance matrix is Hermitian. Considering that the covariance matrix would not differ significantly across multiple Res, the computation of the covariance matrix is in fact about $O\left(N_{UE} \cdot N_{RE}^{data} \cdot (N_{rx} \cdot N_{SF})^2 / 2N_{RE}^{adj}\right)$ where the parameter N_{RE}^{adj} quantifies how many adjacent modulation symbols in the time and frequency domain whose covariance matrix or MMSE weights can be assumed the same. This parameter depends on the mobility speed and the frequency response of the channel.

- The calculation of MMSE weights can be represented as

$$\mathbf{w}_k = \mathbf{h}_k^* \mathbf{R}_{yy}^{-1}$$

The matrix inversion in the weight calculation requires about $(N_{rx} * L)^3$ complex multiplications. The matrix multiplication takes about $(N_{rx} * L)^2$ complex multiplications. Considering that MMSE weights can be reused for multiple Res, the actual computation complexity of MMSE weights is about $O\left(\left(N_{RE}^{data} \cdot (N_{rx} \cdot N_{SF})^3 + N_{iter}^{IC} \cdot N_{RE}^{data} \cdot (N_{rx} \cdot N_{SF})^2\right) / N_{RE}^{adj}\right)$. Here the parameter N_{iter}^{IC} is the number of decoding required for each MMSE hard IC operation.

- The computation of demodulation is

$$\hat{x}_k = \mathbf{w}_k^* \mathbf{y}$$

Demodulation here is essentially to multiply MMSE weights with the received signal. In the case of spreading, the above process requires $N_{rx} * L$ complex multiplications.

- User ordering:

  User ordering primarily involves the following computations which can be expanded as

$$T_k = h_k^* R_y^{-1} h_k == \sum \begin{bmatrix} h_{k,1}^* h_{k,1} r_{1,1} & \cdots & h_{k,1}^* h_{k,N} r_{1,N} \\ \cdots & \cdots & \cdots \\ h_{k,N}^* h_{k,1} r_{N,1} & \cdots & h_{k,N}^* h_{k,N} r_{N,N} \end{bmatrix} := \sum Q$$

Here the summation $\sum$ is over all the elements in the matrix. Considering that when computing the covariance matrix, $h_{k,i}^* h_{k,i}$ is already obtained. The matrix Q is Hermitian. Hence, the only additional computation includes

a. all the diagonal elements, each requiring one real multiplication;

b. off-diagonal elements, each requiring two real multiplications.

Figure 6.11 shows the computation for a 3×3 matrix.

- Symbol reconstruction mainly involves encoding, modulation and possible spreading. Its computation complexity is roughly $O\left(N_{\text{UE}} \cdot N_{\text{RE}}^{\text{data}} \cdot N_{rx}\right)$.
- Interference cancelation just involves subtraction, and its computation complexity is low.
- Estimation of complexity of decoding.

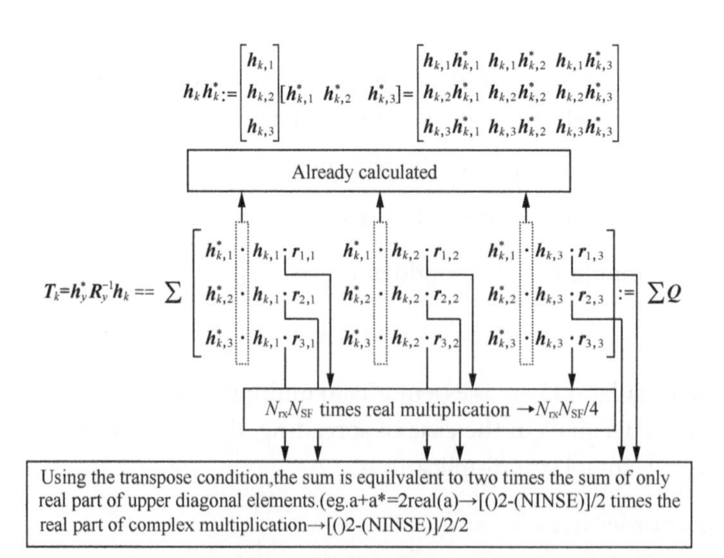

FIGURE 6.11 Computation of metrics for user ordering.

The low-density parity check (LDPC) in NR is irregular, meaning that the number of elements with a value of 1 in each column (also called "column weight") can be different. The number of elements with value 1 in each row (also called "row weight") can also be different. The decoding complexity mainly depends on the average column weight, denoted as d_v, and the average row weight, denoted as d_c. The parity check matrix of LDPC in NR is quite flexible and can be scaled to different block lengths by "lifting" the protomatrix. Code rate can be adjusted by puncturing of protomatrix. In the NOMA study, the size of the LDPC block rarely goes beyond 150 bytes (e.g., 1,200 bits) and protomatrix BG2 is used. Table 6.22 shows the average column weights and average row weights of NR LDPC under different sizes of block and the spreading lengths.

Another factor that determines the complexity of LDPC decoding is the calculation of the marginal probability density function at each check node. This calculation is often called log-belief propagation (log-BP). There are basically two methods: (1) exact calculation, e.g., ideal kernel, which can get the optimal performance, but requires more computation, and (2) approximation, e.g., min-sum with offset, and its performance is about 0.5~0.8 dB worse than the optimal in additive white Gaussian noise (AWGN), but with much less calculation. In the exact calculation, the probability at each check node needs to be calculated by adding up the hyperbolic function of the probabilities of each variable node. The hyperbolic function is highly nonlinear and usually requires a table look-up to calculate. In contrast, for the approximation method, we only need to find the minimum probability of all variable nodes, e.g., only multiple comparisons are required. In Table 6.23, the computation complexities of the NR LDPC decoder are compared between the exact method and the approximation method. In practical systems, the approximation method is often used, so as in the rest of the chapters.

The break-down computation complexities of each detail processing module of the MMSE hard IC receiver are listed in Table 6.24. It is noted that for covariance matrix calculation, demodulation weight calculation and user ordering, the estimated complexities are not exactly the same

TABLE 6.22 Average Column Weights and Average Two Weights of BG2 of NR LDPC

(TBS, Modulation, SF)	d_v	d_c
(10 bytes, QPSK, 2)	3.79	4.69
(20 bytes, QPSK, 2)	3.77	5.28
(20 bytes, QPSK, 4)	3.43	6.55

TABLE 6.23 NR LDPC Decoder's Complexities in the Exact Method and the Approximation Method

Major Processing	Exact Calculation (Optimal) Log-BP + Ideal Kernel	Approximation (Sub-Optimal) Log-BP + Min-Sum + Offset
Check node operation (per code block and per iteration)	#Add : $d_v*N_{bit}+(2d_c-1)*(N_{bit}-K_{bit})$ #LUT : $2d_c*(N_{bit}-K_{bit})$	#Add : $d_v*N_{bit}+2*(N_{bit}-K_{bit})$ #Comp : $(2d_c-1)*(N_{bit}-K_{bit})$
Variable node operation (per code block and per iteration)	#Add : d_v*N_{bit}	#Add : d_v*N_{bit}

between companies. The discussion above takes Option 1 as an example. Table 6.24 also includes slightly different computations, as seen in Option 2.

Typical values for the parameters to calculate the receiver complexity are listed in Table 6.25. Since the enhanced MMSE hard SIC is considered, some code blocks may require more than one decoding attempt before being successfully decoded. Hence, N_{iter}^{IC} is larger than 1.

During NOMA Study Item in 3GPP, a majority of companies use Option 1 in Table 6.24 to calculate the complexity of MMSE hard IC receiver, since MMSE hard SIC is more widely used. For parallel processing, e.g., MMSE hard PIC, Option 2 can be used. For MMSE hard IC, one effective way to reduce the complexity is reflected in the demodulation weight, or covariance matrix calculation where only one weight (vector) needs to be calculated across multiple adjacent Res in the time and frequency domain. There are a few different assumptions of N_{RE}^{adj}, e.g., how many adjacent Res in which the channel is considered as constant.

1. Assumption 1: Res in two adjacent subcarriers and one TTI can share the same weight (vector)

2. Assumption 2: Res within an entire RB can share the same weight (vector)

3. Assumption 3: Res within four adjacent subcarriers and one slot can share the same weight (vector);

Link-level simulation has been carried out under Assumption 1. Comprehensive analysis of the link simulation results can be referred to in Ref. [18], including the case when there is a time and frequency offset. It is observed in Ref. [18] that the performance with Assumption 1 is not

TABLE 6.24 Break-Down Analysis of the Complexity Analysis of MMSE Hard IC Receiver

Key Module	Detailed Processing	O(.) Level Analysis
Detector (complexity measured in number of complex multiplications)	User detection	$O\left(N_{AP}^{DMRS}\cdot N_{RE}^{DMRS}\cdot N_{rx}\right)$
	Channel estimation	$O\left(N_{UE}\cdot N_{RE}^{CE}\cdot N_{RE}^{DMRS}\cdot N_{rx}\right)$
	Calculation of covariance matrix	Option 1: $O\left(N_{UE}\cdot N_{RE}^{data}\cdot(N_{rx}\cdot N_{SF})^2/2N_{RE}^{adj}\right)$Option 2: $N_{itr}\left(\frac{\left(N_{rx}N_{sf}\right)^2 N_{RE}^{DMRS}}{2N_{sf}}+N_{UE}N_{Rx}N_{RE}^{DMRS}\right)$
	Calculation demodulation weight	Option 1: $O\left(\left(N_{RE}^{data}\cdot(N_{rx}\cdot N_{SF})^3+N_{iter}^{IC}\cdot N_{RE}^{data}\cdot(N_{rx}\cdot N_{SF})^2\right)/N_{RE}^{adj}\right)$ Option 2: $\frac{N_{itr}N_{RE}^{Data}}{N_{RE}^{adj}}\left(1.5\left(N_{rx}N_{sf}\right)^2 N_{UE}+\left(N_{rx}N_{sf}\right)^3\right)$, $N_{RE}^{adj}=NN_{sf}$ Option 3: $O\left(\left(N_{RE}^{data}\cdot(N_{rx}\cdot N_{SF})^3+N_{iter}^{IC}\cdot N_{RE}^{data}\cdot(N_{rx}\cdot N_{SF})^2+N_{RE}^{data}\cdot N_{UE}(N_{UE}+1)\cdot(N_{rx}\cdot N_{SF})^2/2\right)/N_{RE}^{adj}\right)$
	User ordering	Option 1: $O\left(N_{RE}^{data}\cdot N_{rx}\cdot N_{SF}\cdot(N_{UE})^2/N_{RE}^{adj,SINR}\right)$ *Option 2:* $O\left(N_{RE}^{data}\cdot N_{rx}\cdot N_{SF}\cdot\frac{(N_{UE})^2}{N_{RE}^{adj,SINR}}+(N_{UE})^2 \log(N_{UE})/2\right)$
	Demodulation	$O\left(N_{iter}^{IC}\cdot N_{RE}^{data}\cdot N_{rx}\right)$
	Generation of soft bits	$O\left(N_{iter}^{IC}\cdot N^{bit}\right)$

(*Continued*)

TABLE 6.24 (*Continued*) Break-Down Analysis of the Complexity Analysis of MMSE Hard IC Receiver

Key Module	Detailed Processing	O(.) Level Analysis
Decoder (complexity measured in number of bit additions and comparisons)	LDPC decoding	$A: N_{iter}^{IC} \cdot N_{iter}^{LDPC} \cdot \left(d_v N^{bit} + 2\left(N^{bit} - K^{bit}\right)\right)$ $C: N_{iter}^{IC} \cdot N_{iter}^{LDPC} \cdot (2d_c - 1) \cdot \left(N^{bit} - K^{bit}\right)$
Interference cancelation (measured in number of complex multiplications)	Symbol reconstruction	$O\left(N_{UE} \cdot N_{RE}^{data} \cdot N_{rx}\right)$
	LDPC encoding	Buffer shifting: $N_{UE} \cdot \left(N^{bit} - K^{bit}\right)/2$ Addition: $N_{UE} \cdot (d_c - 1)\left(N^{bit} - K^{bit}\right)$

TABLE 6.25 Typical Parameter Values for Receiver Complexity Calculation

Type of Parameters	Parameter	Notation	Value
General	Number of Rx antennas	N_{rx}	2 or 4
	Number of REs	$N_{\mathrm{RE}}^{\mathrm{data}}$	864
	Number of users	N_{UE}	12
MMSE or EPA related	Spreading length	N_{SF}	4
MMSE hard IC related	Average number of decoding attempts	$N_{\mathrm{iter}}^{\mathrm{IC}}$	$[1.5-3]\cdot N_{\mathrm{UE}}=[18-36]$
Decoder related	Average column weight of LDPC	d_v	3.43
	Average row weight of LDPC	d_c	6.55
	Number of information bits (including CRC bits)	K^{bit}	176 （对应 20 Bytes）
	Number of coded bits	N^{bit}	432
	Number of internal iterations in LDPC decoder	$N_{\mathrm{iter}}^{\mathrm{LDPC}}$	20
User detection and channel estimation related	Number of DMRS antenna ports	$N_{\mathrm{AP}}^{\mathrm{DMRS}}$	12
	Number of REs of DMRS for channel estimation	$N_{\mathrm{RE}}^{\mathrm{CE}}$	12
	DMRS sequence length (NR Type II)	$N_{\mathrm{RE}}^{\mathrm{DMRS}}$	24

much different from the performance when the channel coefficients are updated for each modulation symbol, e.g., $N_{\mathrm{RE}}^{\mathrm{adj}}=4$. Link simulations with Assumptions 2 and 3 are also carried out in Ref. [19]. It is found that for low mobility, e.g., up to 3 km/h, the performance degradation with respect to $N_{\mathrm{RE}}^{\mathrm{adj}}=1$ is negligible. For medium speed, e.g., 30 km/h, the performance degradation is very small. In Ref. [20], Assumption 3 is used. It is found that even at high speed, e.g., 120 km/h, the performance loss compared to $N_{\mathrm{RE}}^{\mathrm{adj}}=1$ is still quite small.

Assuming multiple modulation symbols to share a covariance matrix or demodulation weight, e.g., $N_{\mathrm{RE}}^{\mathrm{adj}}>4$, has been a common practice in baseband implementation. It has also been used for multi-antenna systems in LTE. Simulations in Ref. [18] assume $N_{\mathrm{RE}}^{\mathrm{adj}}=48$. If even higher $N_{\mathrm{RE}}^{\mathrm{adj}}$ is desirable to further reduce the receiver's complexity, the following processing can be considered, as illustrated in Figure 6.12.

Further complexity reduction can be achieved via aggressive partial covariance matrix sharing. When trying to decode User 1, a common

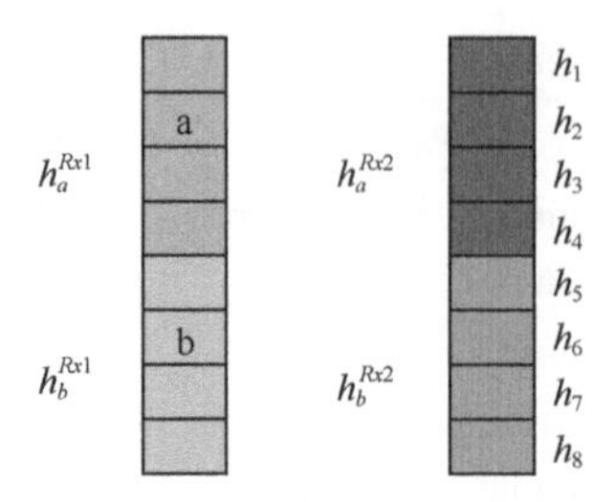

FIGURE 6.12 A demodulation weight shared by multiple symbols.

part of the covariance matrix is shared by the frequency-domain spreading of the element a and b, and across all the time-domain symbols, in order to derive the MMSE demodulation weight, as shown in Figure 6.12. The covariance of the decoding of any given symbol within the spreading unit a or b could be obtained by Eq. (6.3) below, where $\mathrm{Cov}_{b,i}$, is the covariance matrix of the i-th interfering user and $I = 1, 2, \ldots, N_{\mathrm{ue}}$.

$$\mathrm{Cov}_b = \left(\mathrm{Cov}_{b,1} + \sum_{i=2}^{N_{ue}} \mathrm{Cov}_{b,i} \right) \tag{6.3}$$

When calculating the covariance corresponding to a symbol within the range of any given spreading unit c, the following formula is used. Instead of applying the covariance in this particular spreading unit c of the users other than User 1, the covariance in spreading unit b could be used while only the covariance of User 1 is updated as

$$\mathrm{Cov}_c = \left(\mathrm{Cov}_{c,1} + \sum_{i=2}^{N_{ue}} \mathrm{Cov}_{b,i} \right) \tag{6.4}$$

Comparing Eqs. (6.3) and (6.4), it is seen that a bulk of content in the covariance matrix has been reused, e.g., only the covariance of the target user needs to be updated with its own spreading element while calculating the MMSE demodulation weight within a spreading unit other than the spreading unit a or b. This process would not significantly increase the complexity but can benefit the link-level performance when the channel experiences big fluctuations. As seen in Figure 6.13, even when $N_{\mathrm{RE}}^{\mathrm{adj}}$ is increased to 96, the performance degradation compared to $N_{\mathrm{RE}}^{\mathrm{adj}} = 4$ is only about 0.2 dB at BLER = 10%.

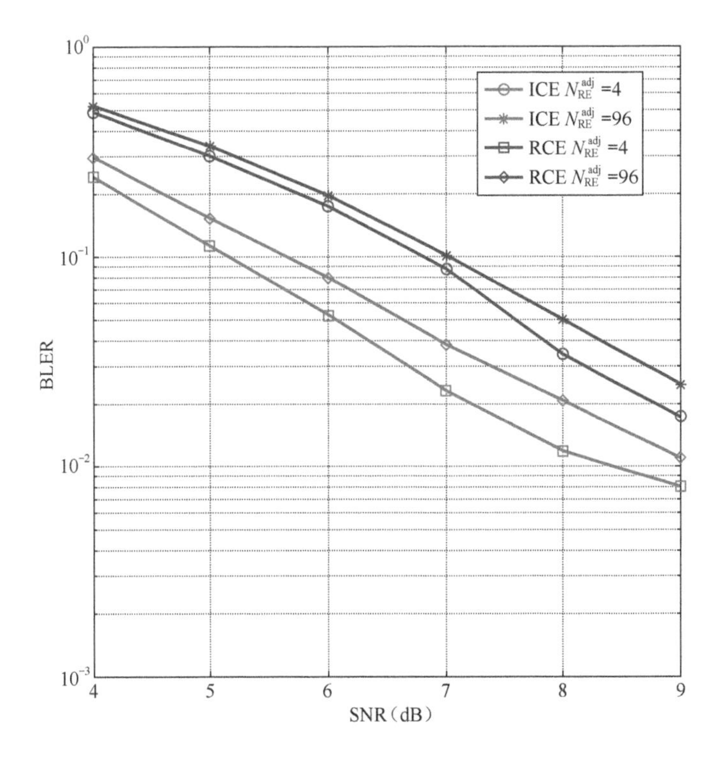

FIGURE 6.13 Link-level performance to verify method of the common covariance matrix.

In summary, for MMSE hard IC, there are plenty of ways to reduce the receiver complexity without degrading the performance. The complexities of the detector module in MMSE hard IC are shown in Figures 6.14 and 6.15 for two Rx antennas and four Rx antennas, respectively. The complexity is compared against that of the MMSE-IRC receiver. Options 1–3 correspond to the three methods to estimate the complexity of demodulation weight determination mentioned earlier.

Most of the calculations in a decoder involve addition operations and table looking-up, and their complexity ratio is roughly 1:6 [21]. Assuming that the computation complexity of a real multiplication is similar to that of a table look-up, the conversion between the complexity of bit processing and the complexity of complex multiplication can be obtained the following way. A complex multiplication usually involves three real multiplications and five real additions. Based on the previously mentioned ratio of 6:1 for the complexity of a real multiplication and a real addition. The ratio between a complex multiplication to a real addition is (3*6 + 5):1, that is, approximately 25:1. With these conversions, the complexities of the entire

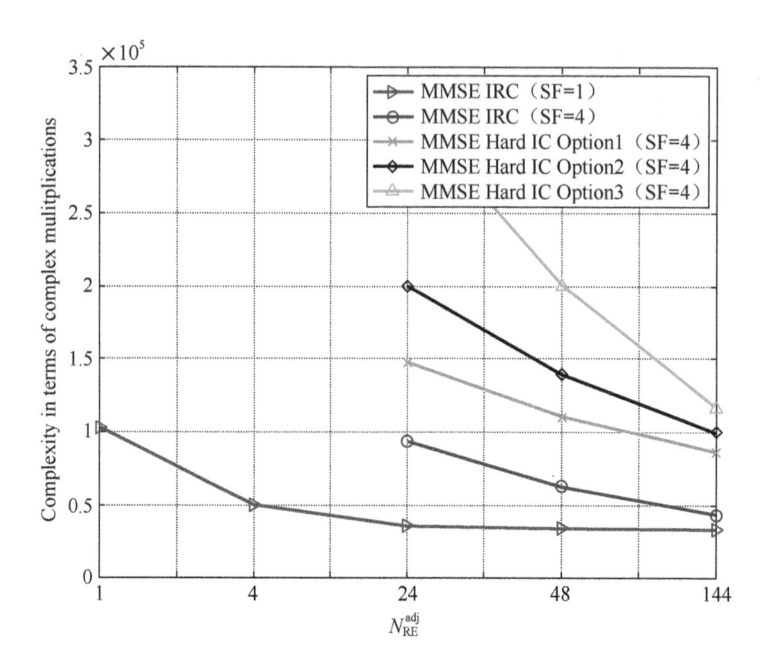

FIGURE 6.14 Complexity of the detector module in MMSE-IRC and MMSE hard IC receivers, 2 Rx antennas.

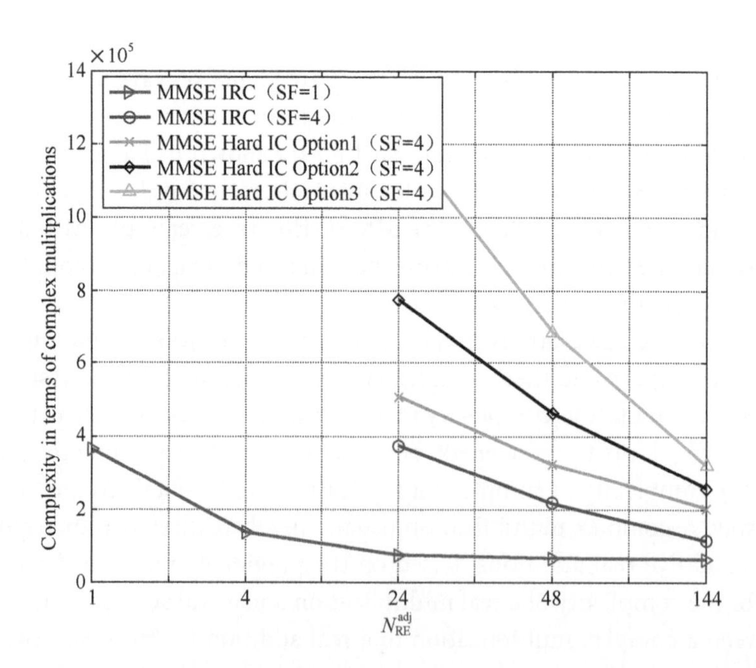

FIGURE 6.15 Complexity of the detector module in MMSE-IRC and MMSE hard IC receivers, 4 Rx antennas.

MMSE hard IC are shown in Figures 6.16 and 6.17 for two Rx antennas and four Rx antennas, respectively. When $N_{\mathrm{RE}}^{\mathrm{adj}} = 24$ and the block size is 20 bytes (176 bits, including CRC bits), the total complexity of MMSE hard IC is between 3×10^5 and 9×10^5 in terms of complex multiplications

MMSE type of receivers can also be hybrid between hard and soft interference cancelation. Different from MMSE hard IC, the soft and hard hybrid relies more on the decoder, in particular the soft bits, or the log-likelihood (LLR), to reconstruct the signal. The complexity analysis is listed in Table 6.26.

The meaning of the notations is as follows:

- $\overline{N_{\mathrm{iter}}^{\mathrm{IC}}}$: average number of interference cancelation.
- $\overline{N_{\mathrm{UE}}}$: the number of users whose interference is to be canceled per round.
- $\overline{N_{\mathrm{UE}}^{+}}$: the number of the remaining users that have not been decoded successfully in the first round.
- $\overline{N_{\mathrm{UE}}^{S}}$: the number of users that have decoded successfully per each round.
- $\overline{N_{\mathrm{UE}}^{F}}$: the number of the remaining users that have not been decoded successfully per each round (excluding the first).

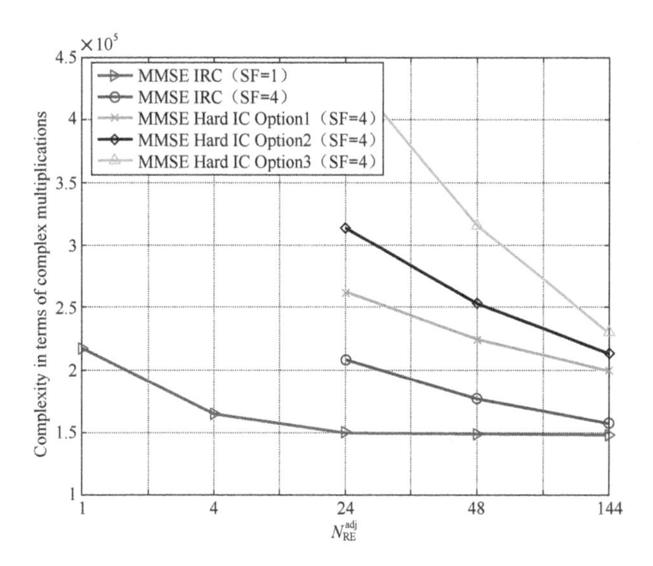

FIGURE 6.16 Complexity of the entire MMSE-IRC and MMSE hard IC receivers, 2 Rx antennas.

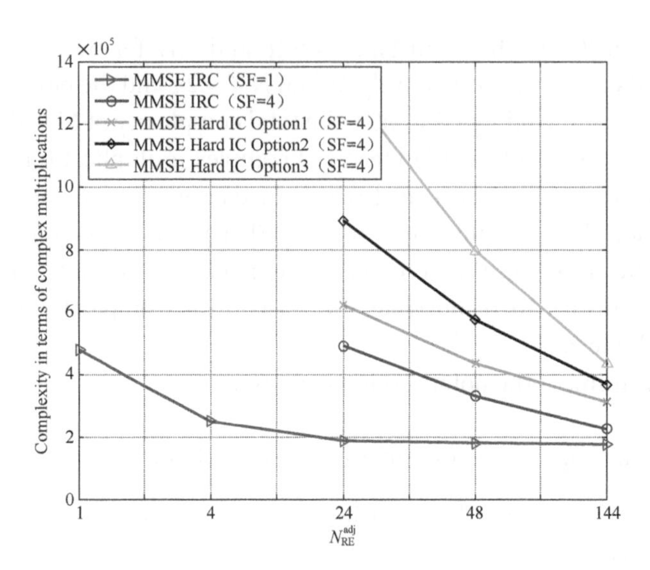

FIGURE 6.17 Complexity of the entire MMSE-IRC and MMSE hard IC receivers, 4 Rx antennas.

In the 3GPP NOMA study item, few companies have thoroughly analyzed MMSE hard-soft hybrid receiver. Some key values of the parameters did not get consensus. Hence its complexity will not be elaborated on in this chapter. However, it is not difficult to see that the hard-soft hybrid receivers require more processing than MMSE hard IC.

6.2 BIT-LEVEL-BASED SCHEMES AND TYPICAL RECEIVERS

6.2.1 Transmitter-Side Schemes

Bit-level-based schemes include interleaver-based schemes and scrambler-based schemes. Either bit interleavers or bit scramblers can be used to differentiate users.

6.2.1.1 Interleaver-Based Bit-Level Processing

Interleaver-division multiple access (IDMA) was first proposed by Prof. Li Ping in [22]. The idea of IDMA was inspired by iterative processing at Turbo receivers. A few related technologies are listed below (Figure 6.18)

- Multi-user detection (MUD) (1998 Verdu and Poor)
- Turbo codes (1993 Berrou, Glavieux and Thitimajshima)
- LDPC codes (Gallager, Mackay, Richardson and Urbanke)
- Iterative MUD (1998, Reed, Schlegel, Alexander and Asenstorfe)

TABLE 6.26 Complexity Analysis of MMSE Soft-Hard Hybrid IC Receiver

Key Module	Detailed Processing	O(.) Level Estimation
Detector (measured in number of complex multiplications)	User detection	$O\left(N_{\text{AP}}^{\text{DMRS}} \cdot N_{\text{RE}}^{\text{DMRS}} \cdot N_{rx}\right)$
	Channel estimation	$O\left(N_{\text{UE}} \cdot N_{\text{RE}}^{\text{CE}} \cdot N_{\text{RE}}^{\text{DMRS}} \cdot N_{rx}\right)$
	Covariance matrix calculation	$O\left(N_{\text{SF}}^2 \cdot N_{rx}^2 \cdot N_{\text{UE}} \cdot N_{\text{RE}}^{\text{data}} / N_{\text{RE}}^{\text{adj}}\right) + O\left(\left(\overline{N_{\text{iter}}^{\text{IC}}} - 1\right) N_{\text{SF}}^2 \cdot N_{rx}^2 \cdot \overline{N_{\text{UE}}^{+}} \cdot N_{\text{RE}}^{\text{data}} / N_{\text{SF}}\right)$
	Calculation of demodulation weight	$O\left(N_{\text{SF}}^2 \cdot N_{rx}^2 \cdot N_{\text{UE}} \cdot N_{\text{RE}}^{\text{data}} / N_{\text{RE}}^{\text{adj}}\right)$
		$+\, O\left(N_{\text{SF}}^3 \cdot N_{rx}^3 \cdot N_{\text{RE}}^{\text{data}} / N_{\text{RE}}^{\text{adj}}\right)$
		$+O\left(\left(\overline{N_{\text{iter}}^{\text{IC}}} - 1\right) \cdot N_{\text{SF}}^2 \cdot N_{rx}^2 \cdot \overline{N_{\text{UE}}^{+}} \cdot N_{\text{RE}}^{\text{data}} / N_{\text{SF}}\right)$
		$+O\left(\left(\overline{N_{\text{iter}}^{\text{IC}}} - 1\right) \cdot N_{\text{SF}}^3 N_{rx}^3 \cdot N_{\text{RE}}^{\text{data}} / N_{\text{SF}}\right)$
	Demodulation	$O\left(N_{\text{SF}} \cdot N_{rx} \cdot N_{\text{UE}} \cdot N_{\text{RE}}^{\text{data}} / N_{\text{RE}}^{\text{adj}}\right)$
		$+O\left(\left(\overline{N_{\text{iter}}^{\text{IC}}} - 1\right) \cdot N_{\text{SF}} \cdot N_{rx} \cdot \overline{N_{\text{UE}}^{+}}^{2} \cdot N_{\text{RE}}^{\text{data}} / N_{\text{SF}}\right)$
		$+O\left(\left(\overline{N_{\text{iter}}^{\text{IC}}} - 1\right) \cdot N_{\text{SF}} \cdot N_{rx} \cdot \overline{N_{\text{UE}}^{+}} \cdot N_{\text{RE}}^{\text{data}} / N_{\text{SF}}\right)$

(*Continued*)

TABLE 6.26 (*Continued*) Complexity Analysis of MMSE Soft-Hard Hybrid IC Receiver

Key Module	Detailed Processing	$O(.)$ Level Estimation
	Soft information generation	$O\left(\overline{N_{\text{iter}}^{\text{IC}}} \cdot \overline{N_{\text{UE}}^{+}} \cdot N^{\text{bit}}\right)$
	Soft-symbol re-generation	$O\left(\overline{N_{\text{iter}}^{\text{IC}}} \cdot \overline{N_{\text{UE}}^{S}} \cdot N_{\text{RE}}^{\text{data}} \cdot N_{rx}\right)$
Decoder (measured in number of binary addition and comparison)	LDPC decoding	$\text{A:}N_{\text{iter}}^{\text{outer}} \cdot N_{\text{UE}} \cdot N_{\text{iter}}^{\text{LDPC}} \cdot \left(d_v N^{\text{bit}} + 2\left(N^{\text{bit}} - K^{\text{bit}}\right)\right)$
		$\text{C}: N_{\text{iter}}^{\text{outer}} \cdot N_{\text{UE}} \cdot N_{\text{iter}}^{\text{LDPC}} \cdot (2d_c - 1) \cdot \left(N^{\text{bit}} - K^{\text{bit}}\right)$
Interference cancelation (measured in number of complex multiplications)	Conversion from LLR to probability	$O\left(\overline{N_{\text{iter}}^{\text{IC}}} \cdot \overline{N_{\text{UE}}^{\text{F}}} \cdot 2^{Q_m} \cdot N_{\text{RE}}^{\text{data}} / N_{\text{SF}}\right)$
	Others	Buffer shifting: $\overline{N_{\text{iter}}^{\text{IC}}} \cdot \overline{N_{\text{UE}}^{\text{S}}} \cdot \left(N^{\text{bit}} - K^{\text{bit}}\right)/2$
		Addition: $\overline{N_{\text{iter}}^{\text{IC}}} \cdot \overline{N_{\text{UE}}^{S}} \cdot (d_c - 1)\left(N^{\text{bit}} - K^{\text{bit}}\right)$

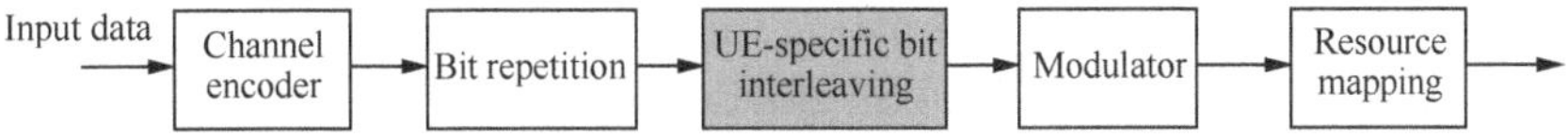

FIGURE 6.18 Transmitter-side processing for interleaver-based schemes.

- Iterative detection for interleaver codes (1998 Moher)
- Trellis code multiple access (2001, Brannstorm, Aulin and Rasmussen)
- Unequal power control for code domain multiple access (CDMA) (1998 Muller, Lampe and Huber)

LDPC as a type of channel code is constructed by sparse matrices. In certain situations, for instance very long block length, LDPC can approach the channel capacity very closely via iterative decoding [23]. Turbo codes can randomize the information bits via interleavers. Together with the iterative decoding between the constituent convolutional codes, Turbo codes can approach the Shannon limit as close as 0.7 dB [24]. CDMA relies on spreading sequences to differentiate users. Via MMSE-based multi-user detection and interference suppression, as well as the soft-output decoder, the iterative receiver for CDMA has the potential to approach the system capacity.

IDMA can be considered as a sparse graph with random inter-connections. Statistically, interleaver can reduce the chance of short loops. This is analogous to LDPC where short loops would negatively affect the decoder's performance. IDMA relies on an iterative receiver. The mean value and the variance of the user signal can be calculated from the soft bit information from the decoder. In the next iteration, other users' signals (as the interference) would be suppressed. The soft information of the target user is calculated and input to the decoder where the mean and the variance of the target user's signal are updated, to be used for the next iteration. Through this iterative process, information bits of each user can be decoded in the end.

Traditional CDMA was widely used in 3G cellular systems such as CDMA2000 and WCDMA. Related key techniques include power control, soft handover, Rake receiver, etc. In CDMA systems, cross-user interference can be suppressed by choosing suitable spreading sequences and zero-forcing (ZF) or MMSE detection. However, the issue with MMSE is the computation complexity of the matrix inversion, especially when the spreading length is long. Compared to CDMA, no matrix inversion is needed in spreading the code domain for IDMA. The complexity of ESE operation for IDMA is lower than that of MMSE hard IC for spreading base schemes [25]. Certainly, due to the multiple iterations required between ESE and the soft-output decoder, when estimating the complexity of the entire receiver for IDMA, multiple rounds of ESE and decoding should be considered.

IDMA was proposed during the 3G era during which the iterative receiver was still too complicated for practical implementation. In the 4G era, OFDMA was the predominant multiple-access scheme where the single-user detection is generally good enough. Hence, over a long period, IDMA did not find much use in the industry.

In IDMA, bit repetition can be carried out to improve the quality of the soft bits by accumulating the statistics. Bit repetition is usually performed before the bit interleaver, e.g., the user-specific interleaving would be applied to the repeated coded bits of each user. Legacy modulations can be used. After the modulation, a zero-value modulation symbol can be inserted to reduce the cross-user interference when the number of users is high and each user's spectral efficiency is also high. Such zero-value insertion can expedite the convergence of the ESE + SISO receiver [26]. In summary, the key ingredients of IDMA include

- User-specific bit interleaver
- Bit repetition before the interleaving
- Zero-value symbol insertion after modulation

The function of the bit interleaver is to randomize the multi-user interference. The superimposed multi-user interference exhibits certain characteristics of AWGN. This is an important assumption of the receiver for IDMA. Based on the mean value and variance per bit, the log-likelihood ratio (LLR) of each bit can be refined and then input to the decoder.

The function of bit repetition resembles spreading. However, the key difference is that spreading is normally operated at the symbol level and each user has its unique spreading sequence. ZF or MMSE inversion is needed for signal detection in the spreading case. The bit repetition in IDMA is common for all the users. To increase the randomness, [1, –1, 1, –1, 1, –1, ...] is often used for repetition. The relation between the input bit and output bit in the repetition is illustrated in Table 6.27.

In the receiver, after the calculation of LLR, de-repetition is carried out. No ZF or MMSE operation is needed. Instead, only summation needs to be performed. Since the elements of repetition sequence can take values of only 1 or –1, no multiplications are needed.

Zero-value symbol insertion is an optional operation. When the number of users is not high and per-user spectral efficiency is low, zero-value

TABLE 6.27 Input and Output Bit Relationship in the Bit Repetition for IDMA

Input Bit	Repetition Value = 1	Repetition Value = −1
1	1	0
0	0	1

insertion is not necessary, e.g., using bit interleaver alone is enough to separate different user's signal. However, when the system loading is high (either for a large number of users or each user having high spectral efficiency), channel coding is often not powerful enough to correct the errors due to the severe cross-user interference. When zero-value symbols are inserted, although the coded bits should shrink to accommodate those zero-value symbols, and the code rate is increased, the cross-user interference is also reduced (certain number of zeros in each user's signal). In NOMA, there are two factors affecting the performance. The first one is multi-user interference and the second one is channel code rate. When the number of users is high and per-user spectral efficiency is high, multi-user interference is the dominant factor. Via zero-value insertion, multi-user interference can be reduced. Even though the code rate is increased, the overall performance can still be improved via iterative decoding.

If the interleaver is uniquely designed for each user, it would increase the complexity of the system. There are several ways to simplify the interleaver design.

Method 1: bit interleaver is used in 5G NR. However, that interleaver is common for all the users and cannot readily be used for IDMA. One way to achieve user-specific interleaving is to define different starting points for the interleaved data. The operation resembles different cyclic shifts for different users. The receiver carries out user signal separation, detection and decoding based on the cyclic shifts

Method 2: cyclic shift can be applied before the bit interleaving as seen in Figure 6.19. If cyclic shifting and interleaver are considered jointly, it would make the interleaver appear as a totally different interleaver to the users.

Method 3: bit interleaver can be implemented in the form of cyclic shifting. This method has the merit of low implementation complexity and no need for interleaving and de-interleaving. Each user is differentiated based on its unique cyclic shift.

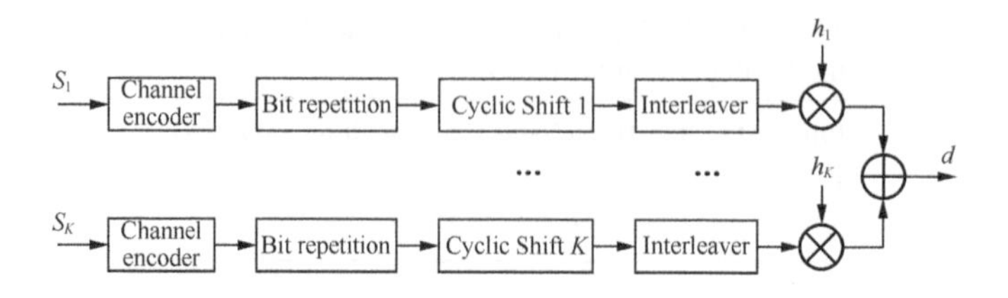

FIGURE 6.19 An example of applying different cyclic shifts before the common bit interleaver.

According to [27], Method 3 can result in a similar performance of IDMA to that of user-specific random bit interleavers, as shown in Figure 6.20. In the simulation, the channel is TDL-C. Each user device has one transmit antenna, and the base station has two receive antennas. The block size is 40 bytes. QPSK is used, with a code rate of 1/2. There are ten users. IDMA can even outperform SCMA (Sparse code multiple access) in this case. SI here stands for shift interleaver. In Methods 1 and 2, not only cyclic shift is used but also the random bit interleaver.

From the perspective of PAPR, since a common random interleaver is used in Methods 1 and 2, the periodicity of the signal after the bit repetition can be destroyed, thus reducing PAPR. However, the periodicity of the repeated signal cannot easily be scrambled by the cyclic shifts in Method 3, leading to the relatively high PAPR.

In IDMA, reference signals and data can be multiplexed in a power superposition fashion with different power ratios [28]. A reference signal

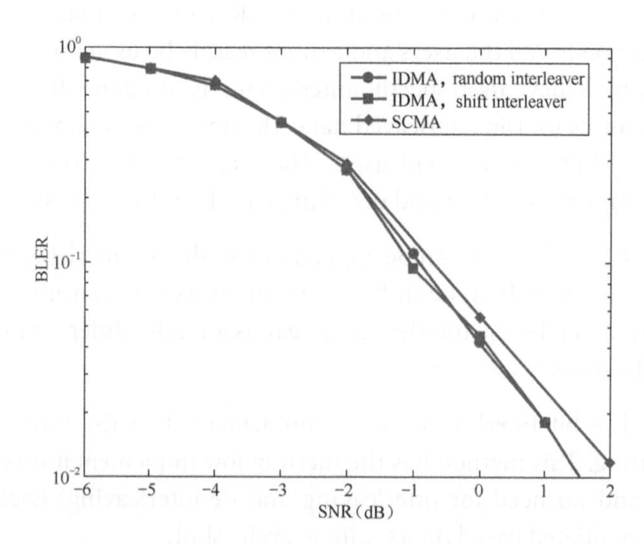

FIGURE 6.20 Link performance of IDMA (random bit interleaver vs. cyclic shift interleaver) and SCMA.

can be used for the initial estimation of the fading channel. After the iterative detection, the detected data can be used to refine the channel estimate. As the iteration continues, more users' data are successfully decoded. So as the accuracy of the channel estimation. The power ratio between the reference signal and data can be adjusted flexibly, as shown in Figure 6.21. This is in contrast to the legacy design of reference signals where the time-frequency resources for reference signals cannot be used to carry data.

One disadvantage of OFDM is the relatively high PAPR. While clipping can be used to reduce PAPR, it brings signal distortion and reduces the system performance. This can be "corrected" by the iterative receiver so that the clipping signal is compensated in a "soft" manner. According to [29], IDMA-OFDM can achieve quite good performance even under very low PAPR scenarios. The performance gain over OFDMA is about 6–7 dB, as shown in Figure 6.22. This is due to the fact that in IDMA-OFDM, with bit repetition, the signal spans over a wide bandwidth and can benefit from multi-user diversity, frequency diversity and spreading diversity. Whereas OFDMA is more sensitive to clipping, which results in significant performance degradation.

When the number of users is large, in order to speed up the convergence of iterative detection, appropriate power allocation may be considered, for instance, to allocate high power to some users to facilitate their decoding. Once these users' signals have been soft canceled, users with lower power can be detected and decoded.

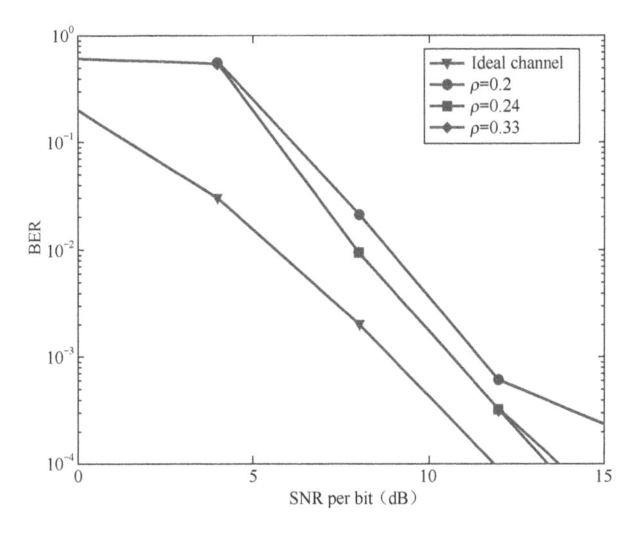

FIGURE 6.21 Performance of IDMA with superposed reference signal under different power ratios, flat channel, 16 users.

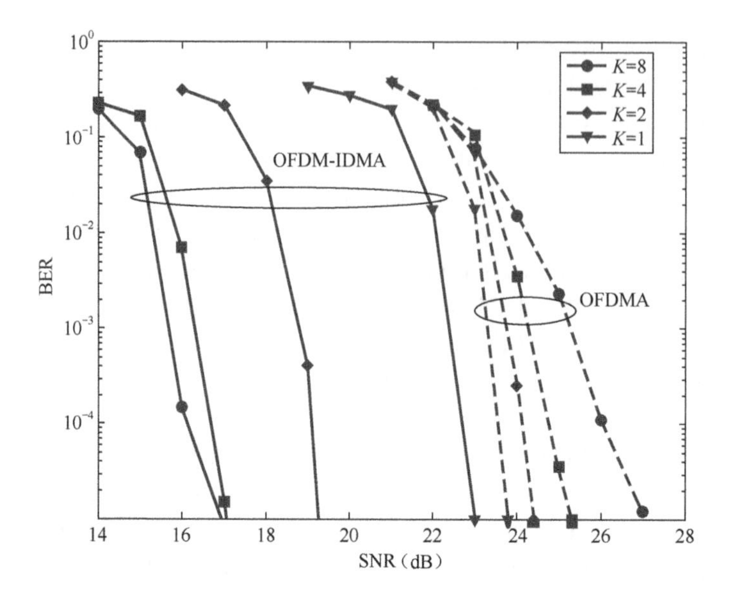

FIGURE 6.22 Performance comparison between IDMA and OFDMA, with clipping ratio = 0 dB, total rate = 3bps/Hz, total 24 layers. The number of layers per user is the same.

A large number of users can share the same time-frequency resources by IDMA. Figure 6.23 shows an example where the channel is AWGN and each user's spectral efficiency is 0.125 bps/Hz. It is seen that 64 users can be accommodated, resulting in total spectral efficiency of 0.125*64 = 8 bps/Hz. The dashed curves are the expected performance via SNR evolution analysis. Different number of iterations are simulated where we can observe the convergence behavior as the iteration process continues.

SNR evolution analysis is a method that can quickly estimate the performance of IDMA in multi-user transmission. Its results are often quite close to the actual simulations as evident in Figure 6.23. There are two functions defined in SNR evolution analysis: $f()$ and $g()$. Both can be computed from the channel codes being used. For instance, if convolutional codes are used, with QPSK modulation, repeated eight times, simulation can be carried out in AWGN for convolutional coding (with soft output) as illustrated in Figure 6.24. After a certain number of iterations, we obtain the soft-output bits under a specific SNR. Then calculate each symbol's variance via the soft bits, and the average of the variance of all the symbols to get the mean squared value f(SNR). Bit error rate (BER) can be calculated by a hard decision on soft bits. That is the BER for the specific SNR.

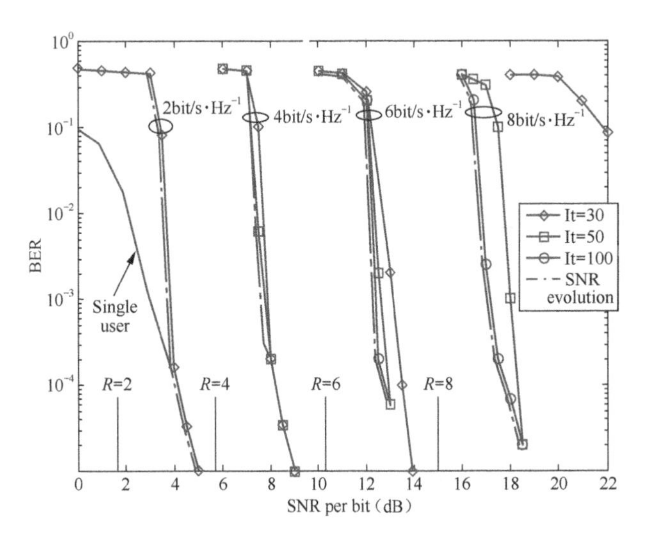

FIGURE 6.23 Performance of IDMA under the different total spectral efficiency, AWGN channel.

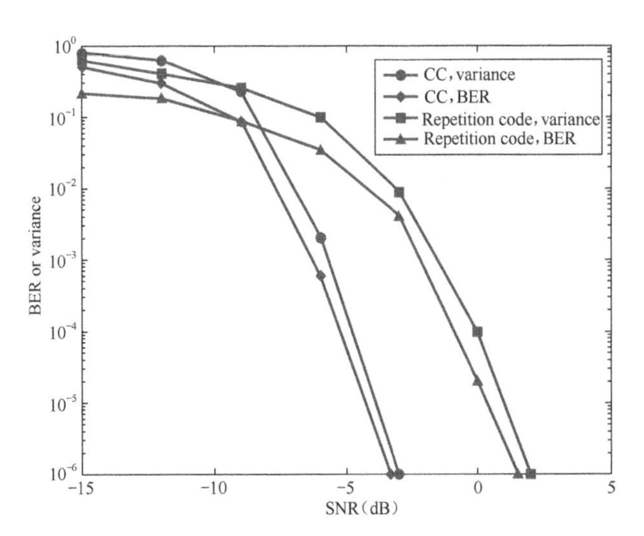

FIGURE 6.24 Functions $f()$ and $g()$ of convolutional codes and repetition code.

Details of SNR evolution analysis can be found in Ref. [25], as shown in Figure 6.25 where SNR is denoted as γ.

1. Initialization, $f\left(\gamma_k^{(0)}\right)=1, \forall k$.
2. SNR update

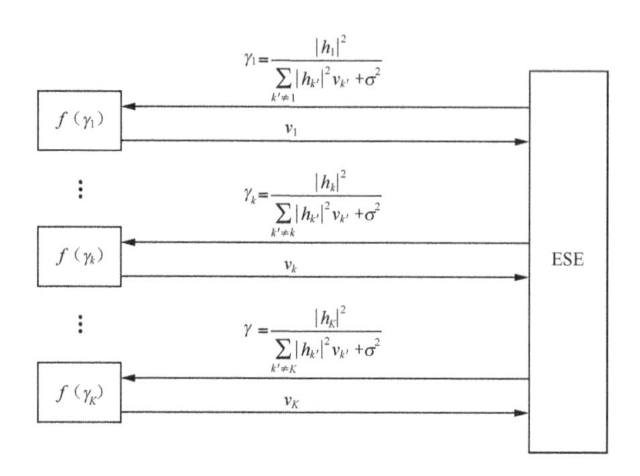

FIGURE 6.25 Block diagram of SNR evolution analysis.

$$\gamma_k^{(q)} = \frac{|h_k|^2}{\sum_{k' \neq k} |h_{k'}|^2 f\left(\gamma_{k'}^{(q-1)}\right) + \sigma^2},$$ for $k = 1$ to K and $q = 1$ to Q, where Q is the number of iterations

3. terminate, BER of the k-th user: $g\left(\gamma_k^{(Q)}\right), \forall k$

It can be observed from the SNR evolution analysis of IDMA that when the received power is the same for all the users, SNR for each user would be relatively low, which may hamper the convergence of iterative detection and decoding. In this situation, unequal power allocation would facilitate the fast convergence of the decoding process of high-power users, thus reducing the interference to other users once high-power users are decoded successfully. Then, the low-power users can be decoded more easily. Table 6.28 shows the numbers of iterations required in the AWGN channel, under different power allocations and number of users (denoted as K). It is found that when K is large, the convergence of ESE can be much faster when the difference in power between users is large.

According to SNR evolution analysis, channel coding has a significant impact on the two functions $f()$ and $g()$, and therefore would affect the performance of IDMA. Powerful channel codes usually help to get good performance overall. For instance, convolutional codes are more powerful than repetition codes. Turbo-Hadamard codes are more powerful than Turbo codes. Figure 6.26 compares IDMA performance between using

TABLE 6.28 Numbers of Iterations Required in AWGN Channel, under Various Power Allocations

Total Number of Users (K)	Number of Iterations	Number of Users in Each Power Level
16	5	16(0 dB)
32	15, 20, 30	25(0 dB) + 7(5.38 dB)
48	20	26(0 dB) + 8(7.45 dB) + 8(10.35 dB) + 6(10.76 dB)
	50	27(0 dB) + 8(7.86 dB) + 4(9.93 dB) + 9(10.34 dB)
64	20	24(0 dB)+ 6(7.86 dB) + 8(8.29 dB) 8(13.66 dB)+ 4(14.07 dB) + 14(19.45 dB)
	30	25(0 dB) + 7(7.86 dB) + 7(8.28 dB) 5(13.25 dB) + 7(13.66 dB) + 13(18.63 dB)
	50	26(0 dB) + 6(7.86 dB) + 9(8.28 dB) 7(12.42 dB) + 13(16.97 dB) + 3(17.39 dB)

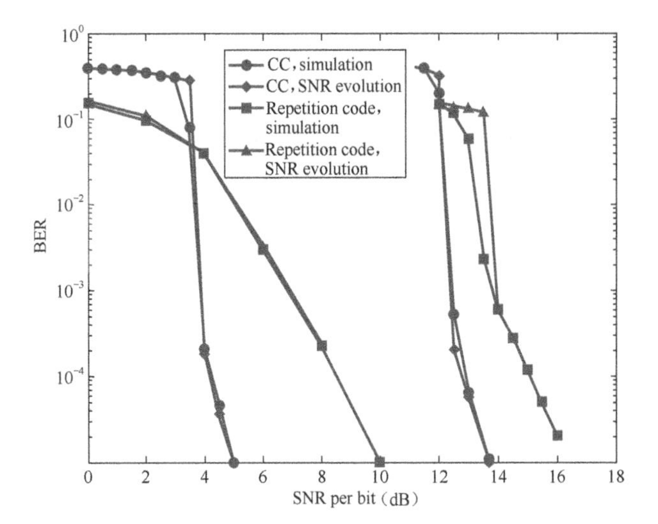

FIGURE 6.26 IDMA performance between using convolutional codes and repletion codes, between actual simulations and SNR evolution analysis.

convolutional codes and using repetition codes. Figure 6.27 compares IDMA performance between using Turbo-Hadamard codes and Turbo codes. In Figure 6.28, the actual simulation results are also compared with the results of SNR evolution analysis.

System throughputs for different channel codes are compared in Figure 6.28 [30]. It is observed that the performance difference between channel codes is not significant when SNR is high, e.g., even repetition codes can deliver quite good performance. However, in the low

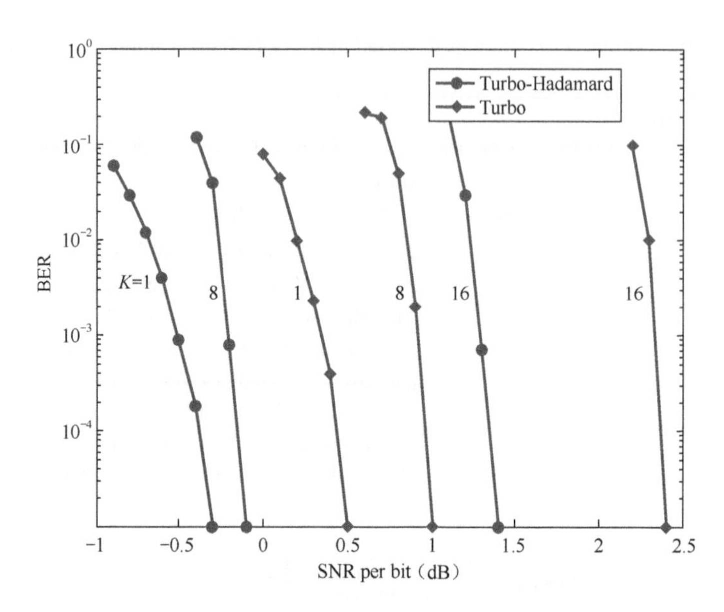

FIGURE 6.27 IDMA performance between using Turbo-Hadamard codes and Turbo codes.

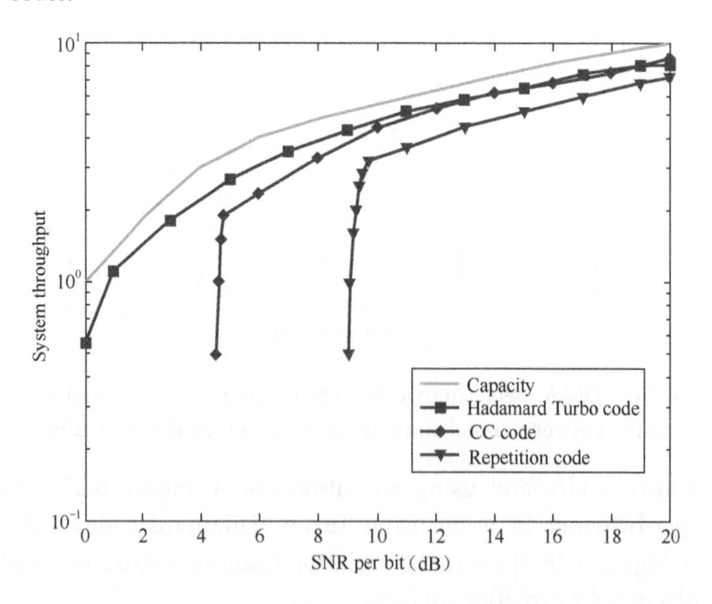

FIGURE 6.28 System throughput of IDMA with different channel codes.

SNR region, convolutional codes are much better than repetition codes, e.g., repetition codes would not work for SNR < 8 dB. Similarly, when SNR is <4 dB, convolutional codes would not work, whereas Turbo-Hadamard can work until SNR is reduced to 1 dB.

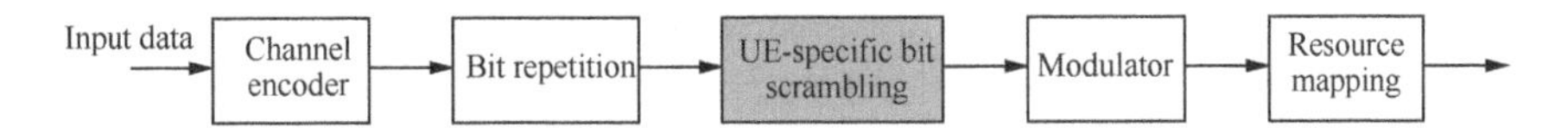

FIGURE 6.29 Transmitter-side processing for scrambler-based schemes.

6.2.1.2 Bit Scrambler-Based Processing

Similar to the interleavers used for mobile communications, scramblers are also widely used, which can be applied to non-orthogonal multiple access, e.g., different users transmit the signals with different scrambling codes. Its typical processing at the transmitter side is illustrated in Figure 6.29.

Although both interleavers and scramblers can achieve bit randomization, there is still a certain performance difference between them. When the number of users is not large and the per-user code rate is not high, interleavers and scramblers perform almost the same. However, in the case of a large number of users, IDMA significantly outperforms scrambler-based NOMA schemes. This is because interleavers can reduce the chance of small loops in the Tanner graph, whereas scramblers cannot. Hence, in theory, interleavers have better performance than scramblers in NOMA.

Similar to IDMA, ESE detector (or MMSE ESE) and MMSE hard IC with a spreading factor of 1 can be used for scrambling-based schemes When detecting the current user, signals of all other users are considered as the interfering signals which can be estimated based on their mean values and variance. After the soft cancelation of the interfering signals, then to calculate the soft information of the current user and input to the channel decoder. The mean value and variance of the current signal can be calculated based on the soft output of the channel decoder, which can be used for detection of other users' signals.

It has been proposed that scrambling and delaying are used together to support NOMA, e.g., different users transmit the signals with different scrambling codes and delays, as shown in Figure 6.30. One of such schemes is called asynchronous coded multiple access (ACMA) [31].

It should be pointed out that when ACMA and OFDM are used together, synchronization should be maintained at OFDM symbol boundaries. However, the indices of OFDM symbols can be different between users, thus allowing asynchronous transmission as illustrated in Figure 6.31.

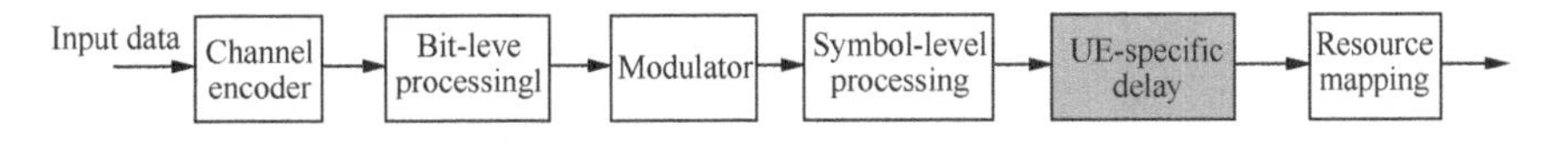

FIGURE 6.30 NOMA schemes based on different scrambling and delay.

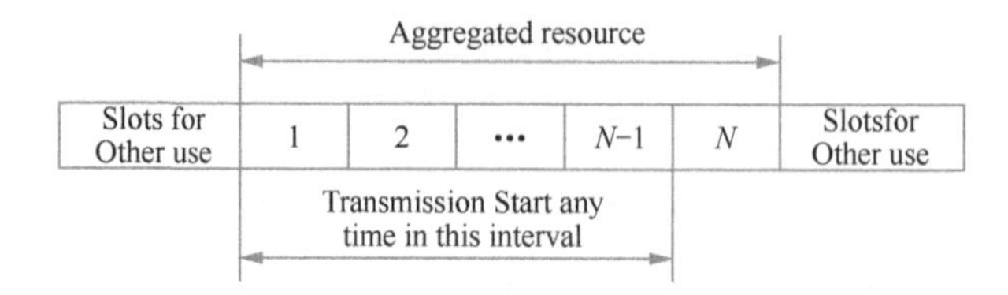

FIGURE 6.31 Time-staggered transmission of ACMA.

6.2.2 ESE + SISO Receiver and Complexity Analysis

6.2.2.1 ESE + SISO Receiver Algorithms

The receiver is an important part of NOMA. Once the transmitter-side scheme is determined, different receiver algorithms would lead to different performances. It is fair to say that transmitter is the key design of NOMA, and the receiver algorithm is the challenging point of the NOMA design. The typical receiver of IDMA is ESE. When there is only one antenna at the receiver, the received signal can be represented as

$$y(j) = h_k x(j) + \xi_k(j) \tag{6.5}$$

where $\xi_k(j) = \sum_{k' \neq k} h_{k'} x(j) + \eta(j)$ and $\eta(j)$ is assumed to be an additive white Gaussian variable. Then the SINR of the k-th user is

$$\text{SINR}_k = \frac{|h_k|^2 P_k}{\sum_{k' \neq k} |h_{k'}|^2 v_{k'} P_{k'} + \sigma^2} \tag{6.6}$$

$v_{k'}$ is the variance of the soft symbol calculated based on the soft bits at the output of the decoder. The essential operation in the ESE receiver is the iterative process between the transfer function of the detector and the transfer function of the decoder, as illustrated in Figure 6.32.

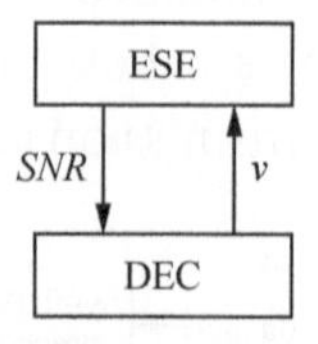

FIGURE 6.32 Information exchange between ESE detector and soft-output decoder.

For simplicity, let us assume the AWGN channel where the SNR can be written as

$$\mathrm{SNR} = \phi(v_k) = \frac{P}{(K-1)Pv+\sigma^2} \tag{6.7}$$

The transfer function of the decoder is

$$v = \psi(\mathrm{SNR}) \tag{6.8}$$

The convergence behavior can be analyzed by observing the trace of the two curves $\phi(v)$ and $\psi(\mathrm{SNR})$[32]. It is seen in Figures 6.2 and 6.16 that since the two curves in the left plot do not intersect each other, the iterative process can converge. By contrast, the two curves cross each other in the right plot. Hence, the iterative process would not cover (Figure 6.33).

ESE was originally proposed for single transmit and single receive antenna. When there are multiple receive antennas, a multi-antenna scenario can be converted to a single receive antenna via maximal ratio combining (MRC) combining. It is known that MRC cannot effectively reduce the cross-user interference and its performance in general is inferior to that of MMSE ESE. Let us use the following equation to represent the received signal for a multi-user case:

$$\mathbf{y}(j) = \sum_{k=1}^{K} \mathbf{h}_k x_k(j) + \mathbf{n}(j), j = 1,2,\ldots,J \tag{6.9}$$

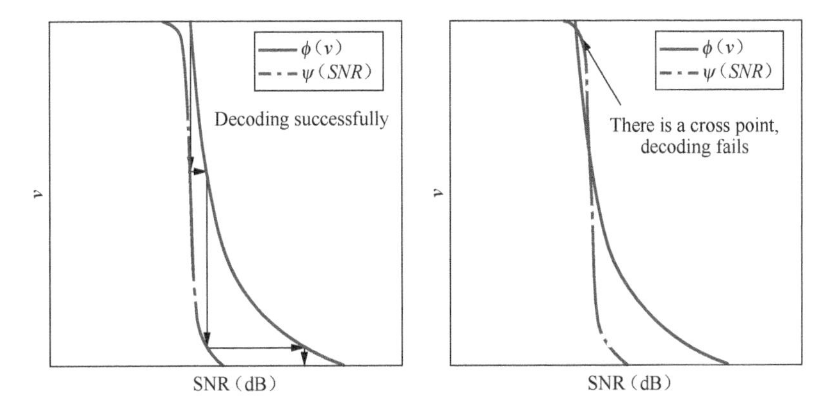

FIGURE 6.33 EXIT Chart analysis of $\phi(v)$ and $\psi(\mathrm{SNR})$.

where $\mathbf{h}_k$ is a $N_r \times 1$ vector, representing the channel of the k-th user. K is the number of users and J is the number of modulated symbols in the block. N_r is the number of receiver antennas. $\mathbf{n}(j)$ is the zero-mean AWGN with a variance of $2\sigma^2$. Equation (6.9) can be written in a more compact form as

$$\mathbf{y}(j) = \mathbf{H}\mathbf{x}(j) + \mathbf{n}(j) \tag{6.10}$$

where $\mathbf{H} = [\mathbf{h}_1, \mathbf{h}_2, \ldots, \mathbf{h}_K]$, $\mathbf{x}(j) = [x_1(j), x_2(j), \ldots, x_K(j)]^T$.

MMSE ESE algorithm can be described in the following steps:

Initialization:

$$\bar{x}_k(j) = 0,\ \bar{v}_k(j) = 1, e_{\text{DEC}}(llr_k) = 0, \forall k, j \tag{6.11}$$

Key operations:

For It = 1 to It_{num} (outer iteration)

{For k = 1 to K (iteration over each user)

{

$$\mathbf{R} = 2\sigma^2\mathbf{I} + \mathbf{H}\mathbf{V}\mathbf{H}^H \tag{6.12}$$

$$\mathbf{W} = \mathbf{V}\mathbf{H}^H\mathbf{R}^{-1} \tag{6.13}$$

Computation of posterior variance

$$\hat{\mathbf{V}} = \mathbf{V} - \mathbf{W}\mathbf{H}\mathbf{V} \tag{6.14}$$

Computation of posterior mean value

$$\hat{\mathbf{x}}(j) = \bar{\mathbf{x}}(j) + \mathbf{W}\left(\mathbf{y}(j) - \mathbf{H}\bar{\mathbf{x}}(j)\right), \forall j. \tag{6.15}$$

Computation of extrinsic variance

$$\frac{1}{v_k^{ex}} = \frac{1}{\hat{v}_k} - \frac{1}{\bar{v}_k}, \forall k. \tag{6.16}$$

Computation of extrinsic mean value

$$x_k^{ex}(j) = v_k^{ex}\left(\frac{\hat{x}_k(j)}{\hat{v}_k} - \frac{\bar{x}_k(j)}{\bar{v}_k}\right), \forall j. \tag{6.17}$$

Computation of posterior LLR

$$e_{\text{ESE}}\left(x_k(j)\right)=2\frac{\text{Re}\left(x_k^{ex}(j)\right)}{v_k^{ex}},\forall j. \tag{6.18}$$

Computation of extrinsic information of the input to the decoder: $e_{\text{ESE}}(llr_k)-e_{\text{DEC}}(llr_k)$

Input the extrinsic information to the decoder and obtain the posterior LLR: $e_{\text{DEC}}(llr_k)$

Computation of extrinsic information based on the output of the decoder: $e_{\text{DEC}}(llr_k)-e_{\text{ESE}}(llr_k)$

Computation of the mean value $\bar{x}_k(j)$ and the variance $\bar{v}_k$ of the soft symbol

}

}

Eq. (6.18) is applicable for QPSK. In the case of high-order modulations, the LLR of each bit can be calculated from the mean $x_k^{ex}(j)$ and the variance v_k^{ex}

In Eq. (6.14), $\mathbf{V}=\text{diag}(\bar{v}_1,\bar{v}_2,\ldots,\bar{v}_K)$, $\bar{\mathbf{x}}(j)=\left[\bar{x}_1(j),\bar{x}_2(j),\ldots,\bar{x}_K(j)\right]^T$, and $\bar{v}_k\equiv\text{Var}\left(x_k(j)\right)$ is the prior variance of $x_k(j)$. To reduce the computation complexity, the following approximation can be used:

$$\bar{v}_k\approx\frac{1}{M}\sum_{j=1}^{M}v_k(j) \tag{6.19}$$

The value of M is about one-tenth of J. $\hat{x}_k(j)$ is the posterior mean of $x_k(j)$ and is also the k-th element of $\hat{\mathbf{x}}(j)$. $\hat{v}_k$ is the posterior mean of $v_k(j)$ and is also the k-th diagonal element of $\hat{\mathbf{V}}$.

The complexity of MMSE ESE mainly comes from signal detection and channel decoding. Equation (6.15) represents the signal detection where other users' signals are canceled. The calculation of W is represented in Eqs. (6.12) and (6.13). There is matrix inversion in Eq. (6.13). Since $\mathbf{R}$ is symmetric, the complexity of inversion of this type of matrices can be reduced to some extent. In the case of higher-order modulation, the calculation of Eq. (6.18) becomes a little more complicated, but there are several simplifications. The complexity of channel decoding can be reduced by using less number of inner iterations inside the LDPC decoder, for instance, 5, instead of 20–25 iterations generally required for LDPC decoding.

In MMSE ESE, the complexity of the matrix inversion depends only on the number of receive antennas and has nothing to do with the number of repetitions. Its contribution to the complexity of the entire MMSE ESE receiver is well controlled.

In the iterations of MMSE ESE, in order to get good performance, the transfer function of the ESE detector $\psi(\text{snr})$ and the transfer function of the decoder $\phi(\text{snr})$ should match. Two transfer functions of ESE $\psi^{(1)}(\text{snr})$ and $\psi^{(2)}(\text{snr})$, representing two schemes are illustrated in Figure 6.34. When $\psi(\text{snr})$ and $\phi(\text{snr})$ cross, it means that the iterative detection would stop the convergence at the intersection. Therefore, the performance would be quite poor. In contrast, when $\psi(\text{snr})$ is always on the left of $\phi(\text{snr})$, the iterative processing between the detector and the decoder would converge, resulting in good performance, as shown in Figure 6.35. More specifically, the transfer function of $\psi^{(2)}(\text{snr})$ is due to the multi-branch transmission with appropriate power allocations between branches. In another word, the shape of $\psi^{(2)}(\text{snr})$ can be changed by adjusting the power allocations. More detailed discussion on multi-branch transmission for NOMA can be found in Section 6.4.

Performances of MRC ESE and MMSE ESE are compared in Figure 6.36 [33]. It is observed that when the number of users is large, MMSE ESE outperforms MMSE MRC since MMSE can try to maximize SINR, whereas MRC cannot. In order to reduce the receiver complexity,

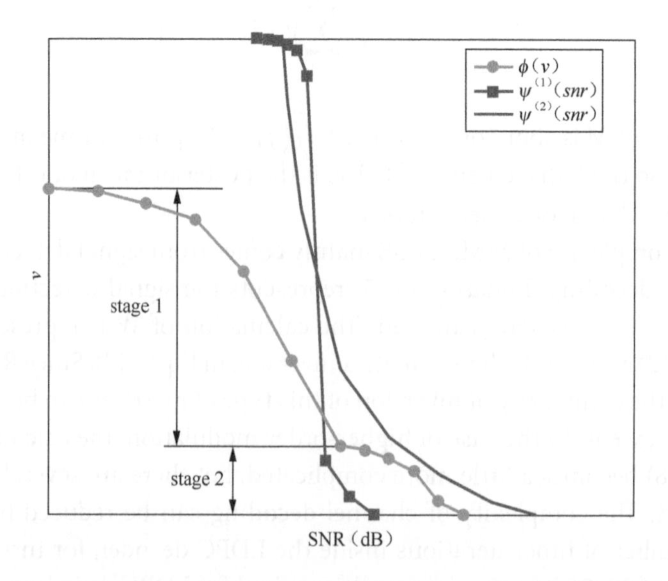

FIGURE 6.34 Transfer functions of ESE detector $\psi(\text{SNR})$ and transfer function of decoder $\phi(v)$.

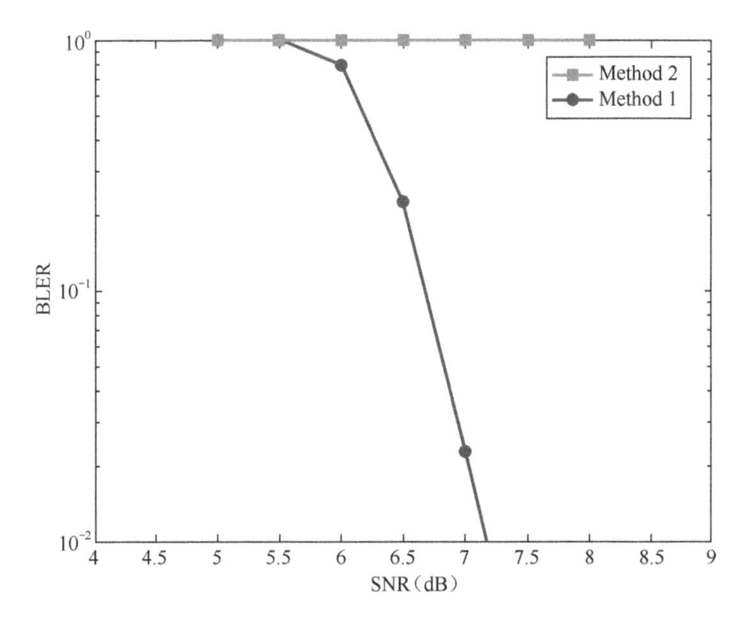

FIGURE 6.35 Comparison of frame error rates between two transfer functions of ESE detector $\psi^{(1)}(\text{SNR})$ vs. $\psi^{(2)}(\text{SNR})$.

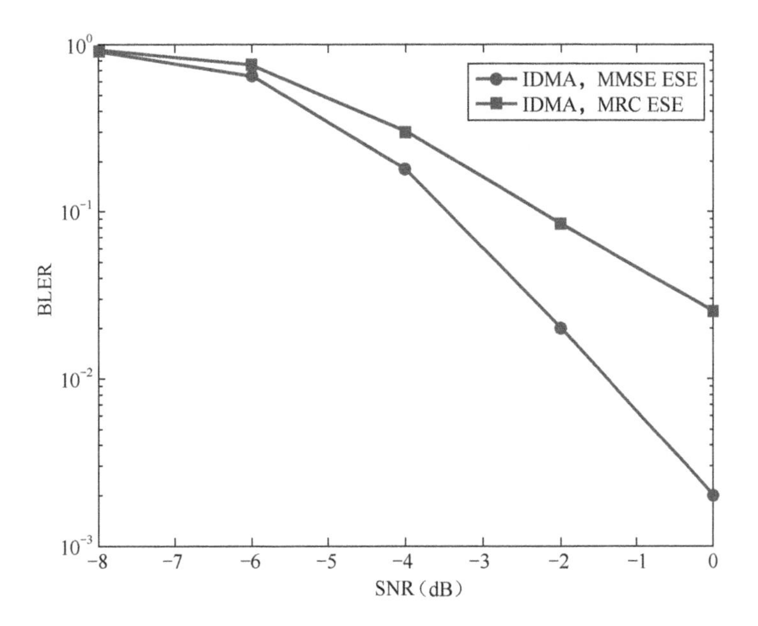

FIGURE 6.36 Performance comparison between MF + ESE and LMMSE + ESE, 6PRB, 20 bytes, 20 users, TDL-C, Nr = 2, QPSK.

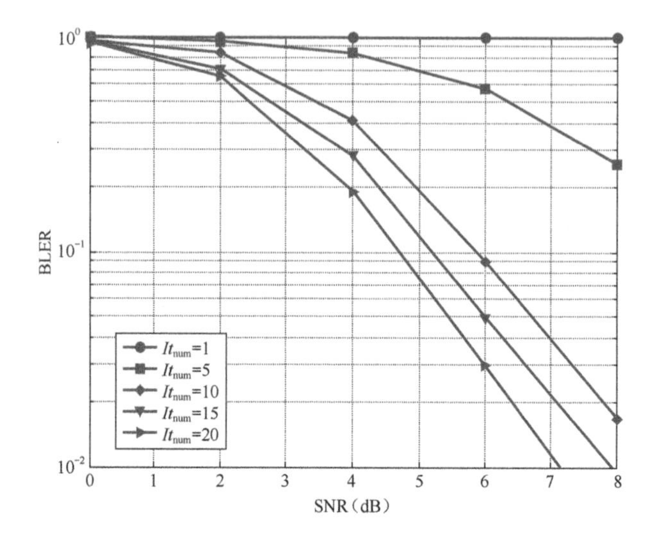

FIGURE 6.37 Performance of iterative receiver, respective to the number of iterations.

when the number of users is not large and per-user spectral efficiency is not high, MRC ESE can be used.

BLER performances of MMSE ESE with a different number of iterations are shown in Figure 6.37. It is seen that the performance quite depends on the number of iterations. The convergence is not reached after 15 iterations. After 20 iterations, required SNR for BLER = 0.1 is reduced by 0.5 dB, compared to 15 iterations.

The performance of the ESE receiver is quite related to the number of iterations. On the other hand, the complexity of the ESE receiver is proportional to the number of iterations. In order to reduce the complexity and processing latency, faster convergence is desirable. One of the ways is to use hard decoded bits to cancel the interference and quickly get cleaner signals.

Channel decoder is an important component of NOMA, and the types of channel decoding impact the performance of channel coding. For Turbo code and LDPC code, log-MAP or maximum log-MAP can be applied as the channel decoder. The latter has inferior performance, albeit with low computational complexity, and is widely used in hardware implementation due to its smaller computational complexity. As we know max-log-MAP usually has about a 0.5 dB penalty compared with log-MAP. Such performance loss may be acceptable for the single-user case due to large savings in the computational complexity. However, when the iterative detector is applied, the NOMA performance will be significantly impacted by the channel decoder. A small performance difference will be greatly amplified

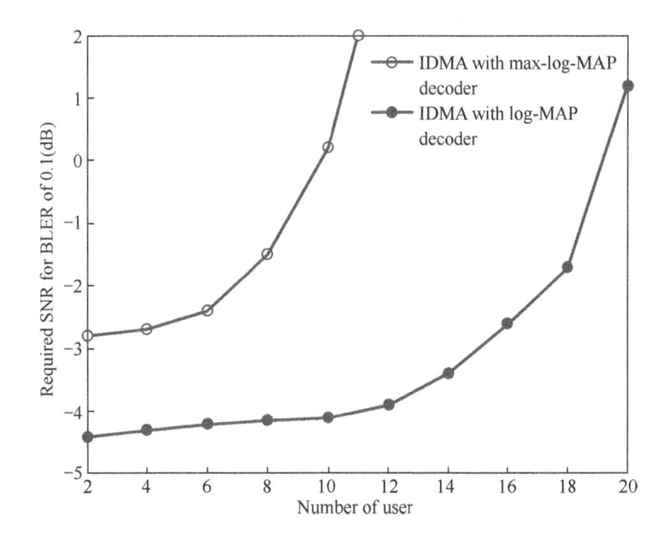

FIGURE 6.38 Performance of IDMA with different channel decoders.

during the iterative process. Figure 6.38 shows the performance of IDMA with different channel decoders [41]. It is observed that when SNR is about −2 dB the supported number of users is 18 for the log-MAP decoder, while it is only 8 for the max log-MAP decoder. The supported number of users is doubled by using a log-MAP decoder. Hence, for mMTC applications where the supported number of users is a crucial requirement, it is recommended to use a log-MAP decoder.

6.2.2.2 Complexity Analysis of the ESE + SISO Receiver

The break-out analysis of the complexity of key modules in ESE + SISO is shown in Table 6.29. In addition to MRC ESE, the complexities of enhanced MRC ESE and enhanced ESE MMSE are also provided. Note that while MRC ESE is simpler, its capability of interference suppression is weaker, thus may increase the burden on the decoder, e.g., more iterations are needed between the detector and the decoder. This would lead to higher complexity in the overall receiver. It is also noted that when a low PAPR waveform such as DFT-s-OFDM is used, multiple FFT and IFFT operations are needed for iterative ESE detection, which would add complexity.

Typical values of the parameters to calculate the complexity of the ESE receiver are listed in Table 6.30. Among them, the parameter $N_{\text{iter}}^{\text{outer}}$ is similar to $N_{\text{iter}}^{\text{IC}}$ in Table 6.30, both referring to the average number of decoding attempts for each code block. Due to the iterative nature of ESE, the typical value of $N_{\text{iter}}^{\text{outer}}$ is greater than $N_{\text{iter}}^{\text{IC}}$, meaning that the ESE receiver relies heavily on the decoder.

TABLE 6.29 Complexity Break-Out of Key Modules of ESE + SISO Receiver

Main Module	Detailed Processing Block	O(.) Complexity Analysis		
		Basic ESE MRC	**Enhanced ESE MRC**	**Enhanced ESE MMSE**
Detector (measured in the number of complex multiplications)	User detection	$O\left(N_{\text{AP}}^{\text{DMRS}} \cdot N_{\text{RE}}^{\text{DMRS}} \cdot N_{rx}\right)$		
	Channel estimation	$O\left(N_{\text{UE}} \cdot N_{\text{RE}}^{\text{CE}} \cdot N_{\text{RE}}^{\text{DMRS}} \cdot N_{rx}\right)$		
	Antenna combining		$O\left(N_{\text{RE}}^{\text{data}} \cdot N_{\text{UE}} \cdot N_{rx}\right)$	$O\left(N_{\text{RE}}^{\text{data}} \cdot N_{\text{UE}} \cdot N_{rx}^{3}\right)$
Decoder (measured in the number of additions and comparisons)	LDPC decoding	$\text{A}{:}N_{\text{iter}}^{\text{outer}} \cdot N_{\text{UE}} \cdot N_{\text{iter}}^{\text{LDPC}} \cdot \left(d_v N^{\text{bit}} + 2\left(N^{\text{bit}} - K^{\text{bit}}\right)\right)$ $\text{C}: N_{\text{iter}}^{\text{outer}} \cdot N_{\text{UE}} \cdot N_{\text{iter}}^{\text{LDPC}} \cdot (2d_c - 1) \cdot \left(N^{\text{bit}} - K^{\text{bit}}\right)$		
Interference cancelation (measured in the number of complex multiplications)	Conversion from LLR to the probability	$O\left(N_{\text{iter}}^{\text{outer}} \cdot N_{\text{UE}} \cdot N^{\text{bit}}\right)$		
	Interference cancelation	$O\left(6 \cdot N_{\text{iter}}^{\text{outer}} \cdot N_{\text{UE}} \cdot N_{\text{RE}}^{\text{data}} \cdot N_{rx}\right)$	$O\left(6 \cdot N_{\text{iter}}^{\text{outer}} \cdot \rho N_{\text{UE}} \cdot N_{\text{RE}}^{\text{data}} \cdot N_{rx}\right)$	$O\left(6 \cdot N_{\text{iter}}^{\text{outer}} \cdot \rho N_{\text{UE}} \cdot N_{\text{RE}}^{\text{data}} \cdot N_{rx}\right)$

TABLE 6.30 Typical Values of Parameters for Complexity Calculation of ESE Receiver

Type of Parameters	Parameter	Notation	Value
General	Number of receive antennas	N_{rx}	2 or 4
	Number of Res	$N_{\mathrm{RE}}^{\mathrm{data}}$	864
	Number of users	N_{UE}	12
Decoding related	Average column weight of LDPC	d_v	3.43
	Average row weight of LDPC	d_c	6.55
	Number of information bits	K^{bit}	176
	Number of coded bits	N^{bit}	432
	Number of iterations within LDPC decoder	$N_{\mathrm{iter}}^{\mathrm{LDPC}}$	20
Soft cancelation related	Number of iterations between ESE detector and decoder	$N_{\mathrm{iter}}^{\mathrm{outer}}$	5
User detection and channel estimation related	Number of DMRS antenna ports	$N_{\mathrm{AP}}^{\mathrm{DMRS}}$	12
	Total Res of DMRS for channel estimation	$N_{\mathrm{RE}}^{\mathrm{CE}}$	12
	Length of DMRS sequence (NR Type II)	$N_{\mathrm{RE}}^{\mathrm{DMRS}}$	24

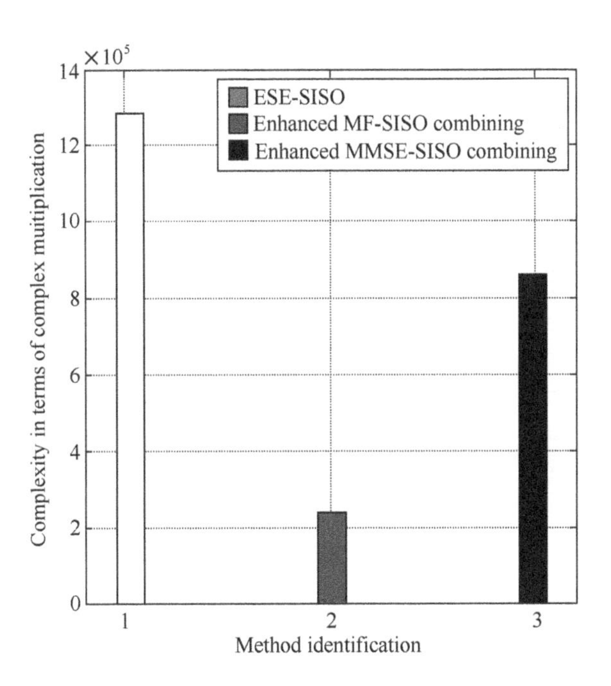

FIGURE 6.39 Complexity of ESE detector, 2 Rx antennas.

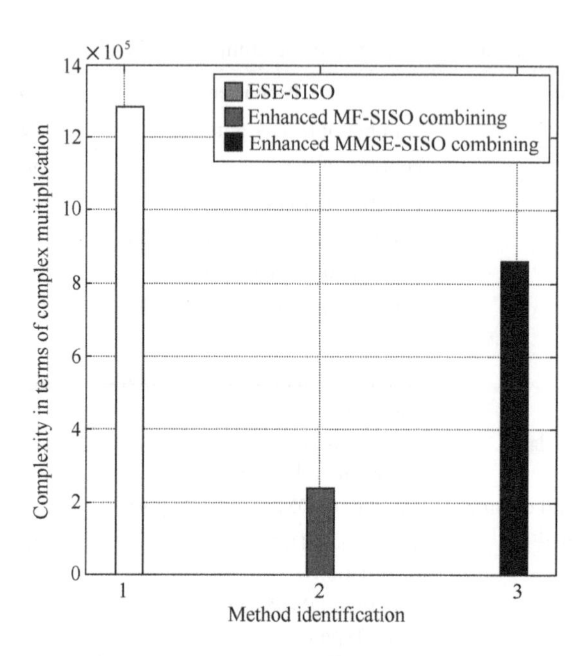

FIGURE 6.40 ESE Complexity of ESE detector, 4 Rx antennas.

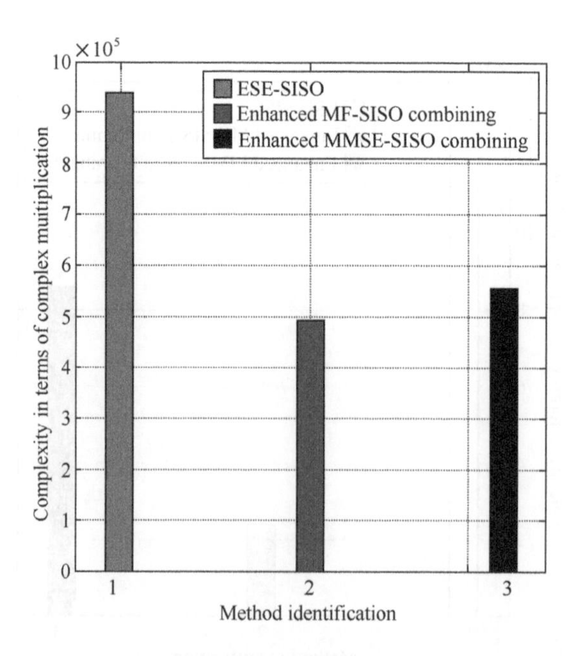

FIGURE 6.41 Total complexity of ESE + SISO receiver, 2 Rx ant.

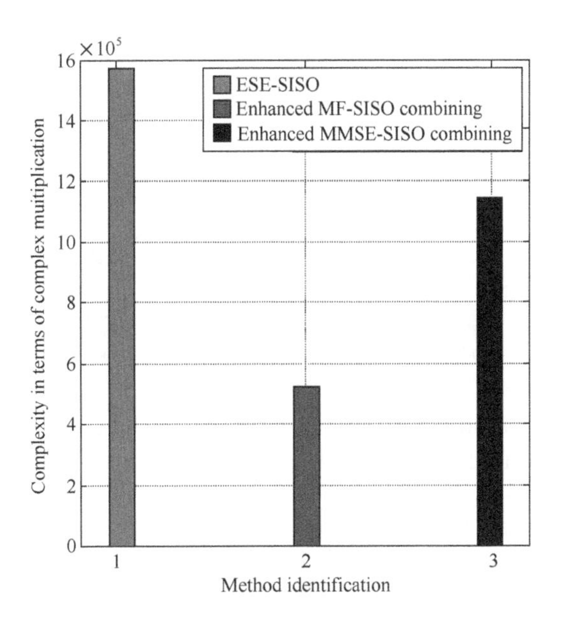

FIGURE 6.42 Total complexity of ESE + SISO receiver, 4 Rx ant.

Figures 6.39 and 6.40 show the complexities of the ESE detector for 2Rx antennas and 4Rx antennas, respectively. The total complexities of the ESE + SISO receiver are illustrated in Figures 6.41 and 6.42, respectively. Take the enhanced ESE MMSE as an example (which can provide a good trade-off between performance and complexity), the ESE+SISO receiver requires roughly 5.6×10^5 and 1.1×10^6 number of complex multiplications to process the multi-user detection and decoding for 2Rx and 4Rx antennas, respectively. Its complexity is higher than that of MMSE hard IC.

6.3 MULTI-DIMENSIONAL MODULATION-BASED SPREADING AND TYPICAL RECEIVERS

6.3.1 Introduction of SCMA

Sparse code multiple access (SCMA) can be considered as an evolution of low-density spreading multiple access (LDSMA) [34]. Different from the traditional CDMA, signals of multiple users are not fully overlapped. Instead, the superposition is of sparsity type, e.g., the number of superposed users is far less than the total number of users, as shown in Figure 6.43 which is a bi-partite graph of six users' signals superposed to four resource nodes. By doing so,

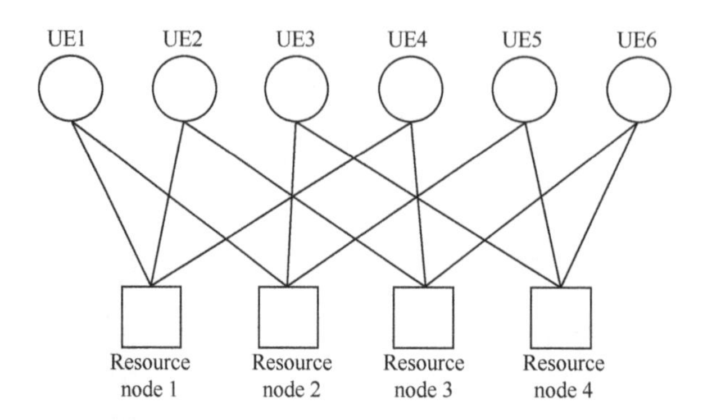

FIGURE 6.43 An example of a bi-partite sparse graph between user nodes and the resource node for LDSMA.

- Searching space on each resource node can be significantly reduced, thus allowing more advanced multi-user detection algorithms such as belief propagation
- To ensure a higher SINR on each resource node and thus to improve the detection performance
- Interference on each resource node for each user comes from different users. Interference would exhibit diversity property, e.g., not all resource nodes would experience strong interference at the same time.

While belief propagation-based multi-user detection can provide optimal performance, the complexity of receivers for LDSMA is significantly higher than the traditional spreading-based system.

Codebook design is a very crucial part of LDSMA and the design involves a number of aspects such as the sparsity mapping, phases of the codeword, power, etc. [35]. Without losing the generality, let us take the AWGN channel as an example. Using x_k as the modulation symbol for the k-th user, the received signal of LDSMA is

$$y(l)=\sum_{k=1} s_k(l)x_k+w(l), \quad l=0,1,\ldots,L-1 \tag{6.20}$$

where L is the spreading factor. The constellation of x_k has total M points. The received signal can be written in matrix form as

$$\mathbf{y}=\mathbf{S}\mathbf{x}+\mathbf{w} \tag{6.21}$$

where $\mathbf{S}$ is the $L \times K$ sparse spreading matrix and each column is a spreading sequence $\mathbf{S} = [\mathbf{s}_1, \mathbf{s}_2, \ldots, \mathbf{s}_K]$, e.g., $\mathbf{s}_k = [s_k(0), \ldots, s_k(l), \ldots, s_k(L-1)]^T$ is the spreading sequence of the k-th user. $\mathbf{x} = [x_1, x_2, \ldots, x_K]^T$ are the modulation symbol of each user. $\mathbf{w}$ is the vector of the AWGN. The optimal maximum a posteriori (MA) receiver can be used:

$$\hat{\mathbf{x}}_{\text{MAP}} = \arg\max_{x \in X^K} \prod_l M_l(\mathbf{x}) \tag{6.22}$$

where $M_l(\mathbf{x})$ is the error probability at the l-th resource node, based on AWGN assumption:

$$M_l(\mathbf{x}) = \exp\left\{ -\frac{1}{\sigma_w^2} \left| y(l) - \sum_{k=1}^{K} s_k(l) x_k \right|^2 \right\} \tag{6.23}$$

In LDSMA, since there are only a small number of resource nodes where the user signal is non-zero, we only need to calculate the error probability of non-zero signals. Let us use $F(l) = \{k : s_k(l) \neq 0\}$ to represent the set of users that have non-zero signals on the l-th resource node, then

$$M_l(\mathbf{x}) = \exp\left\{ -\frac{1}{\sigma_w^2} \left| y(l) - \sum_{k \in F(l)} s_k(l) x_k \right|^2 \right\} \tag{6.24}$$

It can be seen that compared to CDMA, in LDSMA, the computation complexity of the MAP algorithm on each resource node is reduced from $O(L \cdot M^K)$ to $O\left(\sum_{l=0}^{L-1} M^{F(l)}\right)$. If the number of non-zero-value users is the same for all the resource nodes, denoted as d_f, the MAP complexity in LDSMA is reduced to $O(L \cdot M^{d_f})$. MAP algorithm is often called

TABLE 6.31 Similarity and Difference between LDSMA and SCMA

	SCMA	LDSMA
Means to differentiate users	Sparse codebook	Sparse spreading sequence
Shaping Gain	Yes	No
Degree of freedom	Resource mapping: from coded bits to spread symbol, joint mapping	Resource mapping: from modulation symbol to spread symbol

message passing algorithm (MPA) in SCMA since it is based on the belief propagation for the bi-partite graph.

SCMA is based on LDSMA, but with quite significant enhancements. As outlined in Table 6.31, similar to LDSMA, the codeword mapping of SCMA also has sparsity property. Hence, when both use advanced receivers such as belief propagation for multi-user detection, the complexity of SCMA would be lower than non-sparsity-based spreading schemes. The difference between SCMA and LDSMA is that LDSMA is based on symbol spreading, that is, each non-zero spread symbol is the weighted sum of the modulation symbols, whereas SCMA is not the traditional symbol-level spreading, e.g., there is no such modulation process where coded bits are mapped to a modulation symbol. Instead, the coded bits are mapped to a multiple-dimensional codebook. In high SNR scenarios, the SCMA codebook is expected to provide certain shaping gain.

SCMA operation can be further divided into two processes: multi-symbol joint modulation (from coded bits to modulation symbol, the mapping is denoted as $g(*)$) and the sparse mapping to the resources (from modulation symbols to the resource nodes, denoted as $\mathbf{V}$). For the k-th user whose coded bits are $\mathbf{b}_k$, its transmitting signal would be [36]

$$\mathbf{x}_k = \mathbf{V}_k\left(g_k(\mathbf{b}_k)\right) \tag{6.25}$$

In an SCMA system with K users, each with a spreading factor of L. The number of states for the cluster of coded bits is M and the non-zero elements in the spreading block is N. The codebook set can be written as [37]: $\boldsymbol{S}(\boldsymbol{V},\boldsymbol{G};K,L,M,N)$ where $\boldsymbol{V} := \left[\mathbf{V}_k\right]_{k=1}^{K}$ is the sparse resource mapping set of K users. $\boldsymbol{G} := \left[g_k(*)\right]_{k=1}^{K}$ is the multi-user joint modulation mapping for K users. Let us use $f(*)$ as a certain metric function, the optimal SCMA codebook can be designed to

$$\boldsymbol{V}^*,\boldsymbol{G}^* = \arg\max_{V,G} f\left(\boldsymbol{S}(\boldsymbol{V},\boldsymbol{G};K,L,M,N)\right) \tag{6.26}$$

For the optimization of this type of multi-dimensional system, the problem often is disintegrated into several sub-problems to solve. First, for the optimization of sparse resource mapping, exhaustive searching over all the permutations, e.g., $K = \begin{pmatrix} L \\ N \end{pmatrix}$, would result in the optimal solution. Then, multi-symbol joint modulation and mapping would be optimized.

$$G^{+} = \arg\max_{G} f\left(S\left(V^{+}, G; K, L, M, N\right)\right) \tag{6.27}$$

This is a problem in designing a K-layer M dimension constellation. To simplify the optimization, the problem can be further divided into two layers: a constellation of the mother layer and the constellation in other layers, that is, $g_k(*) = (\Delta_g)g$. Then the codebook design problem is simplified as:

$$g^{+}, \left[\Delta_k^{+}\right]_{k=1}^{K} = \arg\max_{g^{+}, \left[\Delta_k^{+}\right]_{k=1}^{K}} f\left(S\left(V^{+}, G = \left[(\Delta_k)g\right]_{k=1}^{K}; K, L, M, N\right)\right) \tag{6.28}$$

This can be considered sub-optimal for the overall constellation design as the mother constellation and the constellation of each layer can be independently optimized. The design of the mother constellation [37] provides the method of rotation of the constellation and re-grouping of real and imaginary components. It is proved in [38] that constellation rotation can achieve the maximization of the product of minimum distances in the constellation as shown in Figure 6.44.

For constellation in each layer, different design principles of multi-symbol joint modulation and mapping should be used for the uplink and the

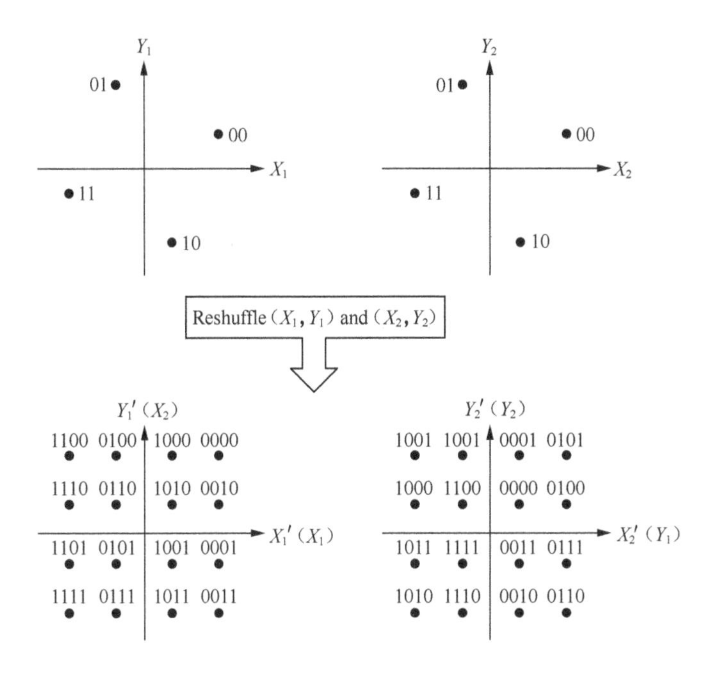

FIGURE 6.44 An example of the mother constellation design of SCMA.

downlink. In the uplink, since different users experience different channels, it is possible to use the same multi-symbol joint modulation and mapping for different users. However, signal superposition is performed at the transmitter. If collision occurs in the composite constellation, the performance would be significantly degraded. Hence, in order to separate different users' data, different multi-symbol joint modulation and mappings are used. In [35], it is proposed that for the normalized power of the mother constellation, the optimal performance can be achieved by using different phases to separate the superposed constellation, similar to LDS.

6.3.1.1 Multi-Symbol Joint Modulation

In multi-symbol joint modulation, a bunch of coded bits are mapped to several symbols. Hence, a certain correlation exists between symbols. Figures 6.45 and 6.46 show some examples of how 8-point and 16-point bits are mapped to two symbols [34].

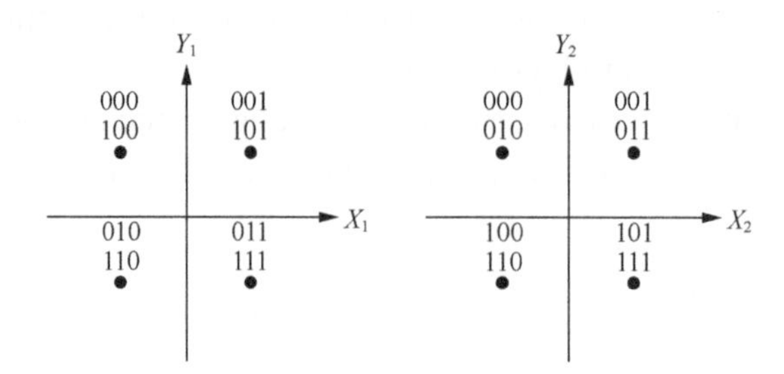

FIGURE 6.45 An example of 8-point bits mapped to two symbols for SCMA.

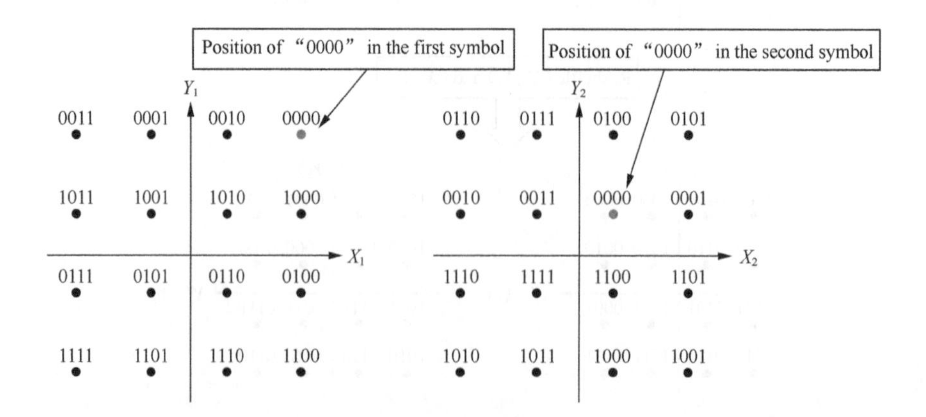

FIGURE 6.46 An example of 16-point bits mapped to two symbols for SCMA.

To further separate the users, it is proposed in Ref. [34] to add a user-specific transformation matrix after multi-symbol joint modulation, that is, multi-symbol joint modulation no longer uses the same mapping, e.g., the ultimate mapping from the coded bits to the modulation symbol can be different for different users. The transformation matrices for two modulation symbols can be

$$\begin{bmatrix} 1 & 0 \\ 0 & 1 \end{bmatrix}, \begin{bmatrix} 1 & 0 \\ 0 & -1 \end{bmatrix}, \begin{bmatrix} 1 & 0 \\ 0 & j \end{bmatrix}, \begin{bmatrix} 1 & 0 \\ 0 & -j \end{bmatrix},$$

$$\begin{bmatrix} 0 & 1 \\ 1 & 0 \end{bmatrix}, \begin{bmatrix} 0 & 1 \\ -1 & 0 \end{bmatrix}, \begin{bmatrix} 0 & 1 \\ j & 0 \end{bmatrix}, \begin{bmatrix} 0 & 1 \\ -j & 0 \end{bmatrix}$$

6.3.1.2 Sparse resource mapping

After the mapping from coded bits to modulation symbols, sparse resource mapping is carried out, that is, to map the modulation symbols to resource elements. Taking two symbols to four resource elements as an example, the mapping is as follows:

TABLE 6.32 SCMA Codebook Resource Pool (Number of Mapped Resource Elements $L = 4$)

Total Pool Size of the Codebooks	**Way to Generate the Overall Codeword**
6	6 codewords of sparse resource mapping + 1 transformation matrix
12	6 codewords of sparse resource mapping + 2 transformation matrix
24	6 codewords of sparse resource mapping + 4 transformation matrix
48	6 codewords of sparse resource mapping + 8 transformation matrix

TABLE 6.33 SCMA Codebook Resource Pool (Number of Mapped Resource Elements $L = 6$)

Total Pool Size of the Codebooks	**Way to Generate the Overall Codeword**
15	15 codewords of sparse resource mapping + 1 transformation matrix
30	15 codewords of sparse resource mapping + 2 transformation matrix
60	15 codewords of sparse resource mapping + 4 transformation matrix
120	15 codewords of sparse resource mapping + 8 transformation matrix

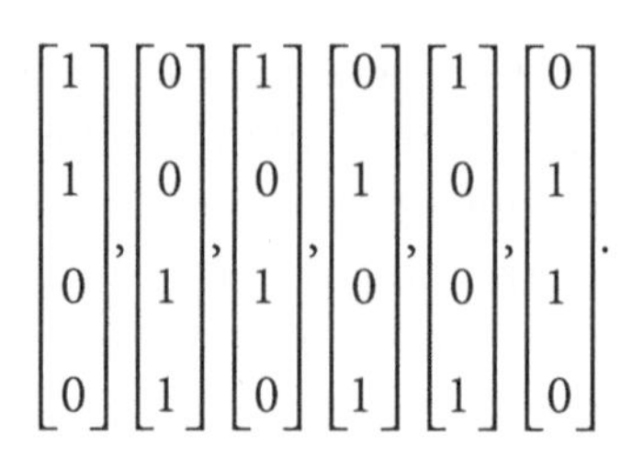

$$\begin{bmatrix}1\\1\\0\\0\end{bmatrix},\begin{bmatrix}0\\0\\1\\1\end{bmatrix},\begin{bmatrix}1\\0\\1\\0\end{bmatrix},\begin{bmatrix}0\\1\\0\\1\end{bmatrix},\begin{bmatrix}1\\0\\0\\1\end{bmatrix},\begin{bmatrix}0\\1\\1\\0\end{bmatrix}.$$

6.3.1.3 Codebook Resource Pool

For SCMA, a codebook resource pool would be used to differentiate the users. The pool consists of two parts: (1) codebook with sparsity resource mapping and (2) codebook based on user-specific transformation matrices. By using these two jointly, a maximum number of codewords can be obtained. Tables 6.32 and 6.33 show the sizes of codebook pools when the number of mapped resource elements L is 4 and 6, respectively. Since transformation matrices have limited capability in separating users, normally the codebook of sparse resource mapping is applied first. After all the codewords of the sparse resource mapping codebook are allocated, the codebook of transformation matrices would be used [39].

6.3.2 EPA + SISO Receiver Algorithm and Complexity Analysis

As mentioned earlier, the receiver for SCMA can be MPA. Because of the sparsity of SCMA, the complexity of MPA is lower compared to non-sparse mapping. Even so, the complexity of MPA exponentially grows with the number of non-zero users. EPA is a simplified version and an approximation of MPA. The complexity of EPA linearly grows with the number of non-zero users.

It should be pointed out that EPA can be applied to all three major categories of NOMA schemes: linear symbol-level spreading, bit-level scrambling/interleaving and joint modulation and spreading.

6.3.2.1 Principle of EPA

EPA has been widely used in machine learning. It is a type of Bayesian interference technique: to use an exponential-like distribution q to approximate the target distribution p. In another word, to project the target distribution p to the set of exponential-like distribution Φ, so that Kullback–Leibler divergence can be minimized, that is

$$\mathrm{Proj}_{\Phi}(Q) = \arg\min_{\Phi} KL(Q\,||\,q) \tag{6.29}$$

It can be proved that the optimal solution to (6.29) can exactly match the statistics of the target Q. For instance, if q is Gaussian distributed, the mean and the variance of q would be equal to the mean and the value of the target distribution Q, respectively. EPA can be used for multi-user detection [34]. In the context of multi-user detection, the target distribution Q is usually the product of a series of factors where D is the normalization constant:

$$Q = \frac{1}{D} \Pi_i f_i(x) \tag{6.30}$$

In EPA, it can be approximated as

$$q = \frac{1}{Z} \Pi_i \tilde{f}_i(x) \tag{6.31}$$

where each factor $\tilde{f}_i(x)$ corresponds to the factor $f_i(x)$ in the target distribution Q. The constant Z is for normalization. If each factor is an exponential-like function, the product of these factors is also exponential-like function. It is usually difficult to get the optimal solution to Eq. (6.29). Instead, each factor is obtained via an iterative process. For example, when we optimize $\tilde{f}_i(x)$, first remove $\tilde{f}_i(x)$ from q, and get $q^{\setminus i} = q / \tilde{f}_i(x)$. Then we get the approximated distribution q^{new} by minimizing $KL\left(\frac{1}{A_i} f_i(x) q^{\setminus i} \,||\, q^{\text{new}}\right)$. Then we get $\tilde{f}_i(x)$ via $q^{\text{new}/q \setminus i}$. After a few iterations, the approximation distribution q can be obtained by calculating the product of $\tilde{f}_i(x)$.

Assuming there are K users (UEs) in a NOMA system with a spreading factor of L. The received signal can be represented as:

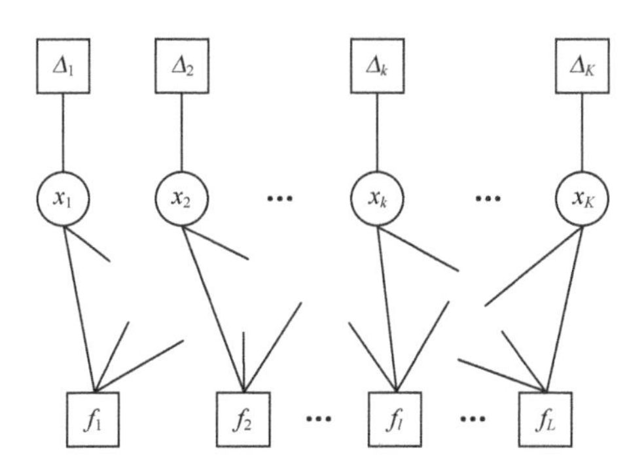

FIGURE 6.47 Factor graph of NOMA.

$$\mathbf{y} = \mathbf{H}\mathbf{x} + \mathbf{n} \tag{6.32}$$

where $\mathbf{H}$ is the channel matrix. $\mathbf{n}$ is the noise vector. The factor graph of the NOMA system is shown in Figure 6.47 which includes K variable nodes (VNs) denoted as $\mathbf{x}_k$, corresponding to the transmitter signal of K users, as well as L factor nodes (FNs), corresponding to the probability $p(\mathbf{y}_l|\mathbf{x})$ on L REs. $\mathbf{y}_l$ is the received signal on l-th RE. There are K nodes to contain K prior information Δ_k.

Let $V(k) = \{l : x_{kl} \neq 0\}$ represent the set of FNs that are connected to the variable node $\mathbf{x}_k$ and $F(l) = \{k : x_{kl} \neq 0\}$ represent the set of VNs that are connected to the factor node f_l. The numbers of elements in $V(k)$ and $F(l)$ are d_v and d_f, respectively

There are two sub-modules in EPA: iteration of VNs, and iteration of FNs. In the t-th iteration of variable nodes, the extrinsic information $I^t_{k\to f}(\mathbf{x}_k)$ which is from the variable node $\mathbf{x}_k$ to the factor node f_l is calculated. This extrinsic information does not include the extrinsic information $I^t_{l\to k}(\mathbf{x}_k)$ which is from the factor node f_l to the variable node $\mathbf{x}_k$. That is, $I^t_{k-l}(\mathbf{x}_k) = I^t_{kl}(\mathbf{x}_k) - I^t_{kl;l\to k}(\mathbf{x}_k)$, and $I^t_{kl;l\to k}(\mathbf{x}_k)$ is the mutual information between $I^t_{kl}(\mathbf{x}_k)$ and $I^t_{l\to k}(\mathbf{x}_k)$.Thus, according to the information theory, the probability $P^t_{k-l}(\mathbf{x}_k)$ from the variable node $\mathbf{x}_k$ to the factor node f_l can be calculated as

$$P^t_{k-l}(\mathbf{x}_k) = \frac{\text{Proj}_\Phi(Q^t(\mathbf{x}_k))}{P^{t-1}_{l\to k}(\mathbf{x}_k)} \tag{6.33}$$

Similarly, in the t-th iteration of FNs, the probability $P^t_{l-k}(\mathbf{x}_k)$ from the factor node f_l to the variable node $\mathbf{x}_k$ can be calculated as

$$P^t_{l-k}(\mathbf{x}_k) = \frac{\text{Proj}_\Phi\left(q^t(\mathbf{x}_k)\right)}{P^{t-1}_{k\to l}(\mathbf{x}_k)} \tag{6.34}$$

where

$$Q^t(\mathbf{x}_k) = P_{\Delta\to k}(\mathbf{x}_k)\Pi_{l\in V(k)} P^{t-1}_{l\to k}(\mathbf{x}_k) \tag{6.35}$$

$$q^t_l(\mathbf{x}_k) = P^t_{k\to l}(\mathbf{x}_k)\sum_{\mathbf{x}_m, m\in F(l), m\neq k} p(\overline{\mathbf{y}}_l|\overline{\mathbf{x}}_l)\Pi_{m\in F(l), m\neq k} P^t_{m\to l}(\mathbf{x}_m) \tag{6.36}$$

If the projection set Φ is Gaussian distributed, the extrinsic information $P_{k-l}^{t}(\mathbf{x}_k)$ and $P_{l\to k}^{t}(\mathbf{x}_k)$ can be sufficiently described by their mean values and variance. In the t-th iteration, the extrinsic information $P_{l\to k}^{t-1}(\mathbf{x}_k)$ which is from the factor node l to the variable node k can be considered as a Gaussian process with mean $u_{l\to k}^{t-1}$ and variance $\xi_{l\to k}^{t-1}$.

$$P_{l\to k}^{t-1} \propto \mathrm{CN}\left(x_{kl};u_{l\to k}^{t-1},\xi_{l\to k}^{t-1}\right) \tag{6.37}$$

Substitute into Eq. (6.35), then we have

$$Q^{t}(x_k) \propto P_{\Delta\to k}(\mathbf{x}_k)\prod_{n\in V(k)}\mathrm{CN}\left(x_{kn};u_{n\to k}^{t-1},\xi_{n\to k}^{t-1}\right) \tag{6.38}$$

By normalizing Eq. (6.38), the following probability can be obtained

$$p^{t}(\mathbf{x}_k=\mathbf{a})=\frac{P_{\Delta\to k}(\mathbf{x}_k=\mathbf{a})\prod_{n\in V(k)}CN\left(x_{kn};u_{n\to k}^{t-1},\xi_{n\to k}^{t-1}\right)}{\sum_{\mathbf{a}\in\chi_k}P_{\Delta\to k}(\mathbf{x}_k=\mathbf{a})\prod_{n\in V(k)}CN\left(x_{kn};u_{n\to k}^{t-1},\xi_{n\to k}^{t-1}\right)} \tag{6.39}$$

where χ_k is the modulation codebook. From Eq. (6.29), we can obtain the mean u_{kl}^{t} and the variance ξ_{kl}^{t} of User k on the resource node l at the t-th iteration:

$$u_{kl}^{t}=\sum_{\mathbf{a}\in\chi_k}p^{t}(\mathbf{x}_k=\mathbf{a})a_l \tag{6.40}$$

$$\xi_{kl}^{t}=\sum_{\mathbf{a}\in\chi_k}p^{t}(\mathbf{x}_k=\mathbf{a})\left|a_l-u_{kl}^{t}\right|^{2} \tag{6.41}$$

From Eq. (6.36), the probability from the variable node k to the factor node l is

$$P_{k\to l}^{t}(\mathbf{x}_k)=\frac{CN(x_{kl};u_{kl}^{t},\xi_{kl}^{t})}{CN(x_{kl};u_{l\to k}^{t-1},\xi_{l\to k}^{t-1})}\propto CN(x_{kl};u_{k\to l}^{t},\xi_{k\to l}^{t}) \tag{6.42}$$

Hence, based on the assumption of Gaussian distribution, the variance and the mean passed from the variable node k to the factor node l can be derived as

$$\frac{1}{\xi_{k\to l}^{t}}=\frac{1}{\xi_{kl}^{t}}-\frac{1}{\xi_{l\to k}^{t-1}}$$

$$\frac{1}{u_{k\to l}^{t}}=\frac{1}{u_{kl}^{t}}-\frac{1}{u_{l\to k}^{t-1}} \tag{6.43}$$

Next, we discuss the message passing from the FNs to the VNs. The mean $\hat{u}_{kl}^{t}$ and the variance $\hat{\xi}_{kl}^{t}$ of the signal on the factor node l can be obtained by MMSE detection:

$$\mathbf{U}_l^{\text{post}}=\mathbf{U}_l^{\text{pri}}+\xi_l^{\text{pri}}\bar{\mathbf{H}}_l^{H}\left(\bar{\mathbf{H}}_l\xi_l^{\text{pri}}\bar{\mathbf{H}}_l^{H}+\sigma^2\mathbf{I}\right)^{-1}\left(\mathbf{y}_l-\bar{\mathbf{H}}_l\mathbf{U}_l^{\text{pri}}\right) \tag{6.44}$$

$$\xi_l^{\text{post}}=\xi_l^{\text{pri}}+\xi_l^{\text{pri}}\bar{\mathbf{H}}_l^{H}\left(\bar{\mathbf{H}}_l\xi_l^{\text{pri}}\bar{\mathbf{H}}_l^{H}+\sigma^2\mathbf{I}\right)^{-1}\bar{\mathbf{H}}_l\left(\xi_l^{\text{pri}}\right)^{H} \tag{6.45}$$

where $\mathbf{U}_{k\to l}^{\text{pri}}=\left[u_{k\to l}^{t}\,|\,k\in F(l)\right]\in C^{d_f\times 1}$ and $\xi_l^{\text{pri}}=\left[\xi_{k\to l}^{t}\,|\,k\in F(l)\right]\in C^{d_f\times 1}$ are the prior information of the mean and the variance, passed from the VNs to the FNs, respectively. $\mathbf{U}_l^{\text{post}}=\left[\hat{u}_{lk}^{t}\,|\,k\in F(l)\right]\in C^{d_f\times 1}$ and $\xi_l^{\text{post}}=\left[\hat{\xi}_{lk}^{t}\,|\,k\in F(l)\right]\in C^{d_f\times 1}$ are the mean and the variance calculated by the factor node.

Based on the assumption of Gaussian distribution and Eq. (6.34), the mean and the variance passed from the factor node to the variable node are derived:

$$\frac{1}{\xi_{l\to k}^{t}}=\frac{1}{\hat{\xi}_{lk}^{t}}-\frac{1}{\xi_{k\to l}^{t}}$$

$$\frac{1}{u_{l\to k}^{t}}=\frac{1}{\hat{u}_{lk}^{t}}-\frac{1}{u_{k\to l}^{t}} \tag{6.46}$$

6.3.2.1.1 Algorithm of EPA + SISO

The procedure is as follows:

- Outer iteration:
  - To compute the prior probability (reset to zero initially)
  - Iterations within EPA:
    - Variable node side:
      - To compute the probability of each of the constellation point

  - To calculate the mean and the variance on VNs
  - To calculate the mean and the variance to be passed from the variable nodes to the factor nodes (VN->FN)
- Factor node side
  - To calculate the mean and the variance on FNs
  - To calculate the mean and the variance to be passed from the factor nodes to the variable nodes (FN->VN)

- To calculate the LLR which is the output of the detector
- To calculate the extrinsic information by decoding

6.3.2.1.2 Computation of Prior Probability

To calculate the prior probability of each constellation point based on the prior information from the decoder (the prior information is initialized to be 0 during the first outer iteration)

$$p^0\left(x_k = a_{l,k}\right) = \Pi_{i=1}^{M}\left(\frac{\exp\left(\frac{1}{2}b_i^l\mathrm{LLR}_{kM+i}\right)}{\exp\left(-\frac{1}{2}b_i^l\mathrm{LLR}_{kM+i}\right)+\exp\left(\frac{1}{2}b_i^l LLR_{kM+i}\right)}\right)$$

$$= \Pi_{i=1}^{M}\left(\frac{\exp\left(b_i^l LLR_{kM+i}\right)}{1+\exp\left(b_i^l LLR_{kM+i}\right)}\right) \tag{6.47}$$

where b_i^l is the value of i-th bit {1, −1} corresponding to the constellation point $a_{l,k}$.

6.3.2.1.3 EPA: Inner Iterations between FNs and VNs

In the following, VNs are numbered in k. The resource elements (e.g., FNs) are numbered in l.

When the number of inner iterations $t \le T_{\max}$

- On the VN side
  - **To update the probabilities on the VN side**

$$p^t(x_k = a_j) = p^0(x_k = a_j) \cdot \Pi_{u \in \phi_k} \frac{1}{\pi \xi_{u->k}^t} \exp\left(-\frac{\left\|a_j - u_{l->k}^t\right\|^2}{\xi_{l->k}^t}\right) \quad (6.48)$$

- To calculate the mean u_{kl}^t and the variance ξ_{kl}^t, based on Eqs. (6.40) and (6.41)
- To calculate the extrinsic information of the mean $u_{k \to l}^t$ and the variance $\xi_{k \to l}^t$ from VNs to FNs

- on FN side
  - To update the mean $\hat{u}_{lk}^t$ and the variance $\hat{\xi}_{lk}^t$ at the FN side, according to Eqs. (6.44) and (6.45). Then take the diagonal elements of ξ_l^{post} as the total variance $\hat{\xi}_{lk}^t$ at the FN side. $\mathbf{U}_l^{\text{post}}$ would be the total mean $\hat{u}_{lk}^t$ at the FN side
  - To update the extrinsic information of mean $u_{l \to k}^t$ and the variance $\xi_{l \to k}^t$ from FNs and VNs, according to Eq. (6.46).

6.3.2.1.4 To Calculate the LLR at Output of the Detector

To calculate the soft information, e.g., LLR, at the output of the detection, based on the symbol probability $p^t(x_k = a_{l,k})$ at the VN side, for each bit

$$\text{LLR}_{\text{EPA}}(i) = \log \frac{\sum_{a_j \in \Phi(b_i=1)} p^t(x_k = a_j)}{\sum_{a_j \in \Phi(b_i=1)} p^t(x_k = a_j)} \quad (6.49)$$

Then, subtract the extrinsic information input to the detector, to get the extrinsic information to the decoder.

6.3.2.1.5 Decoding

To calculate the extrinsic information

- If the maximum number of decoding has not been reached, then go to the step in Eq. (6.46)
- If the maximum number of decoding has been reached, terminate the receiver processing and output the hard decisions

TABLE 6.34 Complexity Analysis of EPA + SISO Receiver

Key Modules of the Receiver	Detailed Processing	$O(.)$ Complexity Estimation		
		Option 1	Option 2	Option 3
Detection (measured in number of complex multiplication)	User detection		$O\left(N_{AP}^{DMRS} \cdot N_{RE}^{DMRS} \cdot N_{rx}\right)$	
	Channel estimation		$O\left(N_{UE} \cdot N_{RE}^{CE} \cdot N_{RE}^{DMRS} \cdot N_{rx}\right)$	
	Covariance matrix calculation		$O\left(\bar{N}_{iter}^{outer} \cdot N_{RE}^{data}\left(\frac{1}{2}N_{rx}^2 d_f + N_{iter}^{det} \cdot \frac{1}{2}N_{rx} d_f\right)\right)$	
	Demodulation weight computation		$O\left(\bar{N}_{iter}^{outer} \cdot N_{RE}^{data} \cdot N_{iter}^{det} \cdot \left(N_{rx}^3 + N_{rx}^2 d_f\right)\right)$	$O\left(N_{iter}^{outer} \cdot N_{RE}^{data} \cdot N_{iter}^{det} \cdot N_{rx}^3\right)$
	Demodulation		$O\left(\bar{N}_{iter}^{outer} \cdot N_{RE}^{data} \cdot N_{iter}^{det} \cdot 3N_{rx} d_f\right)$	
	Soft bit generation			$O\left(N_{iter}^{outer} \cdot N_{UE} \cdot N_{RE}^{data} \cdot Q_m \cdot 2^{Q_m} / N_{SF}\right)$
	LLR generation	$O\left(6 \cdot N_{iter}^{outer} \cdot N_{iter}^{det} \cdot N_{UE} \cdot N_{RE}^{data} \cdot d_u \cdot 2^{Q_m} / N_{SF}\right)$	$O\left(3\bar{N}_{iter}^{outer} \cdot N_{iter}^{det} \cdot N_{UE} \cdot d_u \cdot \frac{Q_m \cdot N_{RE}^{data}}{4 \cdot N_{SF}} + \bar{N}_{iter}^{outer} \cdot N_{UE} \cdot N^{bit} / 4\right)$	$O\left(2 \cdot N_{iter}^{outer} \cdot N_{iter}^{det} \cdot N_{UE} \cdot N_{RE}^{data} \cdot d_f \cdot 2^{Q_m}\right)$
	Message passing	$O\left(8 \cdot N_{iter}^{outer} \cdot N_{iter}^{det} \cdot N_{UE} \cdot N_{RE}^{data} \cdot d_u / N_{SF}\right)$	$O\left(2 \cdot \bar{N}_{iter}^{outer} \cdot N_{iter}^{det} \cdot N_{UE} \cdot d_u \cdot \frac{N_{RE}^{data}}{N_{SF}}\right)$	

(Continued)

TABLE 6.34 (*Continued*) Complexity Analysis of EPA + SISO Receiver

Key Modules of the Receiver	Detailed Processing	$O(.)$ Complexity Estimation		
		Option 1	**Option 2**	**Option 3**
Decoder (measured in the number of binary additions and comparisons)	LDPC decoding		A: $N_{\text{iter}}^{\text{outer}} \cdot N_{\text{UE}} \cdot N_{\text{iter}}^{\text{LDPC}} \cdot \left(d_v N^{\text{bit}} + 2\left(N^{\text{bit}} - K^{\text{bit}}\right)\right)$ C: $N_{\text{iter}}^{\text{outer}} \cdot N_{\text{UE}} \cdot N_{\text{iter}}^{\text{LDPC}} \cdot (2d_c - 1) \cdot \left(N^{\text{bit}} - K^{\text{bit}}\right)$	
Interference cancelation (measured in number of complex multiplications)	Signal reconstruction (including for DFT-S-OFDM)	Extra for DFT-s-OFDM: $O\left(N_{\text{iter}}^{\text{outer}} \cdot N_{\text{iter}}^{\text{det}} \cdot N_{\text{UE}} \cdot N_{\text{RE}}^{\text{data}} \cdot \log_2 N_{\text{FFT}}\right)$	$O\left(N_{\text{UE}} \cdot N_{\text{RE}}^{\text{data}} \cdot N_{rx}\right)$	$O\left(N_{\text{UE}} \cdot N_{\text{RE}}^{\text{data}} \cdot N_{rx}\right)$
	Conversion from LLR to probabilities	$O\left(N_{\text{iter}}^{\text{outer}} \cdot N_{\text{UE}} \cdot N_{\text{RE}}^{\text{data}} \cdot Q_m \cdot 2^{Q_m} / N_{SF}\right)$	$O\left(\bar{N}_{\text{iter}}^{\text{outer}} \cdot N_{\text{UE}} \cdot \frac{2^{Q_m} N_{\text{RE}}^{\text{data}}}{4 \cdot N_{\text{SF}}}\right)$	$O\left(N_{\text{iter}}^{\text{outer}} \cdot N_{\text{UE}} \cdot N_{\text{RE}}^{\text{data}} \cdot Q_m \cdot 2^{Q_m} / N_{SF}\right)$
	LDPC encoding		Buffer shifting: $N_{\text{UE}} \cdot \left(N^{\text{bit}} - K^{\text{bit}}\right)/2$ Addition: $N_{\text{UE}} \cdot (d_c - 1)\left(N^{\text{bit}} - K^{\text{bit}}\right)$	

TABLE 6.35 Typical Values of Parameters to Calculate the Complexity of EPA + SISO Receiver

Group of Parameters	Parameter	Notation	Value
General	Number of receive antennas	N_{rx}	2 or 4
	Number of resource elements for data	$N_{\mathrm{RE}}^{\mathrm{data}}$	864
	Number of users	N_{UE}	12
Decoding related	Average column weight of LDPC	d_v	3.43
	Average row weight of LDPC	d_c	6.55
	Number of information bits	K^{bit}	176
	Number of coded bits	N^{bit}	432
	Number of inner iterations within LDPC decoder	N_{iter}^{LDPC}	20
Soft interference cancelation reated	Number of outer iterations between the detector and the decoder	$N_{\mathrm{iter}}^{\mathrm{outer}}$	3
EPA receiver related	Number of inner iterations within the detector	$N_{\mathrm{iter}}^{\mathrm{det}}$	3
	Number of FNs connected to each variable node	d_u	2
	Number of variable nodes connected to each factor node	d_f	6
	Modulation order	Q_m	3
Detection and channel estimation related	Number of DMRS antenna ports	$N_{\mathrm{AP}}^{\mathrm{DMRS}}$	12
	Total number of REs for DMRS for channel estimation	$N_{\mathrm{RE}}^{\mathrm{CE}}$	12
	Length of DMRS (NR Type II)	$N_{\mathrm{RE}}^{\mathrm{DMRS}}$	24

6.3.2.2 Complexity Analysis of the EPA Receiver

The complexity calculation of the EPA + SISO receiver in the 3GPP NOMA study is listed in Table 6.34. In addition to the above analysis and complexity calculation (corresponding to Option 1 in Table 6.34 with the complexity of some small modules omitted), there are two other analyses, referring to Options 2 and 3 in Table 6.34.

Table 6.35 shows the typical values of the parameters to calculate the complexity of EPA + SISO receiver. Among them, the typical value of $N_{\mathrm{iter}}^{\mathrm{outer}}$ is about three which is lower than that of ESE. This is due to that EPA has inner iterations that can reduce the burden on the decoder. Nevertheless, $N_{\mathrm{iter}}^{\mathrm{outer}}$ is still greater than $N_{\mathrm{iter}}^{\mathrm{IC}}$ which is typically 1.5–3.0.

Figures 6.48 and 6.49 show the complexity of the EPA detector for two Rx antennas and four Rx antennas, respectively. The total complexity of the EPA +SISO receiver for two Rx antennas and four Rx antennas is shown in Figures 6.50 and 6.51, respectively. By averaging over the three

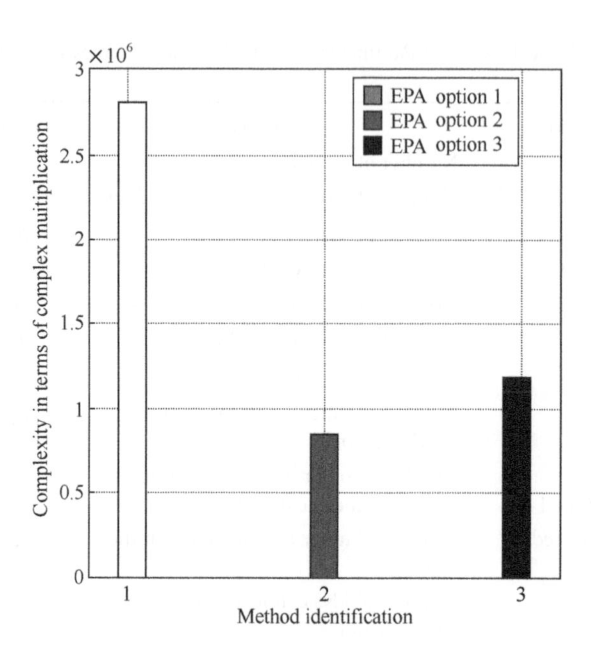

FIGURE 6.48 Complexity of EPA detector, 2 Rx ant.

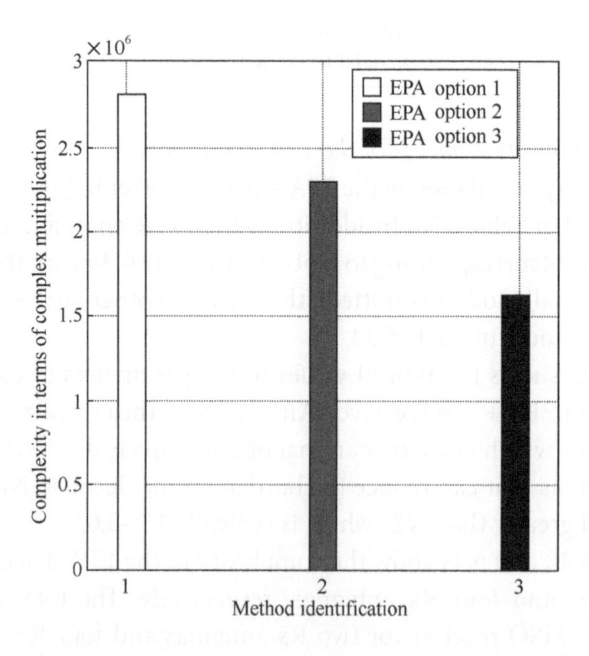

FIGURE 6.49 Complexity of EPA detector, 4 Rx ant.

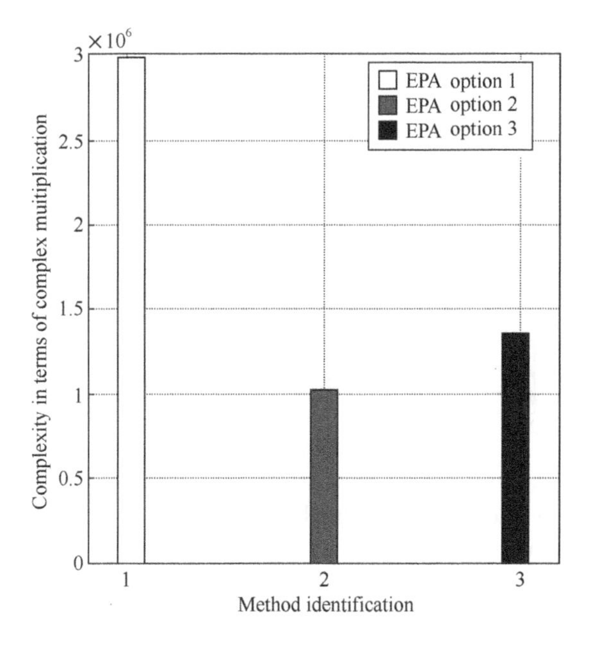

FIGURE 6.50 Total complexity of EPA + SISO receiver, two Rx ant.

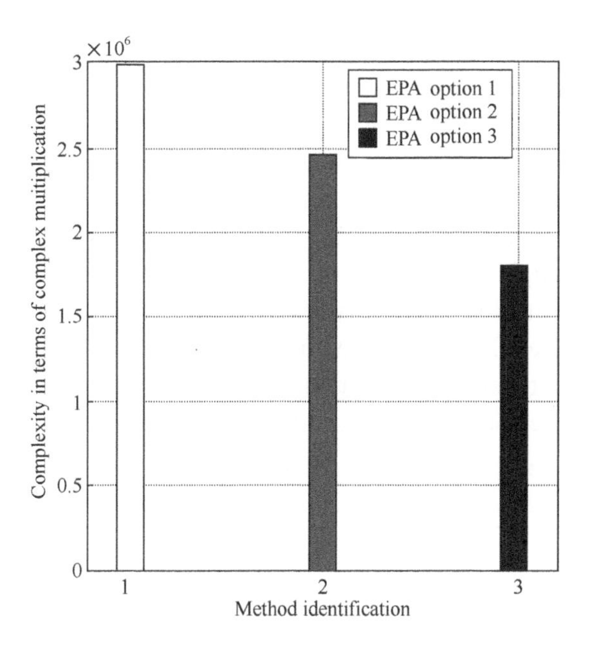

FIGURE 6.51 Total complexity of EPA + SISO, four Rx ant.

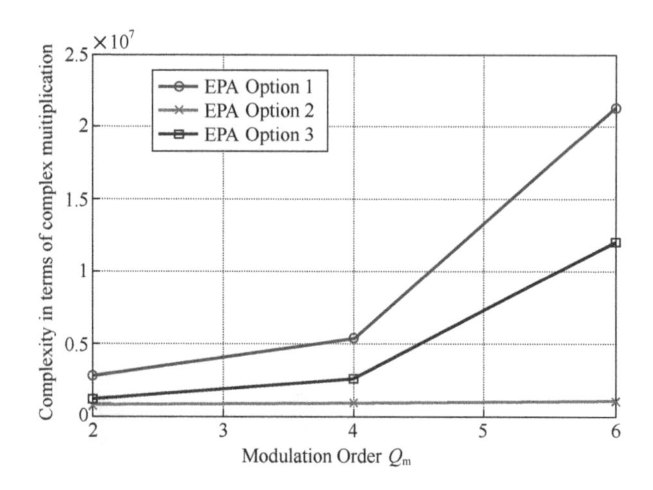

FIGURE 6.52 Relationship between the computation complexity of the EPA detector and the modulation order.

options, the total complexity of EPA + SISO receiver is about 1.8×10^6 and 2.4×10^6 in a number of complex multiplications for 2Rx and 4Rx antennas, respectively, which are significantly higher than those of MMSE hard IC- and ESE-based receivers.

In EPA receiver, the computation-intensive processing is to calculate the probability of each constellation point based on the soft information, e.g., Eq. (6.30). In this step, the mean and variance are the statistical information about the constellation after multiple inner iterations within the EPA detector. The soft output of the decoder should also be considered in this step, that is, to calculate the weight-sum of the probability $I_{\Delta\rightarrow k}(x_k=\alpha)$. Assuming that the number of constellation points is Q, the calculation in Eq. (6.30) requires $(2d_u+1)\cdot Q_m \cdot N_{\text{iter}}^{\text{outer}} \cdot N_{\text{iter}}^{\text{det}} \cdot N_{\text{ue}} \cdot N_{\text{RE}}^{\text{data}} / N_{sf}$ a number of complex multiplications. Since this is proportional to the number of constellation points, the complexity of EPA grows with the modulation order, as shown in Ref. [18] for 2Rx antennas. It is observed from Figure 6.52 that EPA complexity analysis in Options 2 and 3 can accurately reflect the relationship between the complexity and the modulation order.

TABLE 6.36 Computation Complexities of EPA Detector for Different Waveforms

	CP-OFDM Waveform	DFT-S-OFDM Waveform
EPA Option 1	2,800,000	3,375,000
EPA Option 2	1,180,000	1,758,000
EPA Option 3	845,000	1,421,000

It is observed from Table 6.36 that the complexity of EPA for DFT-s-OFDM is about 60% higher than that for CP-OFDM.

6.4 MULTI-BRANCH TRANSMISSION

In some scenarios of NOMA, it is required that per-user spectral efficiency should be high. One of the ways to achieve this is to use high-order modulations such as 16-QAM and 64-QAM. While a high-order modulation symbol can carry multiple bits and thus increase the spectral efficiency, it has some issues. For instance, over multiple bits corresponding to a high-order modulation symbol, some bits have high reliability and some other bits have low reliability. The bits with lower reliability would limit the performance of the link. By contrast, QPSK has the property of a constant module, e.g., the two bits have the same reliability.

Constellations of high-order modulation constellations can be considered as the superposition of multiple QPSK constellations with different powers. For instance, the 16-QAM constellation is the superposition of two QPSK with powers of $\sqrt{1/5}$ and $\sqrt{4/5}$, respectively. 64-QAM constellation is the superposition of three QPSK with powers $\sqrt{1/21}$, $\sqrt{4/21}$ and $\sqrt{16/21}$. Hence, at the transmitter, each QPSK is considered as one branch of the multi-branch transmission.

There are multiple ways to achieve multi-branch transmission. The multi-branch operation can occur before or after the channel encoding. The example shown in Figure 6.53 is the case of the multi-branch before the coding.

In Figure 6.53, the data in each branch is independently encoded, modulated, spread and power-scaled. The modulation is usually QPSK. Multi-branch transmission brings some issues. In particular, not only would it cause cross-user interference but also cross-branch interference. This issue has to be handled properly, otherwise the performance of multi-branch may be worse than single-branch transmission. One of the ways to mitigate

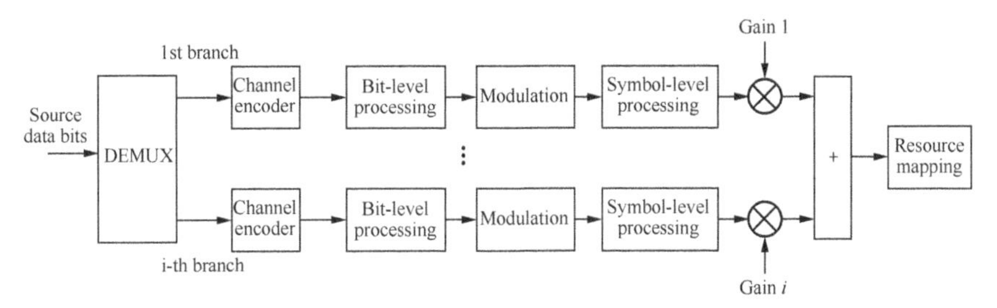

FIGURE 6.53 Pre-encoding multi-branch NOMA.

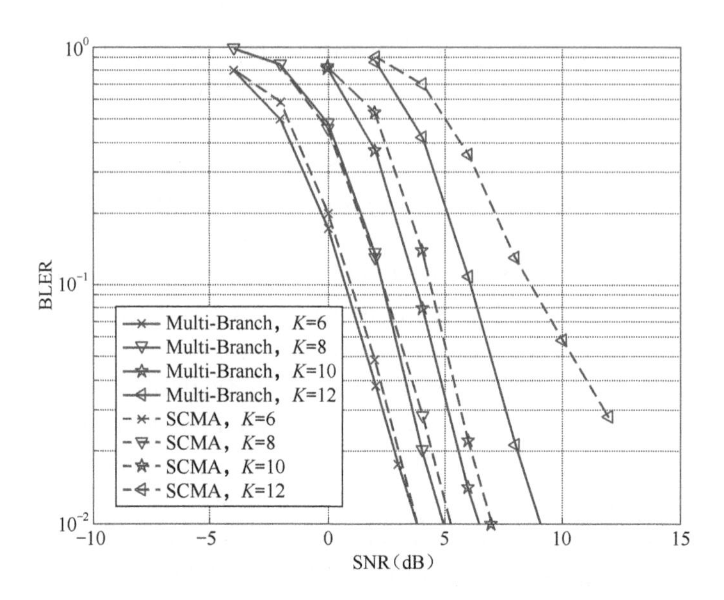

FIGURE 6.54 Performance comparison between multi-branch linear spreading and SCMA under different numbers of users. TDL-C channel, 2 Rx antennas, TBS = 60 bytes, 6 RB resource allocation.

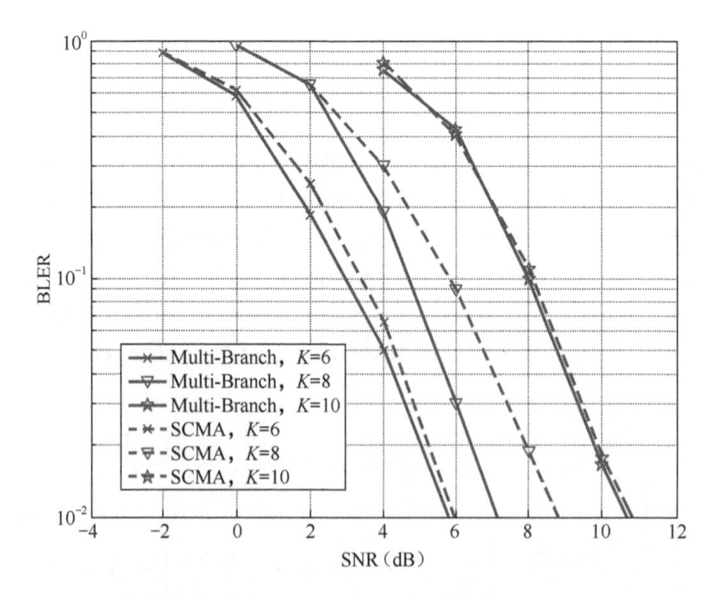

FIGURE 6.55 Performance comparison between multi-branch linear spreading and SCMA under different numbers of users. TDL-C channel, 2 Rx antennas, TBS = 75 bytes, 6 RB resource allocation.

the interference is to use the same spreading code for the two branches of the same user. In this case, the power ratio of these two branches needs to be allocated properly, and the cross-branch interference would be handled by the joint detection at the receiver.

When the power allocation is $\sqrt{0.2}:\sqrt{0.8}$, the composite constellation is 16-QAM. There are four LLRs of each of the four bits corresponding to a 16-QAM symbol. Two bits have higher LLR (corresponding to a higher power) and two other bits have lower LLR (corresponding to lower power). During the iteration, the branch that has higher LLR would quickly converge. The branch that has lower LLR would also converge fast, due to the less interference from the bits of higher LLR. This is in contrast to the single-branch higher-order modulation where the LLRs of the four bits are mingled together during the soft-symbol reconstruction, thus reducing the effectiveness of soft interference cancelation. The essence of multi-branch transmission is the controlled coupling between more reliable bits and less reliable bits, thus allowing efficient soft interference cancelation after more reliable bits are decoded, which is attested in [40].

Figures 6.54 and 6.55 show the performance of the linear spreading type of NOMA with multi-branch, compared with SCMA. It is observed that multi-branch linear spreading performs the same or even better than SCMA in some scenarios, which is due to the good match of the transfer

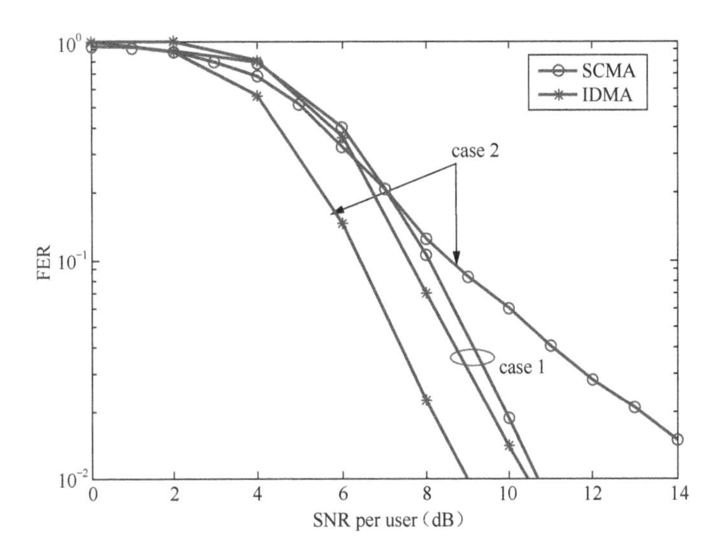

FIGURE 6.56 Performance comparison between multi-branch IDMA and SCMA. TDL-C channel, 2 Rx antennas, 6 RB resource allocation.

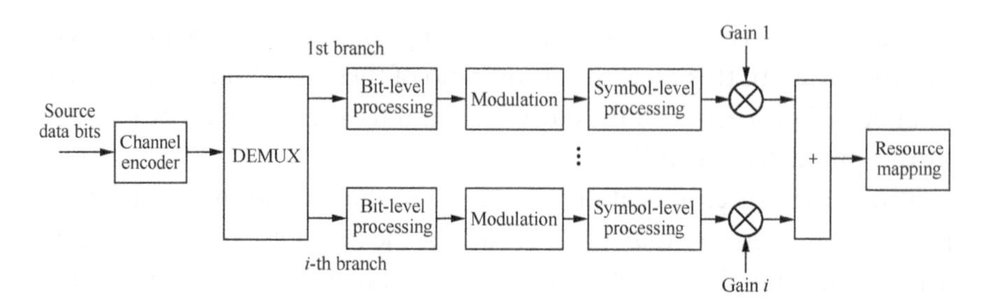

FIGURE 6.57 NOMA scheme based on post-encoder multi-branch spreading.

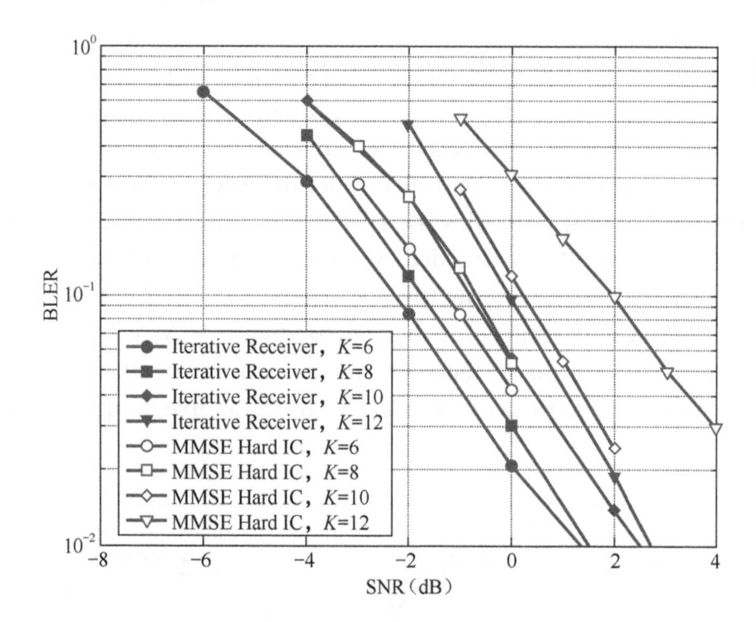

FIGURE 6.58 Performance comparison between the iterative receiver and MMSE hard IC receiver, payload size = 40 bytes.

function of the ESE detector and the transfer function of the soft-output decoder used for the multi-branch receiver.

The performance of multi-branch IDMA is compared with SCMA in Figure 6.56. Here the IDMA has two branches with a power ratio of 0.135:0.865. The transport block size is 75 Bytes in Case 1 with 10 users, and 60 Bytes in Case 2 with 12 users. Bit-level repetition is employed to lower the code rate. Zero-value symbols are inserted to reduce the cross-user interference. More details can be found in Ref. [28]. It is observed that multi-branch IDMA outperforms SCMA, especially in Case 2.

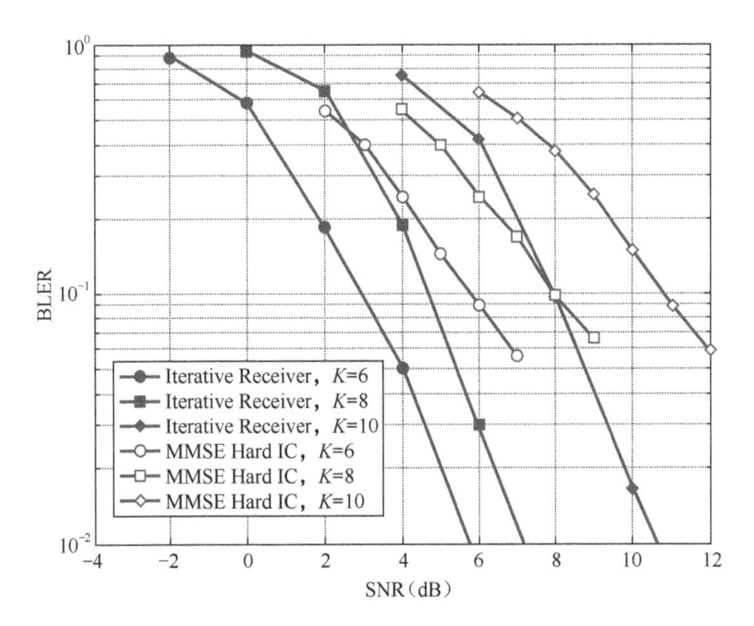

FIGURE 6.59 Performance comparison between the iterative receiver and MMSE hard IC receiver, payload size = 75 bytes.

Multi-branch after channel encoding is illustrated in Figure 6.57.

The data stream is first encoded, and then various bit-level processing, modulation, spreading and power allocation are performed.

Since different branches use different spreading sequences, it is an issue how to handle the cross-branch interference, because users are no longer differentiated by independent channel encoding. One method is to use the Chu sequence for spreading. Different users are assigned different spreading codes. Different branches of the same user are also assigned with different spreading codes. Furthermore, sequences with the least cross-correlations can be allocated to different branches of the same user, in order to reduce the cross-branch interference (they experience the same uplink channel).

Putting multi-branch operation before or after encoder would lead to some difference in receiver implementation. If the multi-branch is a pre-encoder, each branch needs to have a set of CRC bits. If the multi-branch is a post-encoder, only one set of CRC bits is needed. This also affects HARQ retransmission, e.g., in the case of multiple sets of CRC bits, branches of data whose CRC does not pass need to be retransmitted.

When multi-branch is used, the receiver can be MMSE SIC or iterative type. Their performances are compared in Figures 6.58 and 6.59 for payloads of 40 and 75 bytes, respectively.

REFERENCES

1. L. R. Welch, "Lower bounds on the maximum cross correlation of signals," *IEEE Transactions on Information Theory*, Vol. IT-20, 1974, pp. 397–399.
2. D. C. Popescu, O. Popescu, and C. Rose, "Interference avoidance for multiaccess vector channels," in *Proceedings of the International Symposium on Information Theory*, July 2002, pp. 499.
3. S. Ulukus and R. Yates, "Iterative construction of optimum signature sequence sets in synchronous CDMA systems," *IEEE Transactions on Information Theory*, Vol. 47, no. 5, 2001, pp. 1989–1998.
4. J. A. Tropp, "Complex equiangular tight frames", in *Proceedings of SPIE Wavelets XI*, San Diego, August 2005, pp. 590412.01-11.
5. Z. Yuan, C. Yan, Y. Yuan, and W. Li, "Blind multiple user detection for grant-free MUSA without reference signal," *IEEE 86th Vehicular Technology Conference (VTC-Fall)*, Toronto, ON, 2017, pp. 1–5.
6. Z. Yuan, W. Li, Y. Hu, X. Yang, H. Tang, and J. Dai, "Blind receive beamforming for autonomous grant-free high-overloading multiple access," arXiv preprint arXiv:1805.07013, 2018.
7. Y. Yuan, et al. "Non-orthogonal transmission technology in LTE evolution," *IEEE Communications Magazine*, Vol. 54, No. 7, 2016, pp. 68–74.
8. 3GPP, R1-1806930, Considerations on NOMA transmitter, Nokia, RAN1#93, May 2018, Busan, Korea.
9. 3GPP, R1-1806241, Signature design for NoMA, Ericsson, RAN1#93, May 2018, Busan, Korea.
10. 3GPP, R1-1804823, Transmitter side signal processing schemes for NOMA, Qualcomm, RAN1#92bis, April 2018, Sanya, China.
11. 3GPP, R1-1806635, Transmitter side signal processing schemes for NCMA, LGE, RAN1#93, May 2018, Busan, Korea.
12. 3GPP, R1-1811360, Transmitter design for uplink NOMA, NTT DOCOMO, RAN1#94bis, October 2018, Chengdu, China.
13. 3GPP, R1-1810526, NOMA transmitter side signal processing, CATT, RAN1#94bis, October 2018, Chengdu, China.
14. 3GPP, R1-1813309, Receiver complexity reduction by UE-specific power assignment, NTT DoCoMo, RAN1#95, Spokane, USA.
15. 3GPP, R1-1808152, Multi-user advanced receivers for NOMA, ZTE, RAN1#94, Gothenburg, Sweden.
16. 3GPP, R1-1810760, NOMA receiver structure and complexity analysis, Intel, RAN1#94bis, Chengdu, China.
17. 3GPP, R1-1810203, NOMA receiver complexity analysis, ZTE, RAN1#94bis, Chengdu, China.
18. 3GPP, R1-1813858, Complexity analysis of NOMA receivers, ZTE, RAN1#95, Spokane, USA.

19. 3GPP, R1-1812610, Discussion on NOMA receivers, CATT, RAN1#95, Spokane, USA.
20. 3GPP, R1-1813160, Complexity analysis of MMSE-based hard IC receiver, Nokia, RAN1#95, Spokane, USA.
21. 3GPP, R1-060874, Complexity comparison of LDPC codes and Turbo codes, Intel, ITRI, LG, Mitsubishi, Motorola, Samsung, ZTE, RAN1#44bis, Athens, Greece.
22. Li Ping, L. Liu, and W.K. Leung, "A simple approach to near-optimal multiuser detection: Interleave-division multiple access," *IEEE WCNC'03.*
23. S. Y. Chung, On the design of low-density parity-check codes within 0.0045 dB of the Shannon limit. *IEEE Communications Letters*, Vol. 5, No. 2, 2001, pp. 58–60.
24. C. Berrou, A. Glavieux, and P. Thitimajshima, "Near Shannon limit error-correcting coding and decoding: Turbo codes." *Proceedings of IEEE International Conference on Communication*, May 1993, pp. 1064–1070.
25. L. Liu, J. Tong, and Li Ping, "Analysis and optimization of CDMA systems with chip-level interleavers," *IEEE Journal on Selected Areas in Communications*, Vol. 24, No. 1, J2006, pp. 141–150.
26. C. Liang, Y. Hu, L. Liu, C. Yan, Y. Yuan, and L. Ping, "Interleave division multiple access for high overloading applications", *IEEE International Symposium on Turbo Codes and Iterative info Processing Conference (ISTC)*, December 2018.
27. Y. Hu, C. Liang, L. Liu,and Li Ping, "Low-cost implementation techniques for interleave division multiple access." *IEEE Wireless Communication Letters*, Vol. 7, No. 6, December 2018, pp. 1026–10299.
28. L. Ping, L. Liu, K. Wu, and W. K. Leung, "Interleave Division Multiple-Access (IDMA) communication systems," *Proceedings of 3rd International Symposium on Turbo Codes and Related Topics*, 2003, pp. 173–180.
29. J. Tong, L. Ping, and X. Ma, "Superposition coded modulation with peak-power limitation". *IEEE Transactions on Information Theory*, Vol. 55, No. 6, 2009, pp. 2562–2576.
30. L. Ping, "Interleave-division multiple access and chip-by-chip iterative multi-user detection". *IEEE Communications Magazine*, Vol. 43, No. 6, 2005, pp. 19–23.
31. 3GPP, R1-1810623, Transmitter side signal processing of ACMA, Hughes, RAN1#94bis, Chengdu, China.
32. Y. Hu, C. Liang, L. Liu, C. Yan, Y. Yuan, and L. Ping, "Interleave-division multiple access in high rate applications." *IEEE Wireless Communication Letters*, Vol. 8, No. 2, April 2019, pp. 476–473.
33. 3GPP, R1-1808152, Multi-user receivers for NOMA, ZTE, RAN1#94, Gothenburg, Sweden, August 2018.
34. M. A. Imran, M. Al-Imari, and R. Tafazolli, "Low density spreading multiple access." *Information Technology & Software Engineering*, Vol. 2, 2012, p. 4.
35. J. V. De Beek and B. M. Popovic, "Multiple access with low-density signatures," *Proceedings of 2009 IEEE Global Communication Conference*, pp. 1–6.

36. H. Nikopour and H. Baligh, "Sparse code multiple access," *IEEE 24th International Symposium on Personal, Indoor and Mobile Radio Communication*, 2013.
37. M. Taherzadeh, H. Nikopour, and A. Bayesteh, "SCMA codebook design." *IEEE Vehicular Technology Conference*, 2014.
38. J. Boutros and E. Viterbo, "Signal space diversity: A power- and bandwidth-efficient diversity technique for the Rayleigh fading channel." *IEEE Transactions on Information Theory*, Vol. 44, No. 4, 1998, pp. 1453–1467.
39. 3GPP, R1-1810116, Discussion on the design of NOMA transmitter, Huawei, RAN1 #95, Spokane, USA.
40. C. Yan and Y. Yuan, "Spreading based multi-branch non-orthogonal multiple access transmission scheme for 5G". *IEEE Vehicular Technology Conference*, May 2019, pp. 1–5.
41. B. Wang, C. Yan, W. Liu, and H. Zhang, "Multi-user connection performance assessment of NOMA schemes for beyond 5G," *China Communications*, December 2020.

Performance Evaluation of Uplink Contention-free Grant-free NOMA Transmissions

Ziyang Li, Qiujin Guo, Hong Tang, Weimin Li, Jian Li, Yifei Yuan, Chen Huang and Li Tian

As discussed in Chapter 5, there are two types of uplink grant-free transmission: contention-free and content-based. The basic principles and main schemes of uplink non-orthogonal multiple access (NOMA) are elaborated in Chapter 6, together with some link-level evaluations. In the 3GPP Rel-16 NOMA study, there was a big campaign of link- and system-level performance evaluations, participated by many companies, especially for contention-free-based grant-free transmission, which is the main content of this chapter.

7.1 SIMULATION PARAMETERS

Evaluation methodology, key performance indicators, simulation scenarios and traffic model were discussed in Chapter 5. In this chapter, more details on simulation parameter settings will be discussed.

7.1.1 Simulation Parameters for the Link Level

The parameters of link-level simulations are listed in Table 7.1, which are mandatory for all companies in 3GPP that participated in the

DOI: 10.1201/9781003336167-7

TABLE 7.1 Link-Level Simulation Assumptions for Uplink NOMA

Parameter	mMTC	uRLLC	eMBB
Carrier frequency	700 MHz	700 MHz or 4 GHz	4 GHz, 700 MHz (optional)
Waveform (data part)	CP-OFDM and DFT-s-OFDM	CP-OFDM	CP-OFDM
Channel coding	NR LDPC		
Subcarrier spacing (data part)	SCS = 15 kHz, #OS = 14	Case 1: SCS = 60 kHz, #OS = 7 (normal CP), 6 (optional) Case 2: SCS = 30 kHz, #OS = 4	SCS = 15 kHz #OS = 14
Number of RBs	6	6 (for SCS = 60 kHz) 24 (for SCS = 30 kHz)	12
Transport block size (TBS)	{10, 20, 40, 60, 75} bytes	{10, 20, 40, 60, 75} bytes	{20, 40, 80, 120, 150} bytes
Target BLER after initial transmission (%)	10	0.1	10
Antenna configuration at the base station	2 or 4 for 700 MHz 4 or 8 for 4 GHz, where 8 is optional		
Antenna configuration at the terminal	1		
Channel model and mobility	TDL-A 30ns and TDL-C 300ns in TR 38.901 3 km/h, CDL is optional		
Maximum number of HARQ	1	1, or 2 (optional)	1
Channel estimation	Ideal channel estimation Non-ideal channel estimation NR design is reused when the number of DMRS ports is not >12 DMRS overhead should be no less than that of NR DMRS when the number of DRMS ports is larger than 12		
Average SNR	Same or different	Same	Same or different
Timing offset (TO)	Within [0, *y*] for grant-free asynchronous transmission. *y* can take • Case 1: *y* = NCP/2 • Case 2: *y* = 1.5*NCP TO of all the users is independently, identically and uniformly distributed over [0, *y*]. When syn and asyn are mixed, *X*% users with TO = 0, (100 – *X*)% users with TO > 0, where *X* = 80, other *X* values are optional		
Frequency offset (FO) in Hz	700 Hz: 0, or [–70, 70] uniform distribution 4 GHz: [–140, 140] uniform distribution		
Traffic model	Full buffer or non-full buffer (e.g., Poisson arrival of fixed-size packets)		

simulation campaign. The simulation assumptions are for three deployment scenarios: mMTC, eMBB small data, and uRLLC.

Regarding the assumption of channel coding for the mMTC scenario, since mMTC had not been specified in 5G New Radio (NR), there was some debate on which type of channel coding should be used for simulation. One opinion was to reuse the channel codes for eMBB and uRLLC traffic channels, e.g., low-density parity check (LDPC). This can reduce the effort to develop the simulators and maximize code reuse. The second opinion was to use the channel codes in LTE, e.g., Turbo codes, which serve as a neutral position between LDPC and Polar codes. Also, in the Rel-14 NR NOMA study, Turbo codes were used. Reusing the Turbo codes assumption may reduce the effort of implementing LDPC or Polar for mMTC. The third opinion was to enhance LDPC which was briefly discussed in Section 6.2. In theory, channel coding has traditionally been designed assuming single-user transmission which reflects the situation of orthogonal multiple access (OMA). Legacy channel coding, including LDPC for 5G NR, may not be optimal for NOMA. Although the third opinion is more forward-looking and was not included in the scope of study on NOMA. Considering that the second option requires additional work on the implementation of Turbo codes where LDPC has already been developed for eMBB and uRLLC. In the end, it was decided to assume LDPC for NOMA simulations of the mMTC scenario.

There are two waveform assumptions for mMTC: CP-OFDM and DFT-s-OFDM. The reason for considering DFT-s-OFDM is to reduce PAPR and lower the cost and power consumption of the transmitter. The base station antenna configuration depends on the carrier frequency. For 700 MHz bands, the wavelength is relatively long and 2 or 4 Rx antennas are typical. For 4GHz bands, the typical number of antennas at the base station is 4 or 8.

At least five transport block sizes (TBSs) need to be simulated, which represent various operating scenarios with different code rates. For instance, in the mMTC deployment scenario, the number of resource blocks (RBs) is 6 for users to share. The five TBSs are 10, 20, 40, 60, and 75 bytes. Counting in the overhead of demodulation reference signal (DMRS), the total number resource elements in the 6 RBs is $12 * 12 * 6 = 864$. The overhead of cyclic prefix (CP) is about 1/10. If the 16 CRC bits are considered as part of the payload, the corresponding per-user spectral efficiency is 0.10, 0.20, 0.39, 0.57, and 0.71 bps/Hz. It should be pointed out that in the Rel-14 NOMA study [1], the coverage enhancement scenario was included and the per-user spectral efficiency ranges from 0.01 to 0.1

bps/Hz. Considering that in LTE NB-IoT [2] there were a lot of optimizations for these extremely low code rate transmissions, Rel-16 NOMA [3] would not study the scenario with spectral efficiency <0.1 bps/Hz.

For URLLC, the mini-slot configuration is assumed for some settings where there are only four OFDM symbols in a subframe, which is smaller than seven OFDM symbols for regular slots. In the mini-slot configuration, the DMRS overhead is 1/4, significantly higher than 1/7 for the regular slot case. To ensure similar spectral efficiency to mMTC and eMBB, the number of RBs is set to 24. For eMBB, due to the characteristic of its traffic, the payload is often larger than that of mMTC and uRLLC. Hence, TBS = 20, 40, 80, 120 and 150 bytes. In order to maintain a similar spectral efficiency to URLLC and mMTC, the number of RBs for eMBB is 12.

It should be emphasized that in the mandatory simulation parameters, there is no restriction on the exact code rate and modulation order to be assumed. This is because for different NOMA transmission schemes, under the same TBS, the optimal combination of the code rate and modulation may not be the same. Such an optimal combination may also depend on the number of superimposed users.

The signal to noise ratio (SNR) is defined as the average received signal power vs. power of the thermal noise, within a certain transmission bandwidth and per ODM symbol. The average SNR of a user is a long-term measurement. Different users may have equal or unequal average SNRs. Equal average SNR across users often corresponds to ideal open-loop power control without the restriction on maximum transmit power at the terminal. In this case, the power control can fully compensate for the path loss and a large-scale shadow fading. While the assumption of an equal SNR across users may not happen quite often in practical systems, this assumption is kept to facilitate the comparison between companies' results. For the unequal SNR case, the distribution of the average SNR can either be uniform, e.g., 1 dB step size within $[x-a, x+a]$ (dB), where x is the average SNR, the offset a is 3, or Gaussian with the standard deviation of 5 or 9 dB. Figure 7.1 shows the cumulative density functions (CDFs) of the relative receive power in the mMTC scenario. The relative receive power is the difference from the total overall receive power, reflecting the receive power fluctuation (not due to fast fading) between different users. The data is from system-level simulation, and 12 terminals are randomly selected to calculate the relative receive power. It is observed in Figure 7.1 that when the target uplink power control is −100 dBm and the path loss compensation factor alpha is set to be 1, the distribution of the relative receiver power is close to the Gaussian distribution with a standard deviation of

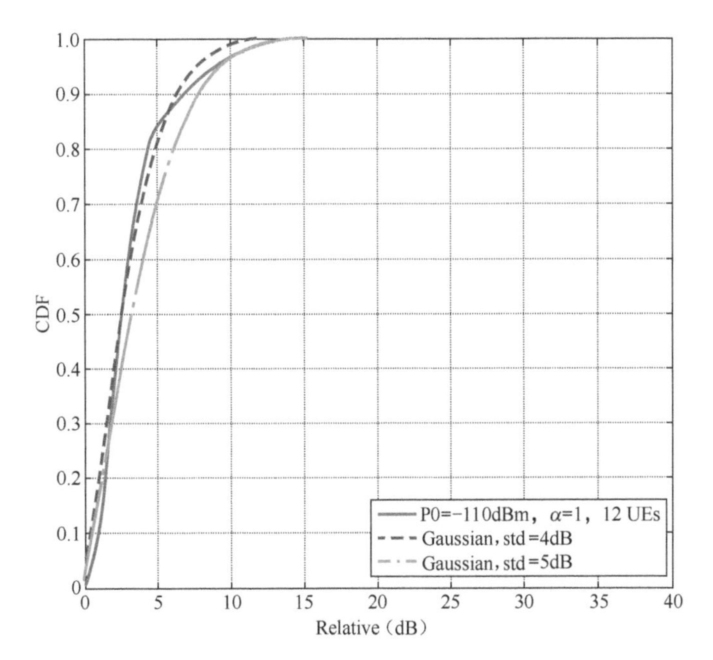

FIGURE 7.1 CDFs of relative power of users from the system level and the curve fitting assuming Gaussian distribution.

8–9 dB. When the target power control is set to −110 dBm and the path loss compensation factor alpha is set to be 1. The distribution of relative receive power is close to the Gaussian distribution with a standard deviation of 4–5 dB. It should be pointed out that (1) the distribution of relative receive power highly depends on the parameter setting of uplink power control. The standard deviation from the curve fitting is small in some settings, whereas large in some other settings; (2) pure Gaussian distribution does not exactly match the distribution of the actual relative power.

Even though it is desirable to assume a more accurate distribution of average SNR across users in link-level simulations, it may not significantly affect the performance comparison between different NOMA schemes. In the case of ideal channel estimation, unequal average SNR would reinforce the near-far effect and be beneficial to all the NOMA schemes in general, regardless of which type of receiver is used. In the case of non-ideal channel estimation, since the channel estimation error is directly related to each user's SNR, a strong near-far effect would degrade the channel estimation quality of some users and in turn would hurt every NOMA scheme. Certainly, if the target is to evaluate the absolute performance of a NOMA scheme, especially at the system level, the average SNR would definitely be unequal between different users and realistic channel estimation should be assumed. In fact, the key factor in system performance evaluation is the link-to-system mapping, e.g., whether the mapping can accurately capture the effect of near-far and channel estimation error. A detailed description of link-to-system mapping can be seen in Section 7.2.

In the block error rate (BLER) vs. SNR curves, the SNR is per-user average SNR, e.g., the mean value of all users' average SNRs in dBs. The reason for not using the sum SNR over users is to reflect the net effect of cross-user interference on the performance.

In addition, to make the simulation setting closer to the practical systems, timing offset and frequency offset can be included in the simulation assumptions. The timing offset is related to the site-to-site distance. Assuming that the site-to-site distance is 1,732 m and users are uniformly distributed in a cell, the timing offset would be uniformly distributed within [0, 1.5*NCP], where NCP stands for the normal CP. The value of the frequency offset is based on carrier frequency and 0.1 ppm of the crystal oscillator. In the case of the 700 MHz band for mMTC, the frequency offset is within [−70, 70] Hz. At the link level, full buffer traffic is normally used. That is, all the users are transmitting in the same time and frequency resources.

From Table 7.1, it is not difficult to find that even for one deployment scenario, for instance, mMTC, there are 5 TBS sizes, two waveforms, two configurations of antennas, two delay spread models, ideal vs. non-ideal channel estimation, equal or unequal average SNR, and two cases of time/frequency errors. The total would be $5 \times 2^6 = 320$ combinations. In addition, there may be multiple receiver types to be simulated for a NOMA scheme, and there are different numbers of users. It would be impractical to simulate all those combinations. Hence in the Rel-16 NOMA study, a limited set of cases are chosen for the link-level simulations and they are mandatory, as listed in Table 7.2. It is required that each company should provide the entire BLER vs. SNR curves, instead of the required SNR at 10% or 1% BLER. Note that only the cases of contention-free transmission are listed in Table 7.2. The cases of contention-based transmission are to be discussed in Chapter 8.

7.1.2 Link-to-System Mapping

Link-to-system mapping is also called physical layer abstraction. It is used to model the various processing in key modules of a receiver, including user identification, channel estimation, minimum mean squared error (MMSE) detection and channel decoding. For uplink NOMA, interference cancellation should also be modeled. Since different NOMA schemes

TABLE 7.2 Cases for Link-Level Simulations of Contention-Free Grant-Free NOMA (Mandatory)

Case Index	Scenario	Carrier Frequency (Hz)	Num Ant	Avg. SNR Distribution	Waveform	Allocation of MA Signature	Channel Model	TBS (Bytes)	#UEs	TO/FO
1	mMTC	700M	2	Equal	CP-OFDM	Fixed	TDL-A	10	12, 24	0
2	mMTC	700M	2	Equal	CP-OFDM	Fixed	TDL-C	20	6, 12	0
3	mMTC	700M	2	Equal	CP-OFDM	Fixed	TDL-A	40	6, 10	0
4	mMTC	700M	2	Equal	CP-OFDM	Fixed	TDL-C	60	6, 8	0
5	mMTC	700M	2	Equal	CP-OFDM	Fixed	TDL-A	75	4, 6	0
6	mMTC	700M	2	Unequal	CP-OFDM	Fixed	TDL-A	20	6, 12	>0
7	mMTC	700M	2	Unequal	CP-OFDM	Fixed	TDL-C	60	6, 8	0
8	mMTC	700M	2	Unequal	DFT-S	Fixed	TDL-C	10	12,24-	>0
9	mMTC	700M	2	Unequal	DFT-S	Fixed	TDL-C	20	6, 12	> 0
Cases 10–13 to be discussed in Chapter 8										
14	uRLLC	700M	4	Equal	CP-OFDM	Fixed	TDL-C	10	6, 12	0
15	uRLLC	700M	4	Equal	CP-OFDM	Fixed	TDL-C	60	4, 6	0
16	uRLLC	4GHz	4	Equal	CP-OFDM	Fixed	TDL-A	10	6, 12	0
17	uRLLC	4GHz	4	Equal	CP-OFDM	Fixed	TDL-A	60	4, 6	0
18	eMBB	4GHz	4	Equal	CP-OFDM	Fixed	TDL-A	20	12, 24	0
19	eMBB	4GHz	4	Equal	CP-OFDM	Fixed	TDL-A	80	8, 16	0
20	eMBB	4GHz	4	Equal	CP-OFDM	Fixed	TDL-A	150	4, 8	0
21	eMBB	4GHz	4	Unequal	CP-OFDM	Fixed	TDL-C	20	12, 24	> 0
22	eMBB	4GHz	4	Unequal	CP-OFDM	Fixed	TDL-C	80	8, 16	> 0
23	eMBB	4GHz	4	Unequal	CP-OFDM	Fixed	TDL-C	150	4, 8	0
Cases 24 and 25 to be discussed in Chapter 8										
26	mMTC	700M	4	Equal	CP-OFDM	Fixed	TDL-C	60	6, 8	0
27	mMTC	700M	4	Equal	CP-OFDM	Fixed	TDL-A	75	4, 6	0
28	mMTC	700M	4	Unequal	CP-OFDM	Fixed	TDL-C	60	6, 8	> 0
29	mMTC	700M	2	Unequal	DFT-S	Fixed	TDL-C	40	6, 10	> 0
30	mMTC	700M	2	Unequal	DFT-S	Fixed	TDL-C	60	6, 8	> 0
31	mMTC	700M	2	Unequal	DFT-S	Fixed	TDL-C	75	4, 6	> 0
32	mMTC	700M	2	5 dB	CP-OFDM	Fixed	TDL-C	20	6, 12	0
33	mMTC	700M	2	4 dB	CP-OFDM	Fixed	TDL-C	60	6, 8	0
34	mMTC	700M	4	4 dB	CP-OFDM	Fixed	TDL-A	60	6, 8	0
35	mMTC	700M	4	5 dB	CP-OFDM	Fixed	TDL-A	20	6, 12	0

may use different types of receivers, multiple links to system mappings or physical layer abstractions would be needed.

Let us first look at the link-to-system mapping for MMSE hard interference cancellation (IC) which is illustrated in Figure 7.2. There are several steps in this physical layer abstraction.

7.1.2.1 User Identification and Channel Estimation

For contention-free-based grant-free transmission, the reference signals and the spreading sequences or interleaver patterns or joint spreading-modulation codebooks are pre-configured without collision. The missed detection probability and false alarm probability are very low during user identification. Hence, the process can be considered ideal, without explicit modeling.

When carrying out the channel estimation, non-ideal channel estimation can be based on the true channel plus certain error terms, as given below:

$$H_R = H_I + H_e \tag{7.1}$$

where H_R is the estimated channel coefficient. H_I is the true channel coefficient (e.g., ideal channel estimate) and H_e represents the estimation error. Since there is no DMRS collision, H_e contains only the interference of users' DMRS in neighboring cells and the thermal noise, which is related to the DMRS design, channel estimation method and smooth filter. Assuming that interference of other cells' DMRS follows Gaussian distribution, H_e can be modeled as a zero-mean Gaussian variable with a variance of σ_e^2. The normalized channel estimation error can be represented as

$$\frac{|H_e|^2}{|H_I|^2} = \frac{\sigma_e^2}{|H_I|^2} = \frac{1}{a * Ns * \text{SNR}} \tag{7.2}$$

where SNR is the ideal SNR of the UE. a is the adjustment factor and can take various values depending on the algorithm for smoothing filter. Ns is the number of resource elements to get one channel estimate.

7.1.2.2 To Calculate the SINR of the Target User Based on the MMSE Criterion

In a multi-cell system, the uplink received signal y can be represented as

$$y = \sum_{k=1}^{K} H_k s_k + \sum_{j=1}^{J} H_j s_j + n \tag{7.3}$$

where K is the number of users in the target cell that are transmitting. s_k is the modulation symbol of the k-th user in the target cell. H_k is the channel containing both spatial-domain and code-domain coefficients of

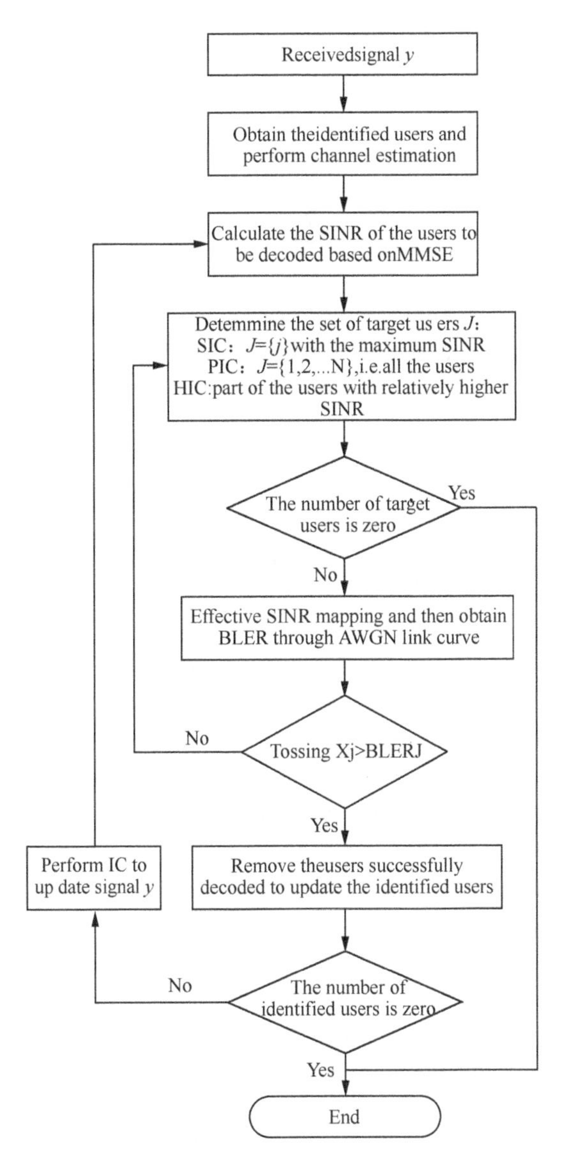

FIGURE 7.2 Link-to-system mapping for the MMSE hard IC receiver.

the k-th user. When the number of transmit antennas is 1 and the number of receive antennas is N and the spreading length is L, H_k would be a NL*1 column vector. J represents the number of users in other cells that are transmitting in the same time-frequency resources of the target user. s_j is the modulation symbol of the j-th other-cell user. H_j is the channel containing both spatial-domain and code-domain coefficients of the j-th other-cell user. H_j would be a NL*1 column vector. n is additive white Gaussian noise (AWGN) with zero mean and variance of σ^2. MMSE weights can be calculated as

$$W_k = H_k^H * R_{yy}^{-1} = H_k^H * \left(\sum_{k=1}^{K} H_k H_k^H + \sum_{j=1}^{J} H_j H_j^H + \sigma^2 I \right)^{-1} \quad (7.4)$$

where $(.)^H$ represents the complex conjugate. R_{yy} is the auto-correlation matrix of the received signal y. I is an identity matrix of size NL*NL. The result of MMSE detection can be written as $\hat{s}_k = W_k * y$. Then the SINR can be calculated as

$$\text{SINR}_k = \frac{|W_k H_k|^2}{\sum_{i=1, i \neq k}^{K} |W_k H_i|^2 + \sum_{j=1}^{J} |W_k H_j|^2 + W_k(\sigma^2 I) W_k^H} \quad (7.5)$$

It should be noted that in ideal channel estimation, true channel coefficients can be used to calculate MMSE weights and SINR. However, in practical systems, non-ideal channel estimation and estimation of R_{yy} should be carried out. Hence, when calculating the MMSE weight, H_k should be the realistically estimated channel of the k-th user in the target cell. R_{yy} can be estimated directly from the received signal. That is, to rearrange the received signal y into a matrix Y of size NL*T, where T is the number of modulation symbols of the user; then $R_{yy} = YY^H / T$.

7.1.2.3 To Obtain the Effective SINR and BLER

According to resource block information rate (RBIR)-SINR mapping, to calculate the effective SINR based on SINRs of the k-th user in the target cell:

$$\text{SINR}_k^{\text{eff}} = \phi^{-1}\left(\frac{1}{M} \sum_{m=1}^{M} \varphi(\text{SINR}_{k,m}) \right) \quad (7.6)$$

where M is the number of resource elements. $\phi(.)$ is the nonlinear invertible function for RBIR mapping. Then, look up the BLER vs. SNR curve of the AWGN channel and determine the BLER. Compare this BLER with a randomly generated number uniformly distributed within (0, 1). If the BLER is less than the number generated at this time, the transmission is claimed successfully. Otherwise, this transmission fails.

As shown in Figure 7.2, when a user or several users do not successfully decode their data, the interference cancellation would continue for the rest of users whose data are not yet decoded. By doing so, the performance can be improved, especially for successive interference cancellation (SIC)

7.1.2.4 To Perform Interference Cancellation

When a user's signal is successfully decoded, its contribution to the received signal can be reconstructed by encoding, modulation, spreading of the information data, and passing through the channel, and then canceled from the received signal. For practical systems, realistic channel estimation should be used. The error model of channel estimation discussed earlier can be considered.

Based on the above procedures, several verifications have been carried out which are based on channel estimation via DMRS.

7.1.2.4.1 Verification of the Channel Estimation Model for DMRS

In 5G NR, two DMRS patterns can be radio resource control (RRC) configured for the CP-OFDM waveform. The first one is called NR DMRS Type 1, as shown in Figure 7.3. The second one is called NR DMRS Type 2, as shown in Figure 7.4. For DMRS Type 1, at most 4 DMRS ports can be supported in each time-domain OFDM symbol, as seen in Figure 7.3a,

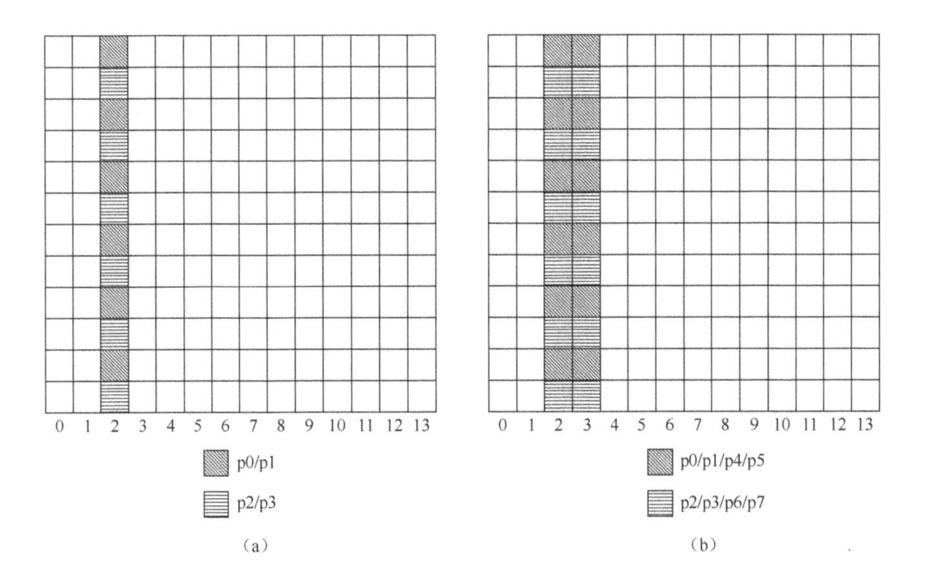

FIGURE 7.3 NR DMRS Type 1. (a) One-OFDM symbol DMRS and (b) Two-OFDM symbol DMRS.

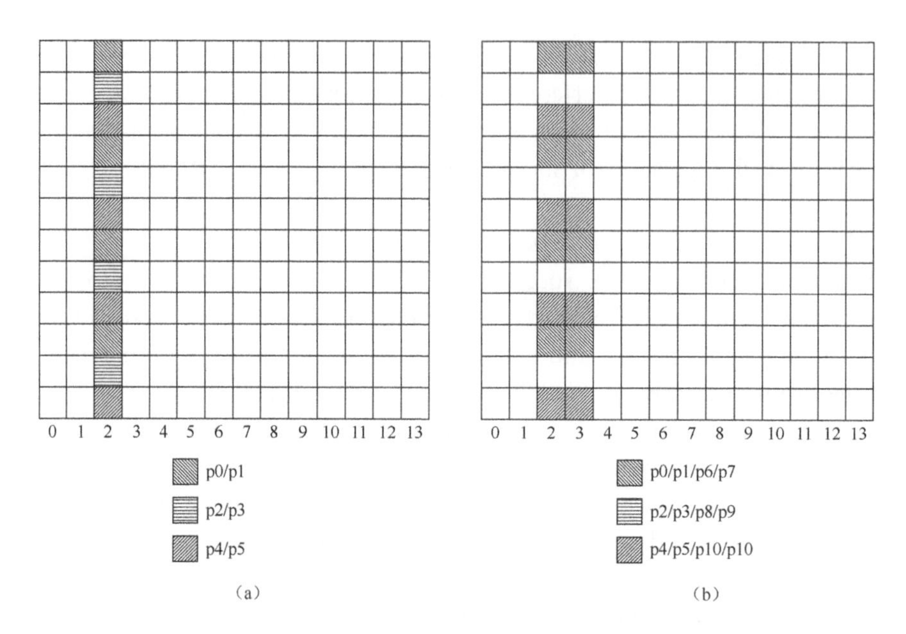

FIGURE 7.4 NR DMRS Type 2. (a) One-OFDM symbol DMRS and (b) Two-OFDM symbol DMRS

different ports can be differentiated in the frequency domain or code domain. At most 8 DMRS ports can be supported with two time-domain OFDM symbols. In Figure 7.3b, 6 DMRS ports are configured within a physical resource block (PRB).

In DMRS Type 2, at most, six DMRS ports can be supported within a time-domain OFDM symbol. As shown in Figure 7.4a, different ports can be differentiated in the frequency domain or code domain. Hence, at most, 12 DMRS ports are supported with 2 OFDM symbols. In Figure 7.4b, only four DMRS sequences are configured within a PRB.

The TDL-C 300 ns channel model is considered in the verification. The least-square channel estimation is based on DMRS Types 1 and 2. Situations with and without smooth filters are considered. The statistic error is measured and compared with the predicted error from the model. The results are shown in Figures 7.5 and 7.6. It is observed that the modeled channel estimation errors are very close to the actual channel estimation errors, no matter whether the smoothing filter is applied.

7.1.2.4.2 Verification of Link-to-System Mapping

For the mMTC scenario, the link-to-system mapping model is verified under TDL-A 30 ns and TDL-C 300 ns, with an equal average SNR, or

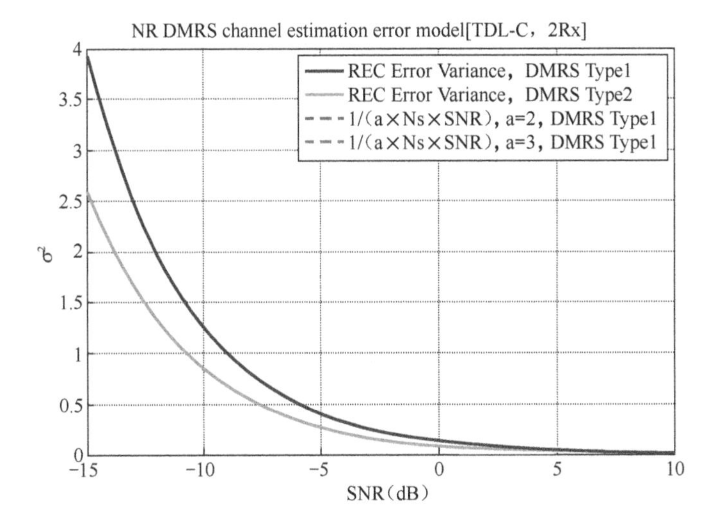

FIGURE 7.5 Comparisons of channel estimation error between the actual measured and the model, without the smoothing filter.

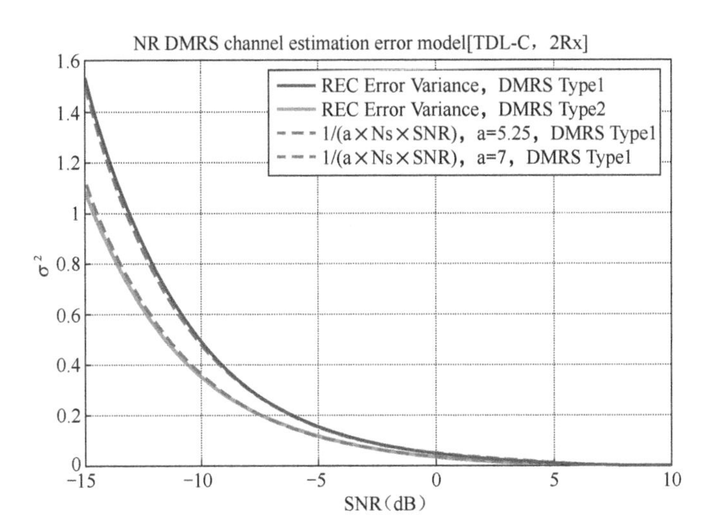

FIGURE 7.6 Comparisons of channel estimation error between the actual measured and the model, with the smoothing filter.

unequal average SNRs, with different loadings. The results are shown in the following (Figure 7.7).

There are two other link-to-system mapping methods listed in [3]. One is suitable for ESE-SISO receivers. In this method, ideal interference cancellation is assumed, e.g., there is no intra-cell interference, when pp-SINR is derived for each resource element. The relationship between the realistic

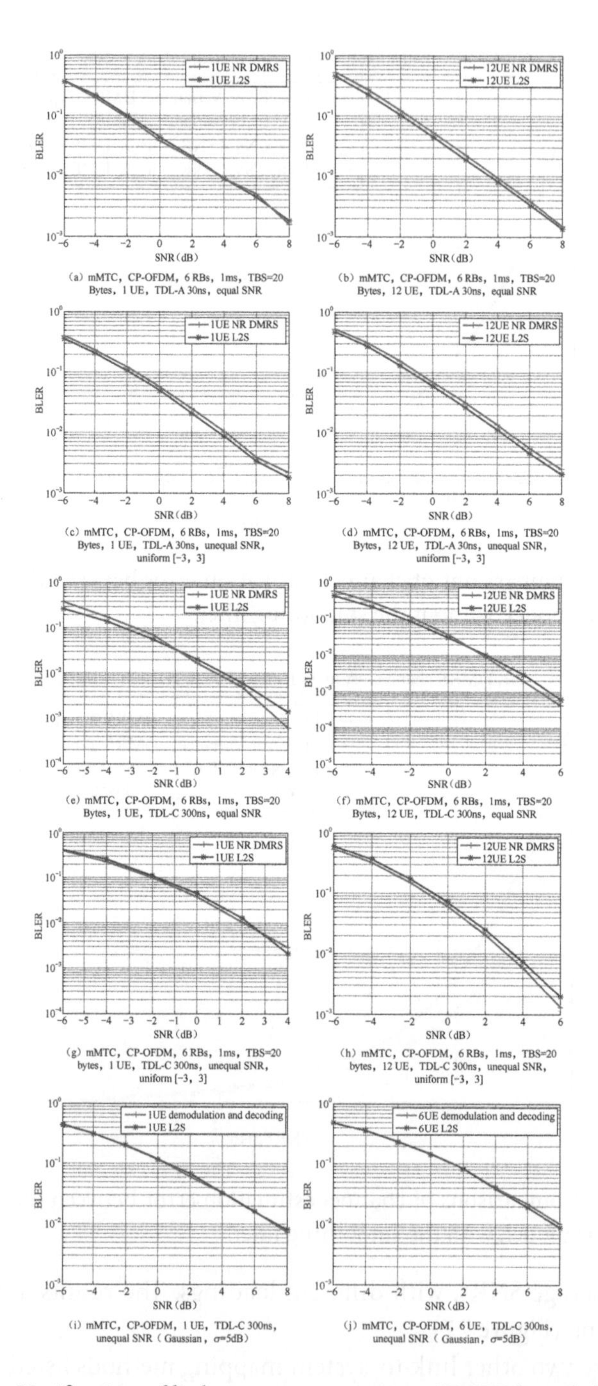

FIGURE 7.7 Verification of link-to-system mapping, DMRS-based channel estimation, contention-free grant-free transmission. (a) mMTC, CP-OFDM, 6 RBs, 1 ms, TBS = 20 bytes, 1 UE, TDL-A 30 ns, equal SNR, (b) mMTC, CP-OFDM, 6 RBs, 1 ms, TBS = 20 bytes, 12 UE, TDL-A 30 ns, equal SNR, (c) mMTC, CP-OFDM,

(*Continued*)

6 RBs, 1ms, TBS = 20 bytes, 1 UE, TDL-A 30 ns, unequal SNR, uniform [–3, 3], (d) mMTC, CP-OFDM, 6 RBs, 1 ms, TBS = 20 bytes, 12 UE, TDL-A 30 ns, unequal SNR, uniform [–3, 3], (e) mMTC, CP-OFDM, 6 RBs, 1 ms, TBS = 20 bytes, 1 UE, TDL-C 300 ns, equal SNR, (f) mMTC, CP-OFDM, 6 RBs, 1 ms, TBS = 20 bytes, 12 UE, TDL-C 300ns, equal SNR, (g) mMTC, CP-OFDM, 6 RBs, 1 ms, TBS = 20 bytes, 1 UE, TDL-C 300 ns, unequal SNR, uniform [–3, 3], (h) mMTC, CP-OFDM, 6 RBs, 1 ms, TBS = 20 bytes, 12 UE, TDL-C 300 ns, unequal SNR, uniform [–3, 3], (i) mMTC, CP-OFDM, 1 UE, TDL-C 300 ns, unequal SNR (Gaussian distribution, σ = 5dB), (j) mMTC, CP-OFDM, 6 UEs, TDL-C 300 ns, unequal SNR (Gaussian distribution, σ = 5dB).

SINR and the SINR (denoted as SINR^{PIC}) assuming ideal interference cancellation can be represented as:

$$\text{SINR} = \left(1 + \text{SINR}^{\text{PIC}}\right)^{\beta} - 1 \tag{7.7}$$

where β is the scaling factor which can be obtained by minimizing the mean squared error with respect to the performance curves of the actual demodulation/decoding. Then the actual SINR goes through RBIR-SINR mapping to get the effective SINR to be looked up in the AWGN curve to get the target BLER.

The third link-to-system mapping can be used for the EPA-hybrid IC receiver or the MMSE hard IC receiver. This method also assumes ideal interference cancellation to get SINR^{PIC} on each resource element and then uses the following formula to get the effective SINR:

$$\text{SINR}_{\text{eff}} = f^{-1}\left(\frac{1}{K}\sum_{k=1}^{K} f\left(\left(\beta \text{SINR}^{\text{PIC}}\right)^{\alpha}\right)\right) \tag{7.8}$$

where β and α are the curve fitting factors that can be obtained by minimizing the mean squared error with the performance curves of the actual demodulation/decoding. The function $f(\cdot)$ represents the RBIR-SINR relationship.

In these two methods for link-to-system mapping, there is no detailed modeling of each key processing module in the receivers. Hence, the scaling factor β or the fitting factor list (β, α) has to be obtained individually for different simulation settings. This work can be very cumbersome for diverse simulation settings as mentioned in Section 7.1.1.

7.1.3 System Simulation Parameters

System simulation parameters of uplink NOMA are listed in Table 7.3 for mMTC, eMBB small data and uRLLC scenarios. The assumptions are particularly for multi-user shared access (MUSA) simulations.

Figures 7.8 and 7.9 show the CDFs of the application layer packet size for the mMTC scenario and eMBB small data scenario, respectively.

7.2 ANALYSIS OF LINK-LEVEL SIMULATION

Among the 35 link-level simulation cases described in Section 7.1, a subset can be selected which may be representative of the scenarios of low-to-medium spectral efficiency and high spectral efficiency.

TABLE 7.3 Uplink System-Level Simulation Assumptions

	Assumption		
Parameter	**mMTC**	**eMBB**	**uRLLC**
Cell layout	Hexagonal macro-cells	Hexagonal macro-cells	Hexagonal macro-cells
Carrier frequency	700 MHz	4 GHz	4 GHz or 700 MHz
Site-to-site distance	1732 m	200 m	4 GHz: 200 m: 700 MHz: 500 m
Simulation bandwidth	6 PRBs	12 PRBs	12 PRBs
Channel model	Based on [4]. UMa model, to use outdoor-to-indoor, O2I) defined in Table 7.4.3-3 for building penetration loss	Based on [4]. UMa model, to use outdoor-to-indoor, O2I) defined in Table 7.4.3-3 for building penetration loss	Based on [4]. UMa model, to use outdoor-to-indoor, O2I) defined in Table 7.4.3-3 for building penetration loss
Number of users per cell	100	100	20
Max transmit power of terminal (dBm)	23	23	23

(*Continued*)

TABLE 7.3 (*Continued*) Uplink System-Level Simulation Assumptions

	Assumption		
Parameter	**mMTC**	**eMBB**	**uRLLC**
Base station antenna configuration	2 Rx; 2 ports: (M, N, P, Mg, Ng) = (10, 1, 2, 1, 1), 2 TXRU; dH = dV = 0.5λ; down-tilt = 92°	4GHz: 4 Rx; 4 ports: (M, N, P, Mg, Ng) = (10, 2, 2, 1, 1), 4 TXRU; dH = 0.5λ, dV = 0.8λ; down-tilt = 102°	4 GHz: 4 Rx; four ports: (M, N, P, Mg, Ng) = (10, 2, 2, 1, 1), 4 TXRU; $dH = 0.5\lambda$, $dV = 0.8\lambda$; 700 MHz: 4 Rx; four ports: (M, N, P, Mg, Ng) = (10, 2, 2, 1, 1), 4 TXRU; $dH = dV = 0.5\lambda$; down-tilt: 98°
Base station antenna height (m)	25	25	25
Base station antenna gain (dBi)	8	8	8
Noise figure of the base station receiver (dB)	5	5	5
Mobile antenna configuration	1 Tx	1 Tx	1 Tx
Mobile antenna height	Refer to [4]	Refer to [4]	Refer to [4]
Mobile antenna gain (dBi)	0	0	0
User distribution	Uniform distribution over an entire cell, 20% outdoor and 80% indoor, mobility speed 3 km/h	Uniform distribution over an entire cell, 20% outdoor and 80% indoor, mobility speed 3 km/h	Uniform distribution over an entire cell, 20% outdoor and 80% indoor, mobility speed 3 km/h
Uplink power control	Open loop: $P_0 = -100$ dBm, alpha = 1	Open loop: $P_0 = -95$ dBm, alpha = 1	Open loop: $P_0 = -90$ dBm, alpha = 1

(*Continued*)

TABLE 7.3 (*Continued*) Uplink System-Level Simulation Assumptions

	Assumption		
Parameter	**mMTC**	**eMBB**	**uRLLC**
Traffic model	Application layer packet size follows Pareto distribution, ranging from 20 to 200 bytes. Shaping coefficient $\alpha = 2.5$. Need to account for 29 bytes of higher layer overhead. Traffic arrival of each user follows the Poisson arrival rate of λ In the case of packet segmentation, additional overhead of five bytes should be considered for the RLC layer and MAC layer header	Application layer packet size follows Pareto distribution, ranging from 50~600 bytes. Shaping coefficient $\alpha = 1.5$. Need to account for 29 bytes of higher layer overhead. Traffic arrival of each user follows Poisson arrival rate of λ In the case of packet segmentation, additional overhead of 5 bytes should be considered for RLC layer and MAC layer header	Application layer packet size can be 60 or 200 bytes, without higher layer overhead. Traffic arrival of each user follows Poisson arrival rate of λ
Transport block size	25 bytes, including 5 bytes overhead in the case of packet segmentation	70 bytes, including 5 bytes overhead in the case of packet segmentation	60 or 200 bytes, same as the application layer packet size
Number of DMRS	24, with proposed enhancement to improve the number of multiplexed users	24, with proposed enhancement to improve the number of multiplexed users	24, with proposed enhancement to improve the number of multiplexed users
HARQ/ repetition	Max. HARQ = 8, non-adaptive retransmission	Max. HARQ = 1	Max. HARQ = 1
ARQ retransmission	Not modeled, e.g., packet drop is declared if a transport block is not successfully decoded when the max number of HARQ transmissions or repetition is reached	Not modeled	Not modeled
Channel estimation	Realistic	Realistic	Realistic
Base station receiver	Baseline: MMSE-IRC or MMSE-PIC Uplink NOMA: MMSE-PIC	Baseline: MMSE-IRC or MMSE-PIC; Uplink NOMA: MMSE-PIC	Baseline: MMSE-IRC or MMSE-PIC; Uplink NOMA: MMSE-PIC

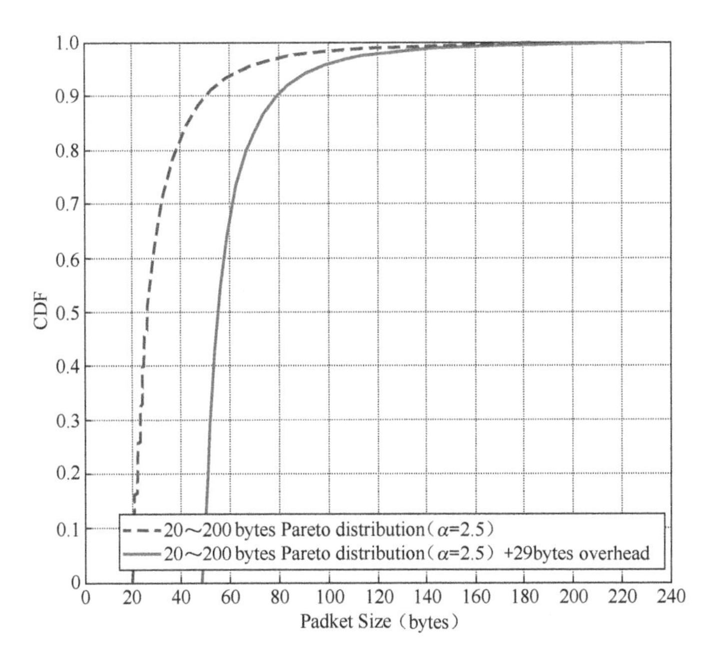

FIGURE 7.8 CDFs of the application layer packet size for the mMTC scenario in system simulation.

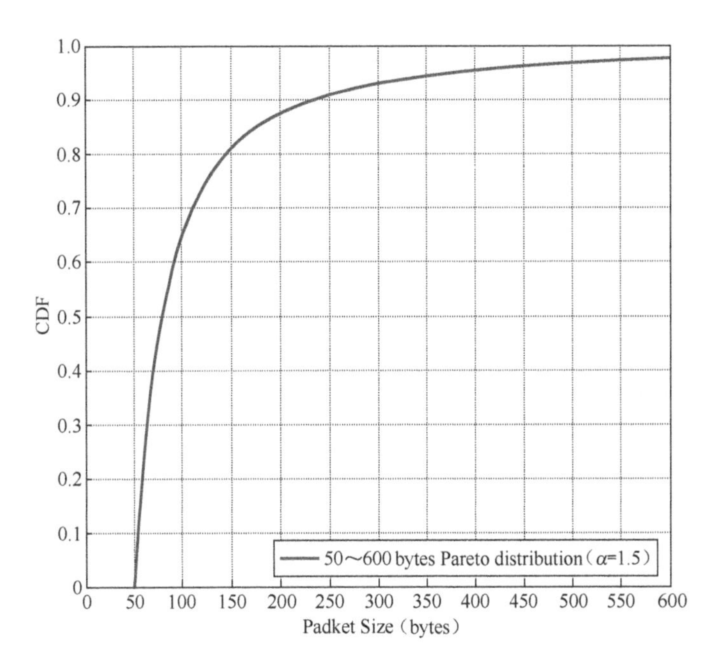

FIGURE 7.9 CDFs of the application layer packet size for the eMBB scenario in system simulation.

For low-to-medium spectral efficiency, typical cases of the mMTC scenario are simulation Cases 1 and 2. Typical cases of the uRLLC scenario are Cases 14 and 16. The typical case of the eMBB scenario is Case 18.

For high spectral efficiency, typical cases of mMTC are simulation Case 3, Case 4 and Case 5. Typical cases of uRLLC are simulation Cases 15 and 17. Typical cases of eMBB are Cases 19 and 20.

For advanced receivers, there are many algorithms for channel estimation with quite different performances of estimation accuracy and computation complexity. To separate the channel estimation and the net performance of the NOMA scheme itself, we in the following example assume ideal channel estimation.

It should be pointed out that the link-level simulation results in this section are from the simulation platforms of different companies. Although a calibration campaign is carried out at the link level, there is still about 0.5 dB difference between the curves of different companies, for either the AWGN channel or fading channels.

7.2.1 Simulation Cases for Low-to-Medium Spectral Efficiency

7.2.1.1 Simulation Case 1

Figure 7.10 shows BLER vs. SNR curves of simulation Case 1 for the mMTC scenario. There are two receiver antennas. The channel model is TDL-A that has a shorter delay spread, meaning that its frequency response is rather flat. An equal average SNR is assumed between superimposed users. It is observed that in the case of ideal channel estimation, regardless of whether there are 12 or 24 users sharing the time-frequency resources, the link performance curves of different NOMA schemes are quite similar. At the target BLER = 10%, the required SNR is about −6.8 and −6.5 dB for 12 and 24 users, respectively. Take a closer look at the 12-user case; the performance of PDMA is slightly worse than other curves, which is because the cross-correlation of PDMA is slightly higher than other linear-spreading-based NOMA schemes.

In simulation Case 1, although per-user spectral efficiency is merely 0.1 bps/Hz, the total spectral efficiency is 0.1 * 24 = 2.4 bps/Hz when 24 users are superimposed. Note that different types of receivers are used in different NOMA schemes in Case 1, for instance, MMSE hard IC, MMSE hybrid IC, ESE + SISO or EPA + SISO. However, no matter which type of receiver is used, similar performance is observed when per-user spectral efficiency is low. Such observation can be explained as follows. While the average SNR between users is the same, each channel experiences independent

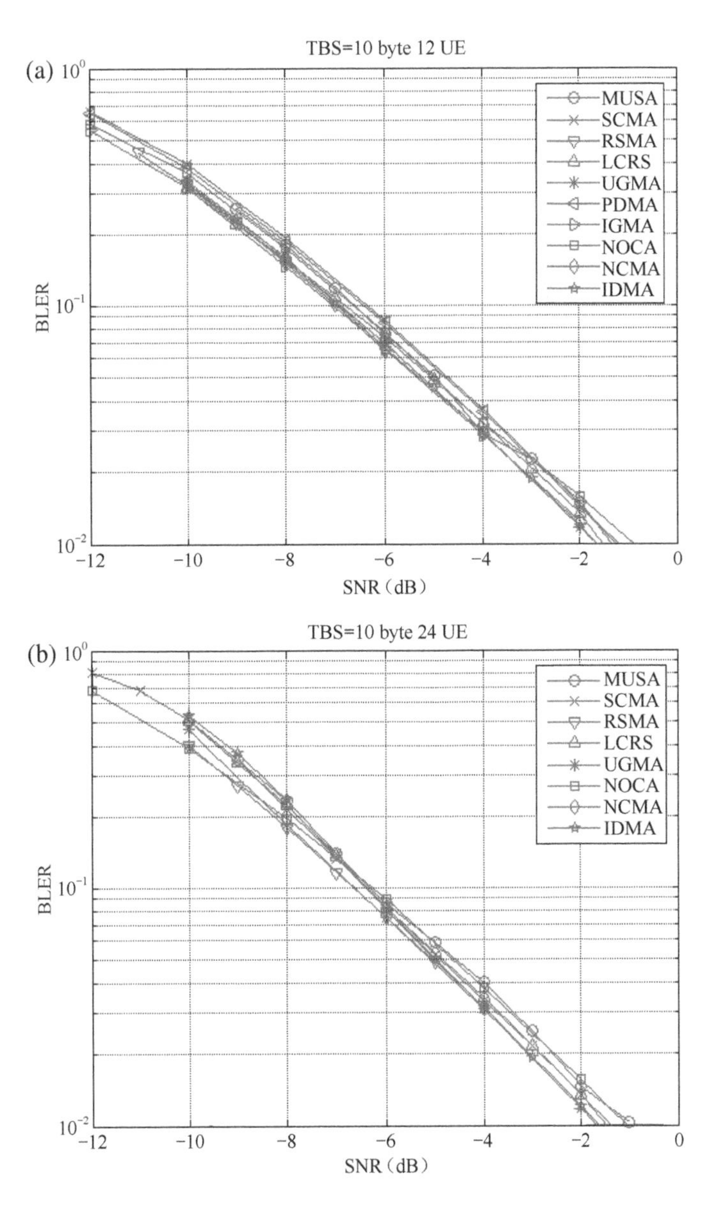

FIGURE 7.10 Link performance curves of NOMA schemes for simulation Case 1 (mMTC). (a) 12 UEs and (b) 24 UEs.

fading in each receive antenna. Hence at each time instant, the SNR differences between users are quite significant. Also, each user's data has a very low code rate and modulation order. Even with relatively simple receivers such as MMSE hard IC, it is highly probable that the stronger user would be first decoded successfully with near-far effect and then the weaker user.

7.2.1.2 Simulation Case 2

Figure 7.11 contains the BLER vs. SNR curves for simulation Case 2 where the number of receiver antennas is 2 and an equal average SNR is assumed between users. The channel model is TDL-C that has longer a multi-path

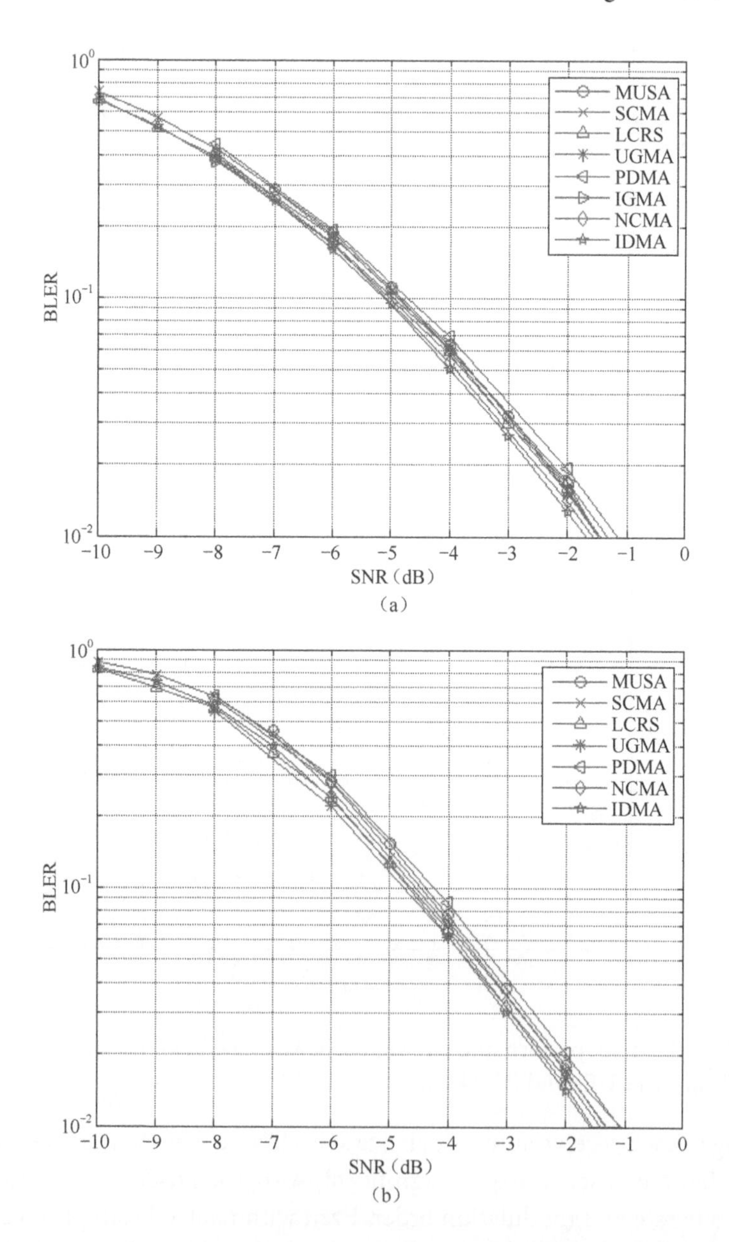

FIGURE 7.11 Link performance curves of NOMA schemes for simulation Case 2 (mMTC). (a) 6 UEs and (b) 12 UEs.

delay, e.g., frequency-selective channel. The per-user spectral efficiency is 0.2 bps/Hz. When 12 users are superimposed, the total spectral efficiency reaches 0.2 * 12 = 2.4 bps/Hz. Different receive types are used, e.g., MMSE hard IC, MMSE hybrid IC, ESE + SISO or EPA + SISO. The general trend is similar to simulation Case 1. That is, under the ideal channel estimation, at BLER = 0.1, no matter whether 6 or 12 users, the link-level performance of different NOMA schemes are close to each other, e.g., the required SNR is −5.0 and −4.5 dB, respectively. The performance of PDMA is slightly worse than others due to the slightly higher cross-correlation of PDMA sequences.

The reason for comparable performance between NOMA schemes is similar to that of simulation Case 1: per-user spectral efficiency is relatively low. The code rate is much lower than that of the mother code rate and quadrature phase-shift keying (QPSK) modulation is used. Even with a relatively simple MMSE hard IC receiver, it is likely that the stronger user's data can be decoded successfully with the near-far effect. Then the weaker users can be decoded.

Note that the per-user spectral efficiency in simulation Case 2 is twice the per-user spectral efficiency in Case 1. The general rule is that when the spectral efficiency is low (as in Cases 1 and 2), the required SNR is proportional to the spectral efficiency, e.g., 3 dB difference between Case 1 and Case 2. However, by comparing Figures 7.10 and 7.11, the required SNR for BLER = 0.1 in Case 2 is only 2 dB higher than that of Case 1. The reason can be explained that TDL-C is a frequency-selective channel, e.g., the SNRs of resource elements are quite different. The near-far effect becomes more significant, which benefits all types of receivers.

Note that the benefit of the frequency-selectivity-induced near-far effect is conditioned on ideal channel estimation and only applicable for low-to-medium spectral efficiency. In general, frequency selectivity would degrade the channel estimation performance and put more strain on the decoders (mostly applicable to high-spectral-efficiency cases).

7.2.1.3 Simulation Case 14

Figure 7.12 shows BLER vs. SNR curves for simulation Case 14 which belongs to the uRLLC scenario. There are four receive antennas and the channel model is assumed to be TDL-C that has longer delay spread. The average SNR of each user is the same. The per-user spectral efficiency is 0.1 bps/Hz. The total spectral efficiency is 0.6 and 1.2 bps/Hz when 6 and 12 users are superimposed, respectively. It is observed that for the ideal channel estimation, at BLER = 0.001, as long as proper configuration is used, the

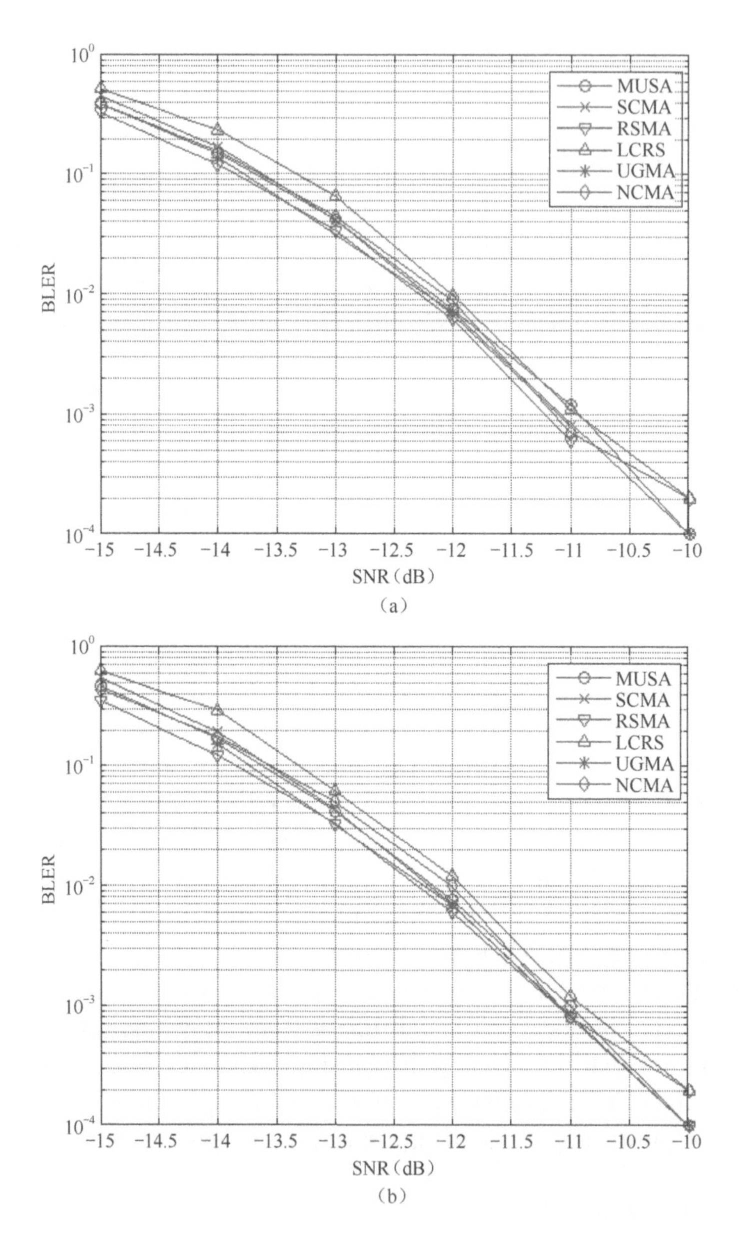

FIGURE 7.12 Link performance curves of NOMA schemes for simulation Case 14 (uRLLC). (a) 6 UEs and (b) 12 UEs.

performances of different schemes are similar, regardless of which receiver type is used. The required SNR is about –11 dB. The performance of LCRS is a little inferior to other NOMA schemes. The reason is that in simulation Case 14 there are four receive antennas. As the number of receive antennas increases, symbol spreading-based schemes, either for linear spreading like MUSA, resource shared multiple access (RSMA), UGMA, NCMA, or multi-dimensional modulation like SCMA, can be benefited from their receiver where code-domain and spatial-domain discrimination can be done jointly to suppress the cross-user interference. This would outperform the LCRS receiver where only spatial-domain discrimination can be carried out.

7.2.1.4 Simulation Case 16

Figure 7.13 shows the BLER vs. SNR curves for simulation Case 16. Many simulation assumptions are the same as those of Case 16. However, there are two differences: (1) carrier frequency, 700 MHz in Case 14 and 4 GHz in Case 16; (2) channel model, TDL-C in Case 14 and TDL-A in Case 16. It is observed that with ideal channel estimation, for the target BLER = 0.001, as long as the configurations are appropriate, most of the schemes perform similarly, e.g., the required SNR is about −8 dB, for both 6- and 12-user cases. As in Case 14, the performance of LCRS is slightly inferior to other schemes.

The TDL-A channel assumed in Case 16 is relatively flat in frequency, meaning that the SNR does not fluctuate significantly within a code block. In contrast, TDL-C assumed in Case 14 is rather frequency-selective. In uRLLC, the subcarrier spacing is 60 kHz, much wider than 15 kHz for mMTC. Hence, the transmission bandwidth of uRLLC is 4 times wider than that of mMTC, leading to a more pronounced frequency selectivity and near-far effect. Hence, the required SNR for BLER = 0.001 in Case 14 is about 3 dB lower than that of Case 16 when the per-user spectral efficiency is kept the same.

7.2.1.5 Simulation Case 18

Figure 7.14 shows the BLER vs. SNR curves for simulation Case 18 which belongs to the eMBB scenario. The number of receive antennas is 4 and the channel model is TDL-A. The carrier frequency is 4 GHz, and the average SNR of each user is assumed the same. The per-user spectral efficiency is 0.1 bps/Hz, translating into a total spectral efficiency of 1.2 and 2.4 bps/Hz for 12- and 24-user cases, respectively. It is observed that with the ideal channel estimation and appropriate configuration setting, most of the schemes have similar performance, e.g., around −12.5 dB for target BLER = 0.1, regardless of the receiver types. The performance of LCRS is a little inferior due to the only spatial-domain discrimination carried out in LCRS receiver, compared to code-domain and spatial-domain joint discrimination in receivers of other NOMA schemes.

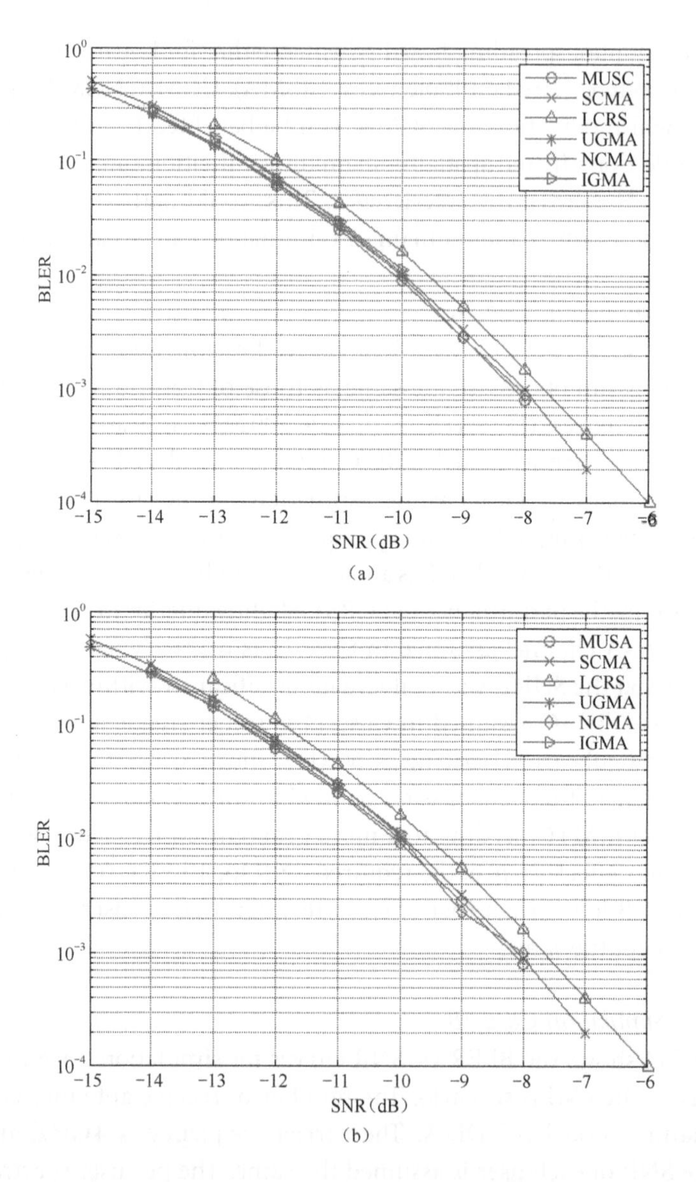

FIGURE 7.13 Link performance curves of NOMA schemes for simulation Case 16 (uRLLC). (a) 6 UEs and (b) 12 UEs.

In summary, for low-to-medium spectral efficiency operation, the simulation results indicate that for mMTC/eMBB/uRLLC scenarios, with ideal channel estimation, equal average SNR, frequency/timing offset = 0, pre-allocated MA signatures, the performance difference between NOMA schemes is not significant, regardless of the receiver types and code rate.

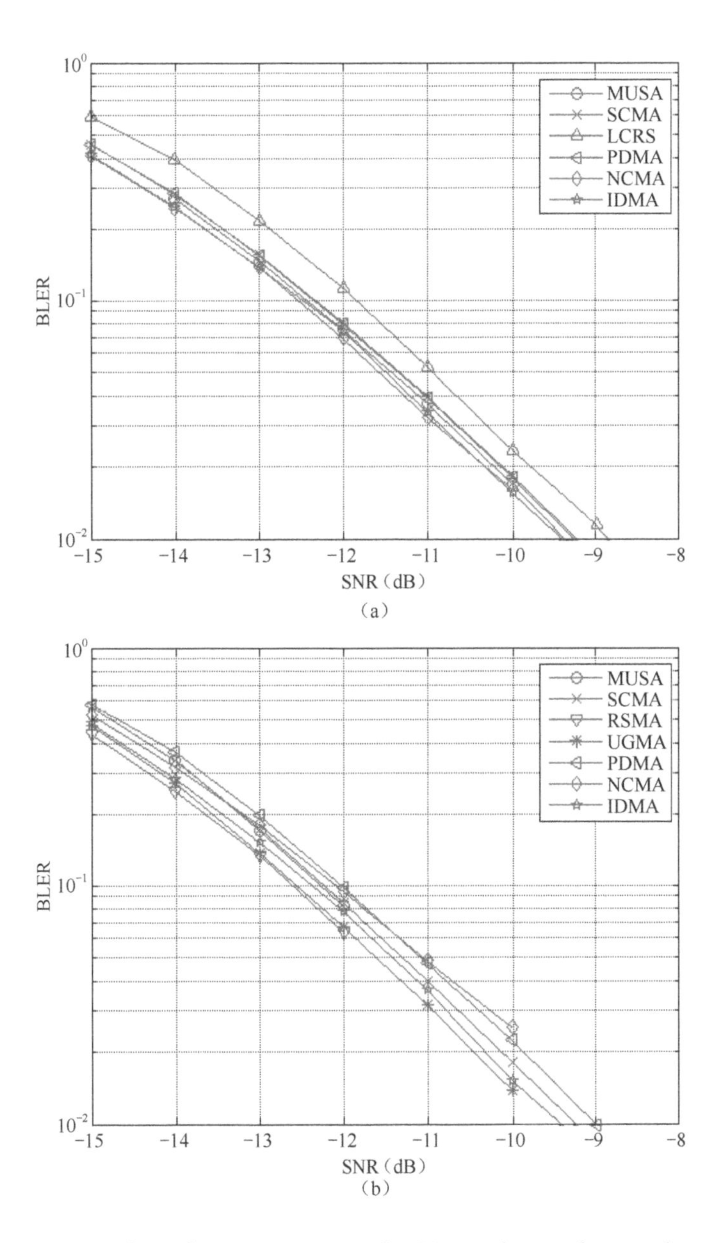

FIGURE 7.14 Link performance curves of NOMA schemes for simulation Case 18 (eMBB). (a) 12 UEs and (b) 24 UEs.

7.2.2 High-Spectral-Efficiency Operation

7.2.2.1 Simulation Case 3

Figure 7.15 shows BLER vs. SNR curves of simulation Case 3 for the mMTC scenario. The TBS is 40 bytes, and the per-user spectral efficiency is 0.39

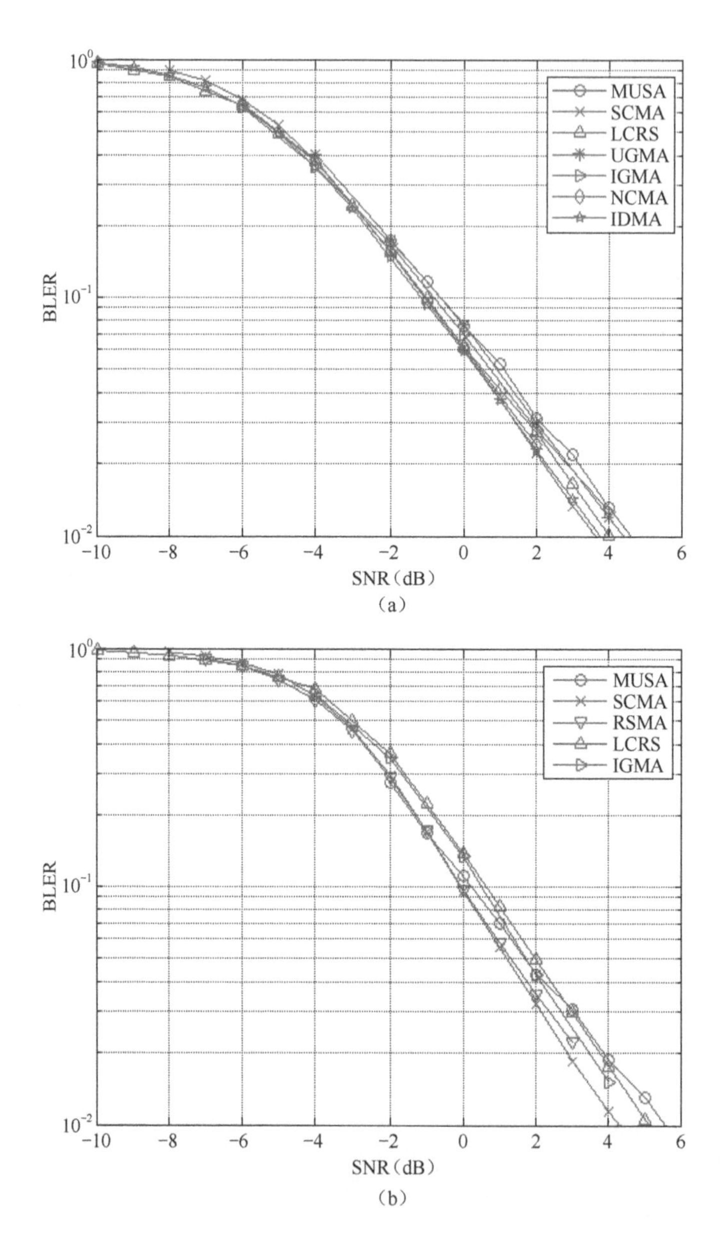

FIGURE 7.15 Link performance curves of NOMA schemes for simulation Case 3 (mMTC). (a) 6 UEs and (b) 10 UEs.

bps/Hz. TDL-A channel model is assumed which is relatively frequency flat. The number of receiver antennas is 2. It is observed that with the ideal channel estimation and appropriate configuration setting, most curves are overlapped and the required SNR for target BLER = 0.1 is about −0.9 and

0.2 dB, respectively, for 6- and 10-user cases. Here, the code rate is kept within 0.4. The performance difference is within 0.5 dB, meaning that as long as the code rate is not high, different NOMA schemes perform more or less the same. Case 3 can be considered as a transition from low spectral efficiency operation to high-spectral-efficiency operation.

7.2.2.2 Simulation Case 4

Figure 7.16 shows the BLER vs. SNR curves for simulation Case 4. The TBS is 60 bytes, and the per-user spectral efficiency is 0.57 bps/Hz. The total spectral efficiency is 0.57 * 6 = 3.42 bps/Hz and 0.57 * 8 = 4.56 bps/Hz, for 6- and 8-user cases, respectively. A frequency-selective TDL-C channel is assumed and the number of receive antennas is 2. Two types of receivers are considered.

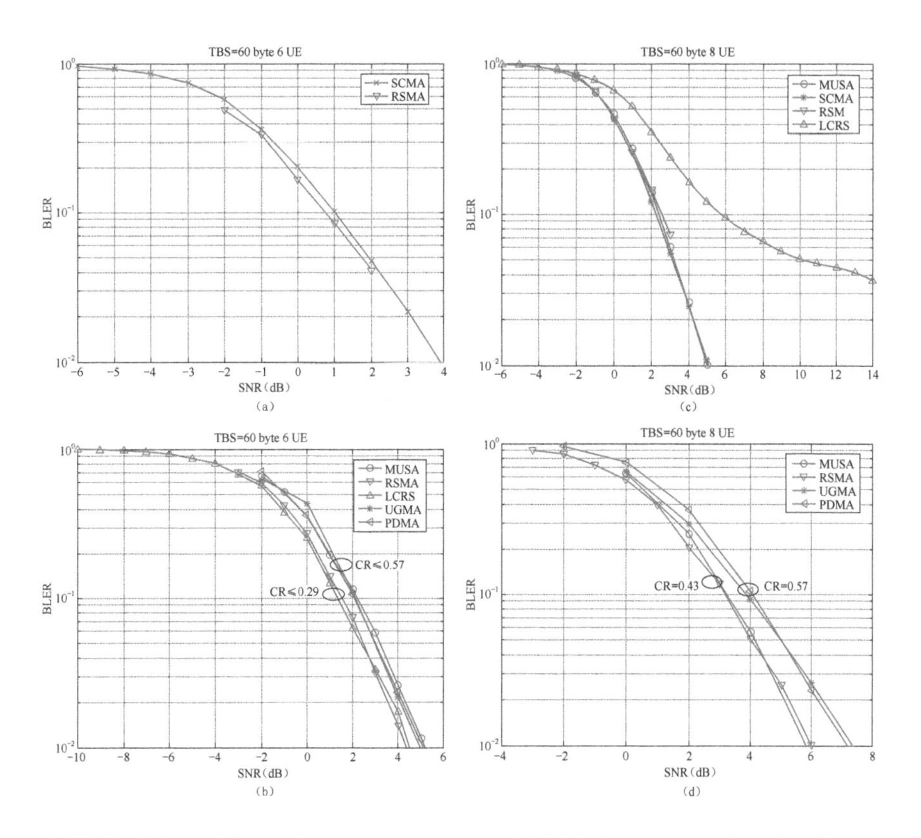

FIGURE 7.16 Link performance curves of NOMA schemes for simulation Case 4 (mMTC). (a) Hybrid IC/EPA + SISO, 6 UEs, (b) MMSE Hard IC, 6 UEs, (c) Hybrid IC/EPA + SISO, 8 UEs and (d) MMSE Hard IC, 8 UEs.

1. Hybrid IC and EPA + SISO: soft bits, e.g., LLR should be provided by the decoder and fed back to the detector of iterative detection. Computation complexity is relatively high.
2. MMSE hard IC: decoder only needs to provide the hard bits. Interference is canceled in the "hard" manner. No iterative detection is needed. Computation complexity is relatively low.

Figure 7.16a contains the results for six users in which RSMA relies on a hybrid IC receiver and SCMA uses EPA + SISO receiver. It is observed that with ideal channel estimation, RSMA and SCMA have similar performance for the target BLER = 0.1. Figure 7.16b shows the results of six users with MMSE hard IC receiver for MUSA, UGMA, PDMA, RSMA and LCRS. Two configurations of code rate are considered: 0.57 and 0.29. It is found that when the code rate is the same, these NOMA schemes show a similar performance. Certainly, using higher code rate would sacrifice certain coding gain, resulting in higher required SNR. The comparison between Figure 7.16a and b shows that the performance of the relatively complex receiver like hybrid IC or EPA + SISO is better than that of less complex receivers like MMSE hard IC.

Similarly, Figure 7.16c contains the results for eight users where RSMA uses a hybrid IC receiver and SCMA/MUSA uses an EPA + SISO receiver. It is observed that RSMA, SCMA and MUSA perform almost the same, whereas LCRS degrades significantly. In Figure 7.16d, MMSE hard IC is used. The code rate of PDMA and UGMA is 0.57, and the code rate of MUSA and RSMA is 0.43. When the code rate is the same, these NOMA schemes show a similar performance.

7.2.2.3 Simulation Case 5

Figure 7.17 shows BLER vs. SNR curves for Case 5. The TBS is 75 bytes, and the per-user spectral efficiency is 0.71 bps/Hz. The total spectral efficiency is 0.71 * 4 = 2.84 bps/Hz and 0.71 * 6 = 4.26 bps/Hz for four and six users, respectively. Frequency flat TDL-A channel is assumed and the number of receive antennas is 2. Compared to simulation Case 4, although the total spectral efficiency of Case 5 is slightly lower, the per-user spectral efficiency in Case 5 is the highest among all the simulation cases. Hence, Cases 4 and 5 represent the extreme operation environment for the NOMA design and receiver implementation. Similar to the practice in Case 4, receivers are categorized into two groups: relatively complex receivers like hybrid IC and EPA + SISO vs. relatively simple receivers like MMSE hard IC. Also, the code rate is differentiated to be 0.71 and 0.36.

Figure 7.17a contains the results of four users with more advanced receivers. It is observed that with ideal channel estimation, the performance of SCMA and RSMA is the same at BLER = 0.1. The results of four users using less advanced receivers, e.g., MMSE hard IC, are shown in Figure 7.17b. Among them, the code rate of 0.71 is assumed in MUSA, PDMA and UGMA. Their performance is quite similar. LCRS uses the code rate of 0.36 and its performance is about 1 dB better than with 0.71 code rate. From the comparison between Figure 7.17a and b, it is found that the performance of the hybrid IC/EPA + SISO receiver is better than that of the MMSE hard IC receiver.

Similarly, Figure 7.17c contains the results of six users with more advanced receivers: RSMA uses hybrid IC and MUSA/SCMA uses EPA + SISO. These three schemes perform the same at BLER = 0.1.

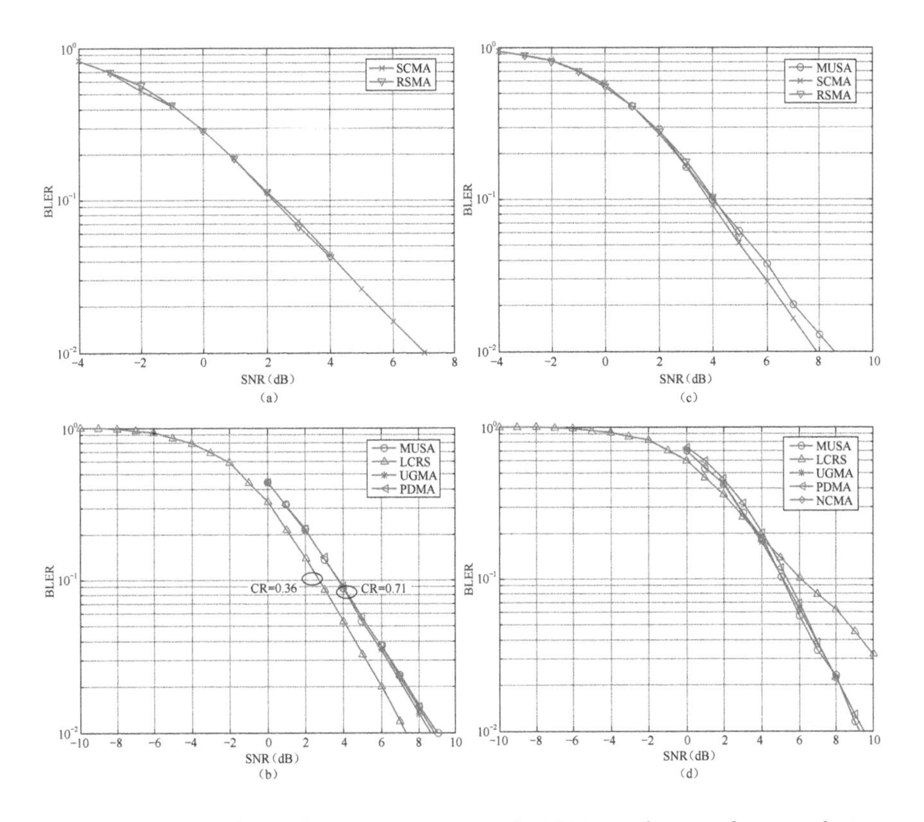

FIGURE 7.17 Link performance curves of NOMA schemes for simulation Case 5 (mMTC). (a) Hybrid IC/EPA + SISO, 4 UEs, (b) MMSE Hard IC, 4 UEs, (c) Hybrid IC/EPA + SISO, 6 UEs and (d) MMSE Hard IC, 6 UEs.

A less advanced receiver, e.g., MMSE hard IC, is used for the simulation in Figure 7.17d. The performances of MUSA, NCMA, PDMA and UGMA are quite close. However, significant performance degradation is seen in LCRS, compared to 4-user case. This is because that the code rate of LCRS cannot be further reduced to match the interference suppression capability of spreading-based schemes. As the number of users increases, the performance of LCRS gradually degrades. Comparing Figure 7.17c and d, it is found that a 1 dB gain is observed by using more advanced receivers compared to the MMSE hard IC receiver.

7.2.2.4 Simulation Case 15

Figure 7.18 shows the BLER vs. SNR curves of Case 15 which belongs to the uRLLC scenario. The number of receiver antennas is 4, and the frequency-selective TDL-C channel is assumed. The TBS is 60 bytes and per-user spectral efficiency is 0.57 bps/Hz, similar to that in Case 3 with the difference in the number of users, e.g., only 4 and 6, as uRLLC does not require a large number of connections.

Figure 7.18a contains the results of four users where the code rate of 0.57 is only simulated for NCMA and UGMA, both using MMSE hard IC and showing similar performance. For the code rate of 0.29, SCMA uses the EPA + SISO receiver. RSMA uses the hybrid IC receiver, and LCRS/MUSA uses the MMSE hard IC receiver. Interleaver-division multiple access (IDMA) uses the ESE + SISO receiver. It is seen that the coding gain (from 0.57 to 0.29) is about 1 dB. For the code rate of 0.29, the choice of receivers would bring <0.5 dB performance difference. The reason is that due to four receiver antennas, there is already ample space in the spatial domain to differentiate different users, which reduces the burden of multi-user detection and coding. Even with less advanced receivers, the performance degradation is insignificant. This is very important to uRLLC as less complicated processing at the receivers would reduce the latency of detection and decoding. The performance of LCRS is slightly inferior to other NOMA schemes, with a similar reason as in Case 14: LCRS cannot benefit from the spreading code-domain interference suppression.

Figure 7.18b contains the results for six users where the code rate is 0.29 for all the schemes simulated. Among them, MUSA and LCRS use the less advanced receiver, e.g., MMSE hard IC, whereas SCMA uses the EPA + SISO receiver and RSMA uses the hybrid IC receiver. It is observed that a more advanced receiver can bring about 0.5 dB performance gain

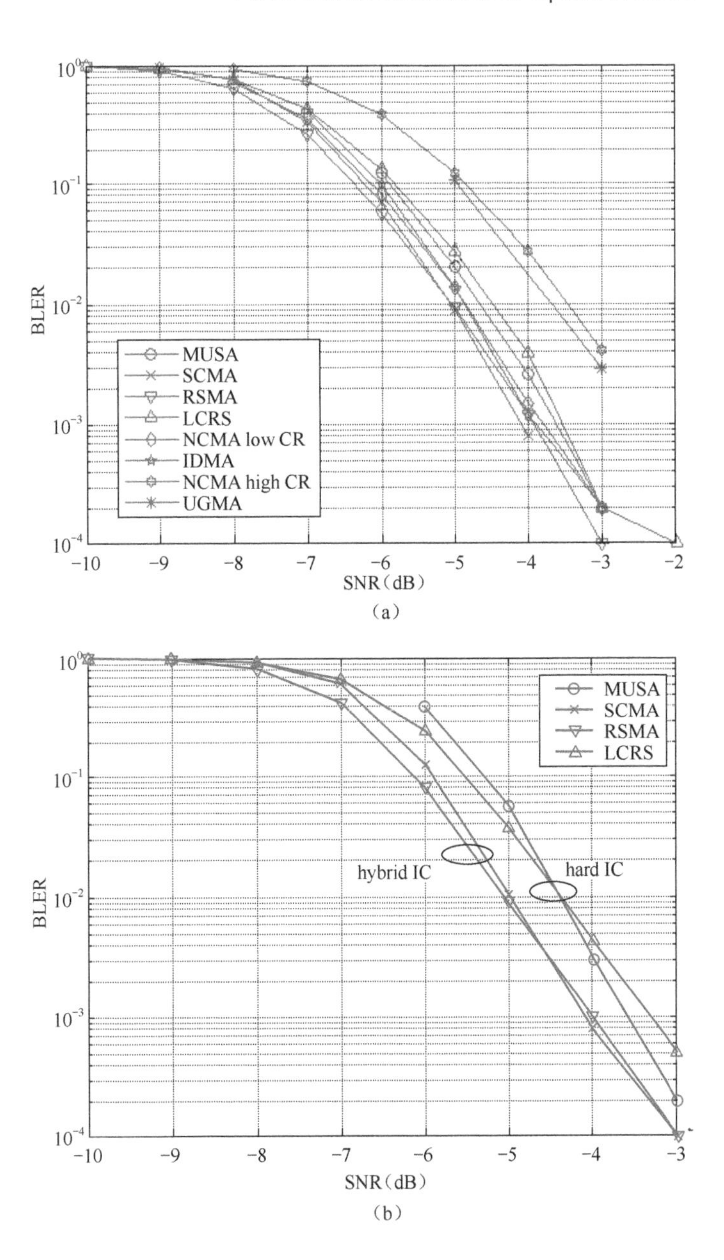

FIGURE 7.18 Link performance curves of NOMA schemes for simulation Case 15 (uRLLC). (a) TBS = 60 bytes, 4 UEs and (b) TBS = 60 bytes, 6 UEs.

compared to the less advanced receiver. Also, the performance of LCRS degrades significantly as the SNR increases. The reason is similar to that for Case 14.

7.2.2.5 Simulation Case 17

Figure 7.19 shows BLER vs. SNR curves for simulation Case 17 which also belongs to uRLLC scenario. The number of receive antennas is 4 and the TBS is 60 bytes. The per-user spectral efficiency is 0.57 bps/Hz and the

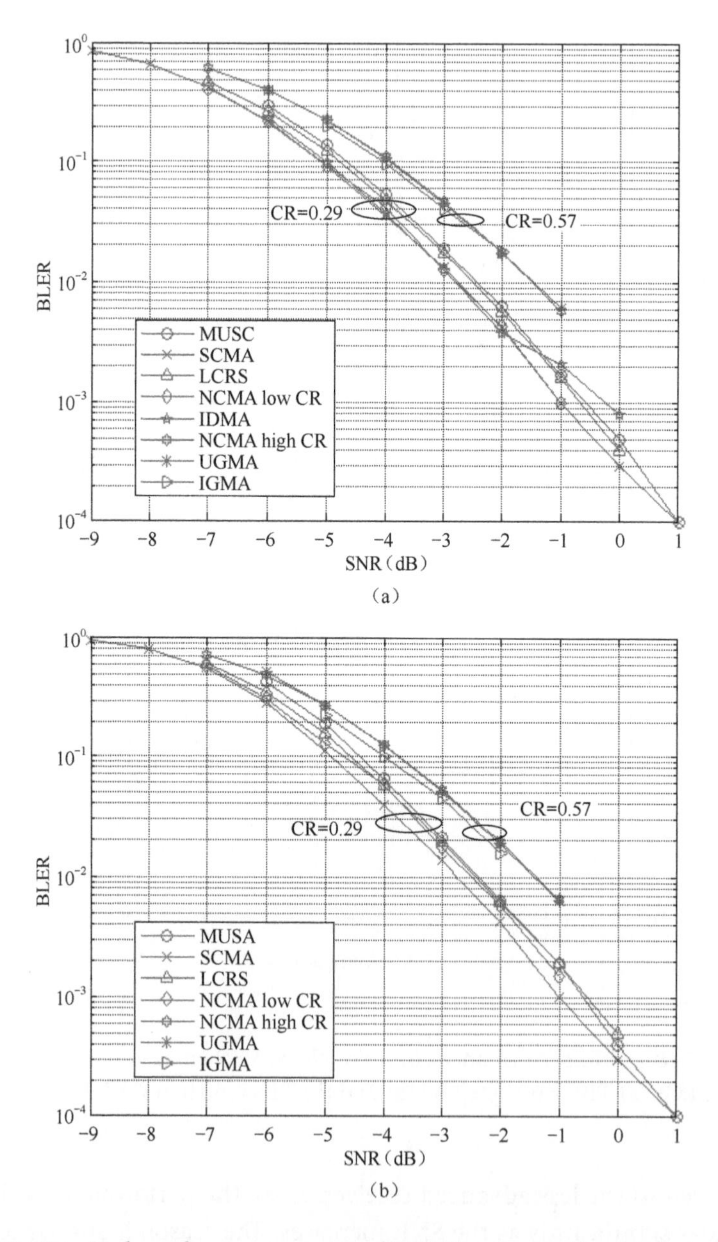

FIGURE 7.19 Link performance curves of NOMA schemes for simulation Case 17 (uRLLC). (a) TBS = 60 bytes, 4 UEs and (b) TBS = 60 bytes, 6 UEs.

number of superimposed users can be 4 or 6. The differences from Case 15 are channel model (flat TDL-A in Case 17 vs. TDL-C in Case 15) and carrier frequency (4 GHz in Case 17 vs. 700 MHz in Case 15).

Figure 7.19a contains the results for four users. Non-orthogonal code multiple access (NCMA), interleaved grid multiple access (IGMA) and user-grouped multiple access (UGMA) are simulated with the code rate of 0.57, NCMA and UGMA use MMSE hard IC receiver, IGMA uses ESE + SISO receiver, leading to similar performance. For the code rate of 0.29, IDMA uses ESE + SISO receiver. SCMA uses EPA + SISO receiver. RSMA uses a hybrid IC receiver. LCRS/MUSA uses a MMSE hard IC receiver. Their performance difference is within 0.5 dB. The coding gain (from 0.57 to 0.29) is about 1 dB.

Figure 7.19b contains the results for six users. The code rate configuration is similar to that in Figure 7.19a. The coding gain from 0.57 to 0.29 is about 1 dB. For the code rate of 0.29, the choice of receiver causes <0.5 dB performance difference.

7.2.2.6 Simulation Case 20

Figure 7.20 shows BLER vs. SNR for simulation Case 20 which belongs to the eMBB scenario. The number of receive antennas is 4 and the TBS is 150 bytes. The transmission bandwidth is 12 RBs. The per-user spectral efficiency is 0.7 bps/Hz. The number of superimposed users can be 4 or 8. Frequency flat channel TDL-A is assumed. The carrier frequency is 4 GHz.

Figure 7.20a contains the results for four users. NCMA, UGMA, PDMA and IGMA are simulated with a code rate of 0.7. NCMA, PDMA and UGMA use the MMSE hard IC receiver, and IGMA uses ESE + SISO receiver. Their performance is quite similar. For the code rate of 0.35, SCMA uses the EPA + SISO receiver. RSMA uses a hybrid IC. LCRS and MUSA use MMSE hard IC. Their performance difference is within 0.4 dB. The coding gain from 0.7 to 0.35 is about 1 dB.

Figure 7.20b contains the results for eight users. The code rate configuration is similar to that in Figure 7.20a. The coding gain (from 0.7 to 0.35) is about 1–1.5 dB. For the code rate of 0.35, the choice of receiver causes <0.8 dB performance difference.

The simulation results for high-spectral-efficiency operation can be summarized as follows:

- Under the ideal channel estimation, for mMTC/eMBB/uRLLC scenario with equal average SNR, time/frequency offset = 0 and pre-allocated MA signatures, as long as the configuration setting is appropriate.

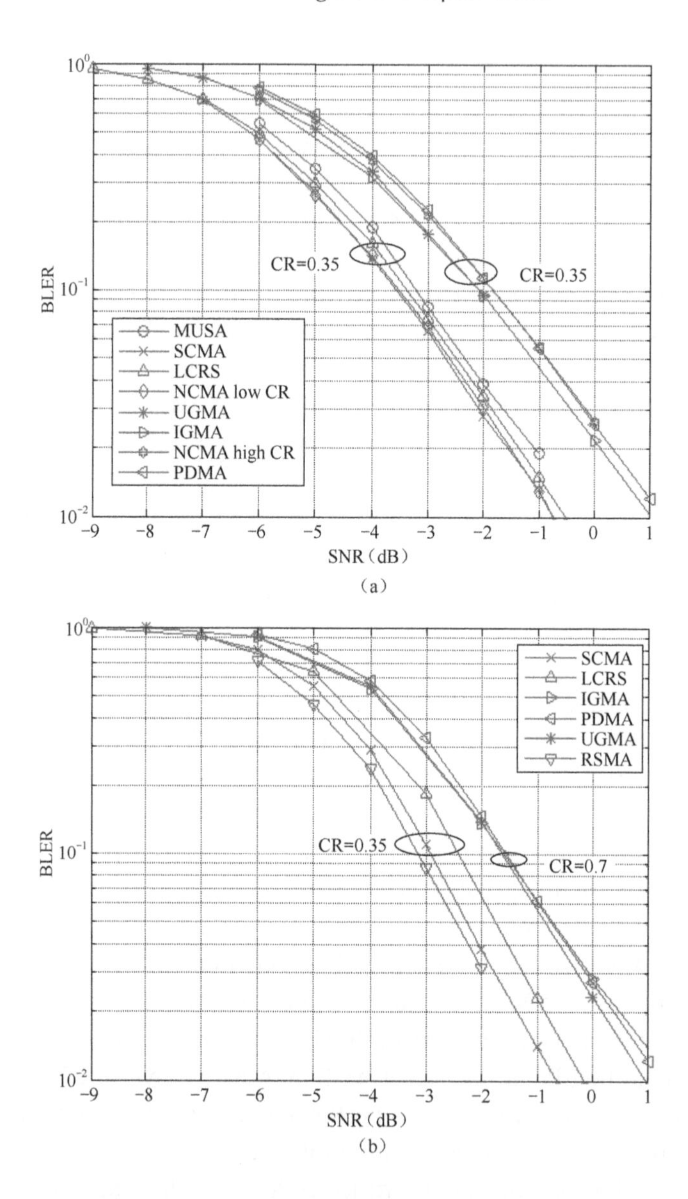

FIGURE 7.20 Link performance curves of NOMA schemes for simulation Case 20 (eMBB). (a) TBS = 150 bytes, 4 UEs and (b) TBS = 150 bytes, 8 UEs.

- Under the same receiver type and the same code rate, the performance difference is not very significant.
- Under the same code rate, using more advanced receivers, e.g., hybrid IC or EPA + SISO would result in better performance than using a less advanced receiver such as MMSE hard IC.

- When the receiver complexity is comparable, using a lower code rate would bring performance gain over using a higher code rate.
- As the TBS and/or the number of users increases, the performance of LCRS degrades rapidly.

In the following, we like to focus on the performance comparison between LCRS and symbol-level spreading. As discussed in Chapter 6, LCRS does not require changes in physical layer specification, whereas symbol-level spreading does. If spreading-based schemes do not provide performance gain, there is no strong motivation to standardize this technology. While the link-level simulations cannot be used to comprehensively evaluate the performance of a solution, the findings from the link level would be useful to explain the system-level performance.

Essentially, LCRS relies on using more redundancy bits to improve the error correction capability of the channel coding to be more robust to the cross-user interference in NOMA transmission. The systematic (info) bits and the redundancy (parity) bits are shown in Figure 7.21. Two users are shown as an example to represent multi-user transmission. The same code rate is assumed for each user.

Normally, LCRS would use the MMSE hard IC receiver. Its performance may be improved by using a more advanced receiver such as ESE + SISO.

Symbol-level spreading relies on spreading sequences with low cross-correlations. The cross-user interference is suppressed during MMSE detection. Its performance is less dependable with the channel coding as in the case of LCRS. Given the same spectral efficiency, the code rate should be increased when the symbol-level spreading is applied, which can be illustrated in Figure 7.22 compared to Figure 7.21. In Figure 7.22, the shadow represents the symbol-level spreading. Different directions of shadow mean different spreading sequences. Generally speaking, each spreading sequence can be considered as a vector in the multi-dimensional space of the spreading domain. The lower the

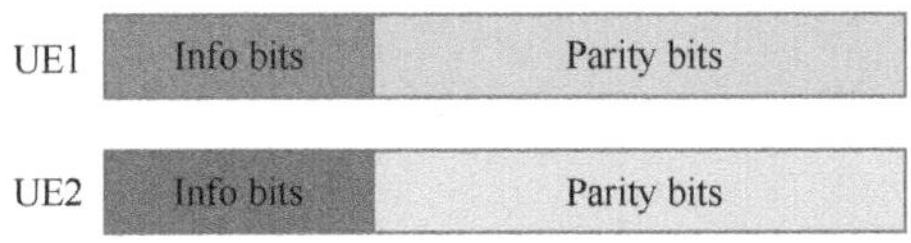

UE1 | Info bits | Parity bits
UE2 | Info bits | Parity bits

FIGURE 7.21 An illustration of info bits and parity bits for LCRS.

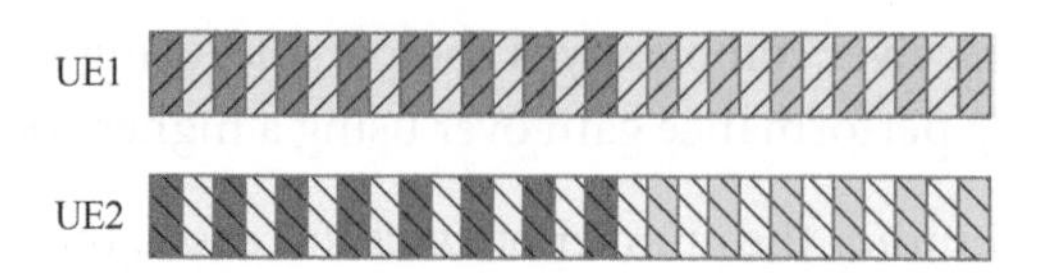

FIGURE 7.22 An illustration of information bits, parity bits and spreading of symbol spreading-based schemes.

cross-correlation, the more difference is between the directions of these spreading sequences.

The typical receiver for symbol-level spreading-based schemes is MMSE hard IC. Different from the case of LCRS, MMSE for symbol-level spreading is joint spreading code domain and spatial domain (if there are multiple receive antennas), whereas the MMSE for LCRS is only in the spatial domain. Of course, the performance of symbol-level spreading can be improved by using a more advanced receiver such as EPA + SISO, to compensate for the negative effect of high code rate operation.

Table 7.4 lists some simulation cases and configurations where significant gains over LCRS are shown [5]. Since there are multiple NOMA schemes in the category of symbol-level spreading, the link-level performance listed in Table 7.4 reflects the average performance of spreading-based schemes. Note that in Table 7.4, only the cases of mMTC scenarios are included, because mMTC scenarios are simulated by many companies and the results are relatively thorough. Also, the number of receive antennas in mMTC is 2 where the spatial-domain discrimination plays less role – so that we can

TABLE 7.4 Link-Level Performance Gains of Symbol-Level Spreading over LCRS

Transmitter Side NOMA Schemes/Categories	Receiver Type	Gain over LCRS (with MMSE Hard IC) TBS (Bytes), #Users: Gain (dB)
Symbol-level linear spreading: for example, MUSA、RSMA、NOCA、NCMA、UGMA、PDMA、WSMA	MMSE hard IC	60 bytes, 6 UEs: **1 dB** 75 bytes, 6 UEs: **2 dB**
SCMA	EPA + SISO	40 bytes, 10 UEs: **0.5 dB** 60 bytes, 6 UEs: **0.5 dB** 60 bytes, 8 UEs: **3 dB** 75 bytes, 6 UEs: **2 dB**
MUSA	EPA + SISO	40 bytes, 10 UEs: **0.5 dB** 60 bytes, 6 UEs: **0.5 dB** 60 bytes, 8 UEs: **3 dB** 75 bytes, 6 UEs: **2 dB**

focus on the coding gain and the spreading induced interference rejection. It should be pointed out that at link-level simulation, even if the simulation is multi-user, other-cell interference is usually modeled as AWGN, meaning that the other-cell interference at the link level is white in both the spatial domain and spreading code domain. In this case, MMSE does not have any interference rejection capability on other-cell interference.

It is seen in Table 7.4 that when the MMSE hard IC receiver is used, only when TBS reaches 60 or 75 bytes, symbol-level spreading would show significant gain over LCRS. This implies that at low-to-medium spectral efficiency, for instance, TBS = 10, 20 or 40 bytes, the benefit of the low code rate of LCRS is comparable to the benefit of low cross-correlation sequences of spreading-based schemes (assuming other-cell interference is AWGN). When the spectral efficiency is increased, low cross-correlation shows more advantage over code rate reduction.

If a more advanced receiver, e.g., EPA + SISO, is used, the multi-dimensional modulation scheme (e.g., SCMA) and linear-spreading-based (e.g., MUSA) one would show better performance than LCRS with MMSE hard IC receiver, when TBS = 40 bytes. The higher the spectral efficiency, the wider the performance gap is.

7.3 SYSTEM-LEVEL PERFORMANCE

In this section, the system-level performance of uplink grant-free NOMA is discussed for the scenarios of mMTC, eMBB small data and uRLLC. Contention-free transmission is assumed where the DMRS is pre-allocated without collision.

7.3.1 mMTC Scenario

For the mMTC scenario, the following three cases are simulated.

7.3.1.1 Case 1: Each User Is Allocated 1 PRB + 1 ms of Time-Frequency Resources in the Baseline; for MUSA, Each Use Transmits in 1 PRB + 4 ms of Time-Frequency Resources

In the baseline, resources are divided into frequency and time domains where each user is pre-allocated to one of them, as well as the DMRS. For uplink MUSA, resources are divided into the frequency domain where each user is pre-allocated to one of them as well as the DMRS. Once the DMRS is allocated, the spreading sequence of this user is also determined. For fair comparison, the total power of the spreading sequence of each user is normalized to be 1.

Figure 7.23 shows the system simulation results of Case 1. In Figure 7.23a, system performance in terms of packet drop rate (PDR) vs. packet arrival rate (PAR) is compared between the baseline with MMSE-IRC receiver and MMSE-PIC (a form of MMSE hard IC), and MUSA with MMSE-PIC receiver. It is observed that at the target PDR = 1%, MUSA can accommodate twice of traffic load of the baseline with MMSE-IRC receiver, and about 1.5 of the traffic load of the baseline with MMSE-PIC receiver.

Figure 7.23b and c shows the maximum numbers of users in a resource and the resource utilization, respectively, where the baseline and MUSA are compared. It is found that the maximum number of users in a resource and the resource utilization increase with the system loading. Compared to the baseline, a higher max number of users and resource utilization are seen in MUSA simulations. This is due to the spreading factor of 4 configured for MUSA, resulting in that a user occupies more (time-domain) resources, thus more chances that multiple users would share the same time-frequency resources and higher resource utilization. It is also noticed that due to the limited resources for pre-configuration in the baseline case, the maximum number of users in a resource saturates once it reaches a certain threshold.

7.3.1.2 Case 2: Each User Occupies 6 PRBs + 1 ms Time-Frequency Resource for Both the Baseline and MUSA

In this case, a user occupies the entire simulation bandwidth with the same time duration. Hence, the code rate of the baseline is relatively low. Note here the baseline is LCRS, e.g., without any change in Rel-15 NR physical layer specification, except that the receiver can be MMSE-IRC or MMSE-PIC.

Simulation results for Case 2 are shown in Figure 7.24. Among them, PDR vs. PAR performance is shown in Figure 7.24a where the baseline with MMSE-IRC receiver and with MMSE-PIC receiver, and MUSA with MMSE-PIC receiver are compared. It is observed that for the target PDR = 1%, MUSA can support twice of traffic load of the baseline with MMSE-IRC or MMSE-PIC receiver.

Figure 7.24b shows the maximum number of users on a resource for the baseline and MUSA. Since the same amount of resource is assumed between the baseline and MUSA. The maximum number of users is more or less the same. Figure 7.24c shows the CDF of the number of users on a resource. As the system loading increases, the CDF moves to the right.

Resource utilizations for the baseline and MUSA are compared in Figure 7.24d. It is observed that the resource utilization of the baseline

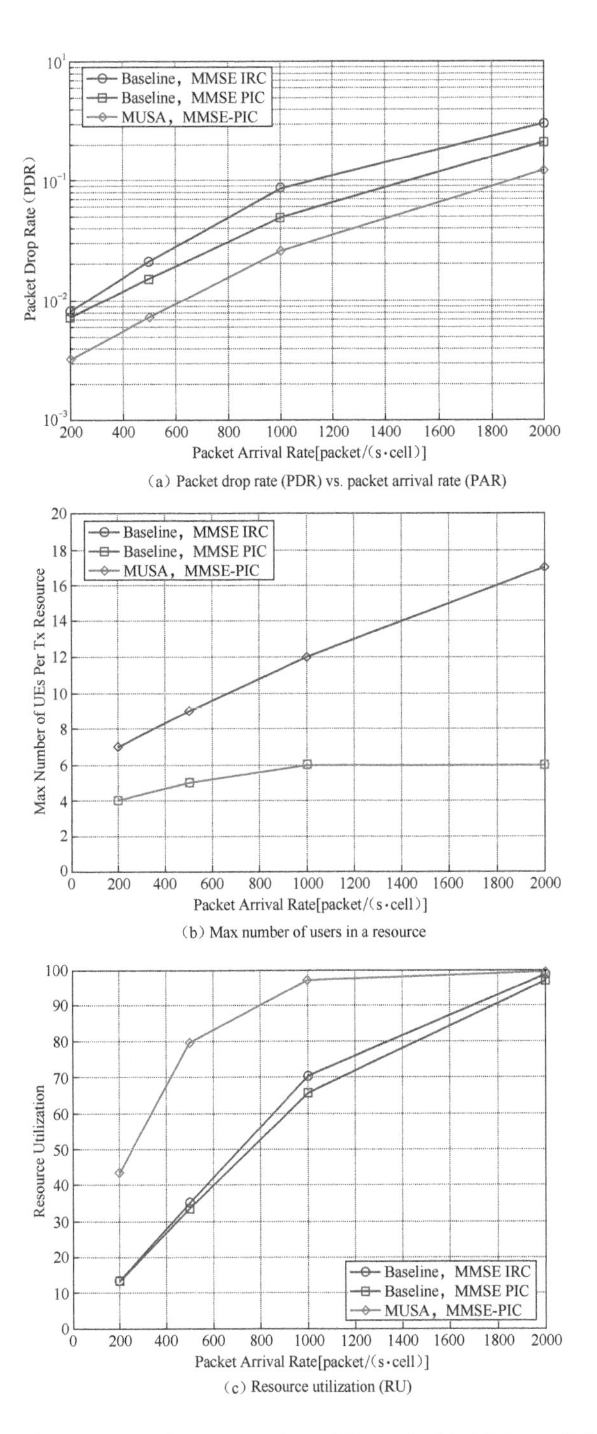

FIGURE 7.23 System simulation results of Case 1. (a) PDR vs. PAR, (b) max number of users in a resource and (c) RU.

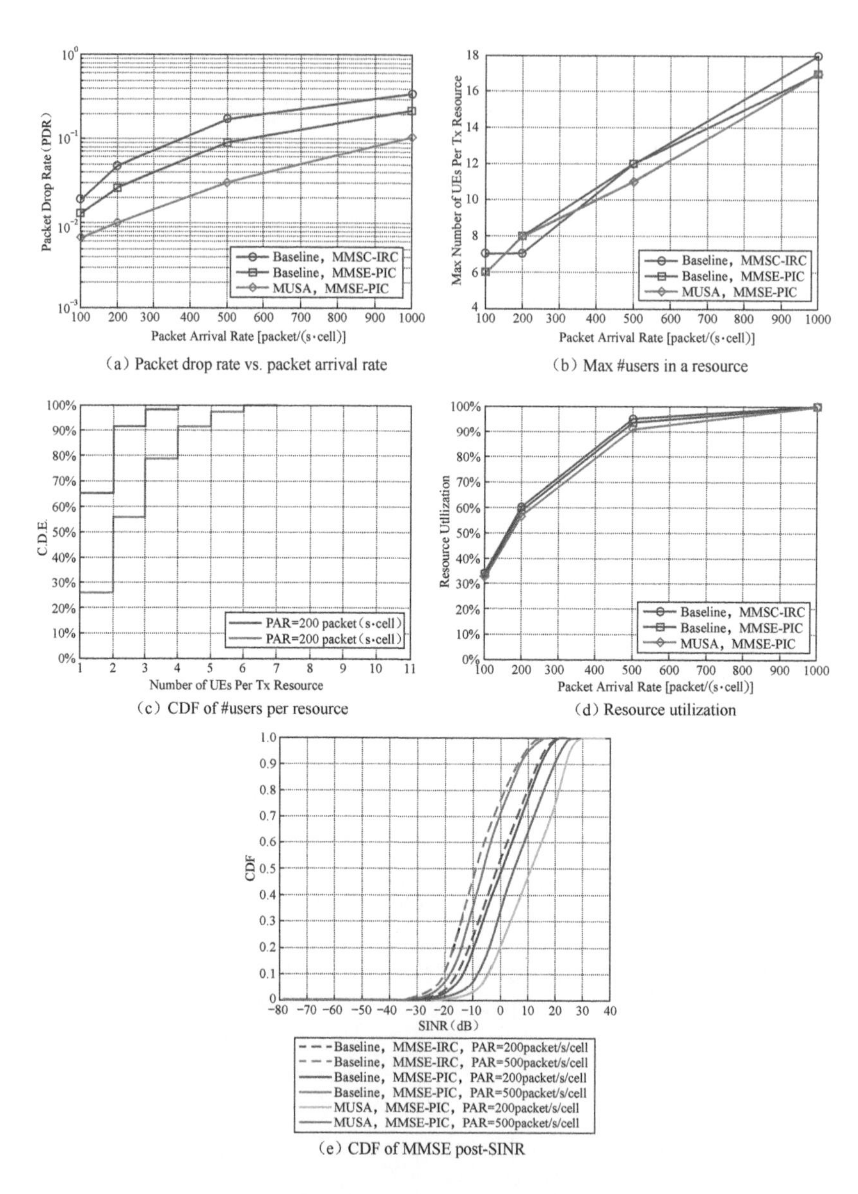

FIGURE 7.24 System simulation results for Case 2. (a) PDR vs. PAR, (b) max #users in a resource, (c) CDF of #users per resource, (d) RU and (e) CDF of MMSE post-SINR.

and MUSA are similar. The slightly lower resource utilization (RU) seen in MUSA is due to the more efficient transmission.

Figure 7.24e shows the CDFs of MMSE post-SINR for the baseline and MUSA. It is observed that the difference is as high as 10 dB. Note that the effective code rate of the baseline is 1/8 whereas the effective code rate of MUSA is 1/2, which translates to about a 7 dB difference in detection threshold. Subtracting this, there is still about a 3 dB net gain which may explain the better performance of MUSA. The root cause would be the inter-cell interference suppression capability of joint spatial-domain and spreading code-domain MMSE detection.

Let us examine whether there are other factors. According to the link-level simulation summary in Table 7.4, symbol-level spreading schemes (including MUSA) would outperform LCRS only when TBS is >40 bytes. However, in the system-level simulation of mMTC, after packet segmentation at the media access control (MAC) layer, each TBS is only about 25 bytes. Obviously, if all the simulation assumptions for system-level simulations are the same as those of link-level simulations, we would not see a significant performance gain of MUSA at the system level. Hence, it is certain that the gain of MUSA at the system level is mainly due to the assumption of inter-cell interference. At the system level, users in the neighboring cells transmit in the same set of time and frequency resources. Although there are many users in the neighboring cells, not many of them play a major role in contributing to inter-cell interference, e.g., only those users that are close to the edge of the target cell would have a significant impact on the interference. Since the number of dominant users (in terms of contributing inter-cell interference) is limited, inter-cell interference exhibits a certain structure, not only in the spatial domain (if there are multiple uplink receive antennas) but also in the spreading code domain (for symbol-level linear spreading). MMSE is capable of suppressing interferences with the structure. By contrast, there is no symbol-level spreading in LCRS. Hence, MMSE cannot provide such interference suppression benefits for LCRS.

7.3.1.3 Case 3: Each User Occupies 1 PRB + 6 ms Time-Frequency Resource for Both the Baseline and MUSA

In this case, a user occupies 1 PRB + 6 ms time and frequency resource. The code rate of the baseline is relatively low. The baseline is LCRS. Based on the simulation assumptions for the mMTC scenario, the maximum couple loss between a user and the serving base station can be 144 dB when the transmit power of a user is concentrated in one PRB. This would improve the cell coverage.

Figure 7.25 shows the system simulation results of Case 3. Among them, Figure 7.25a shows PDR vs. PAR for the baseline and MUSA under different spreading lengths. It is found that at the target PDR = 1%, MUSA with

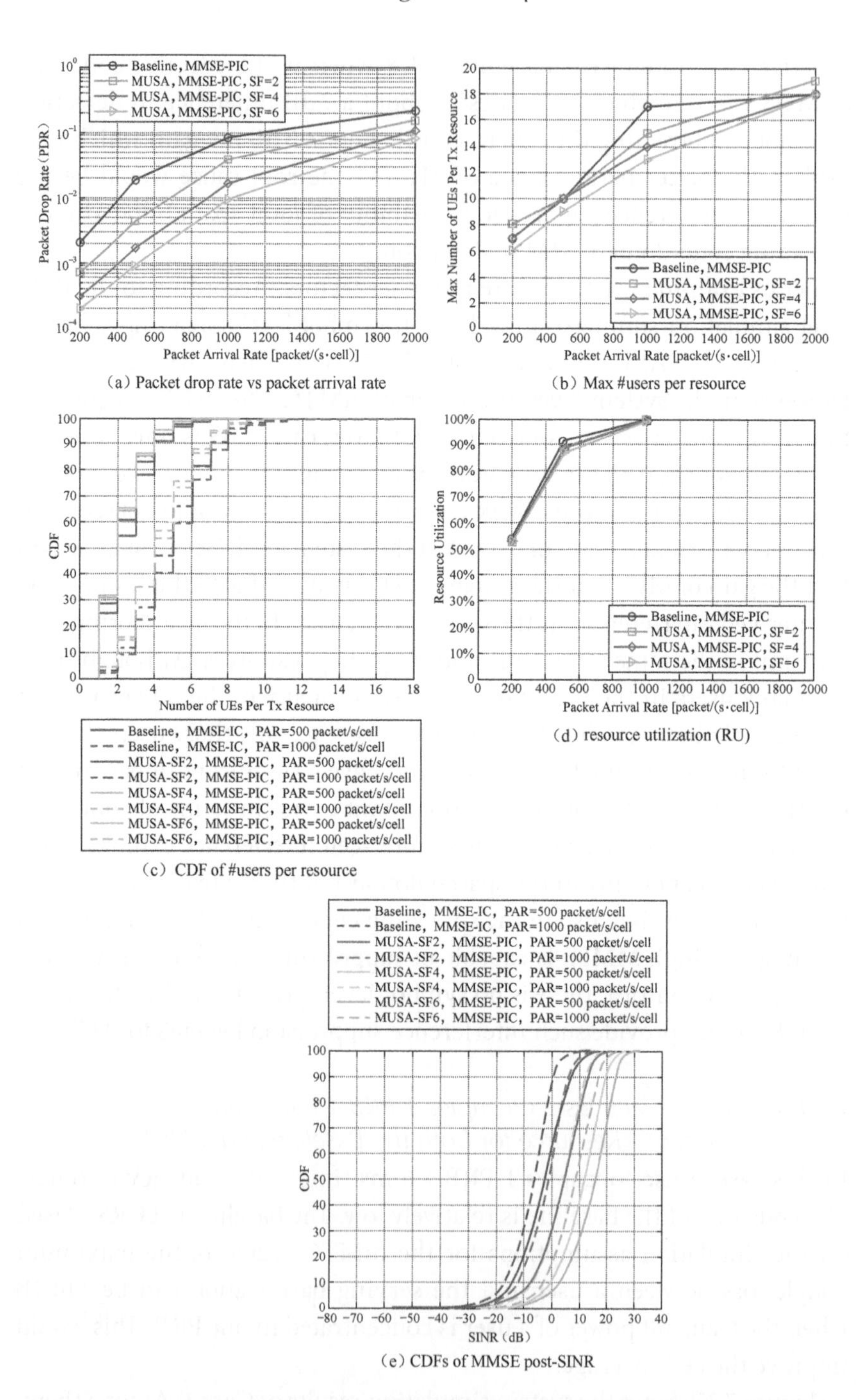

FIGURE 7.25 System simulation results for Case 3. (a) PDR vs PAR, (b) max #users per resource, (c) CDF of #users per resource, (d) RU and (e) CDFs of MMSE post-SINR.

a spreading length of 2 can support 1.5 of the traffic load of the baseline. When the spreading length is increased to 4 or 6, MUSA can double the system loading compared to the baseline since longer spreading can more effectively suppress the inter-cell interference.

Figure 7.25b shows the maximum number of users in a resource for the baseline and MUSA. It is observed that since each user occupies the same amount of resource for the baseline and MUSA, the maximum number of users is similar. Because MUSA transmission is more efficient, the number of multiplexed users is slightly lower in MUSA compared to the baseline. Figure 7.25c shows the CDFs of the number of users in a resource, which increases with the system loading.

Figure 7.25d shows the RU of the baseline and MUSA. The RU statistics are similar. The RU of MUSA is slightly lower due to the more efficient transmission of MUSA.

Figure 7.25e shows the CDFs of MMSE post-SINR for the baseline and MUSA. Similar to Case 2, significant CDFs are observed. Here, the effective code rate of the baseline is 1/8, whereas the code rate of MUSA is 1/4, 1/2 and 3/4 for the spreading lengths of 2, 4 and 6, respectively. There are certain differences in the threshold for the detection. Even with this, MUSA still outperforms the baseline, primarily due to the inter-cell interference suppression capability of joint spatial and code-domain MMSE.

In the above simulations, the covariance matrix of inter-cell interference is assumed to be ideally known. Hence, the interference suppression by MMSE detection is quite effective. In the following, we evaluate the system performance with ideally known covariance matrix or non-ideally known covariance matrix of inter-cell interference. There are two situations of non-ideally known covariance matrices:

1. The MMSE weight is calculated from R_{yy}. Here, it is assumed that there is either no symbol-level scrambling or the same symbol-level scrambling. There are certain estimation errors of covariance
2. The MMSE weight is calculated from the white inter-cell interference

In the latter situation, user-specific symbol-level scrambling is assumed and can reduce PAPR. However, this would affect the covariance matrix estimation for inter-cell interference. That covariance matrix is essentially white. In the simulations, the spreading length is 4. The exact modeling of the receivers can be found in link-to-system mapping discussed in the previous sections. PDRs vs. PAR are compared in Figure 7.26a. It is observed that when MMSE weight is calculated from R_{yy} for both the baseline and MUSA, the supported system loading can be 1.5 times that of the baseline when using MUSA. Even when the inter-cell interference is assumed white when calculating MMSE weight, MUSA can still deliver a 15% gain over the baseline. Figure 7.26b provides the CDFs of MMSE post-SINR when the PAR is 500 packet/s/cell.

7.3.2 eMBB Small Data Scenario

For eMBB small data scenario, two cases are simulated:

7.3.2.1 Case 1: Each User in the Baseline Occupies 3 PRBs + 1 ms Time and Frequency Resource; Each User in MUSA Occupies 12 PRB + 1 ms Time and Frequency Resource

In this case, the resources are divided in frequency for each user in the baseline, as well as the DMRS. For MUSA, the resources in the frequency domain are shared by multiple users and each user is allocated a DMRS.

Figure 7.27 shows the system results for Case 1 in eMBB small data. Among them, Figure 7.27a shows the PDR vs. PAR curves for the baseline with MMSE-IRC receiver and MMSE-PIC receiver, as well as MUSA with

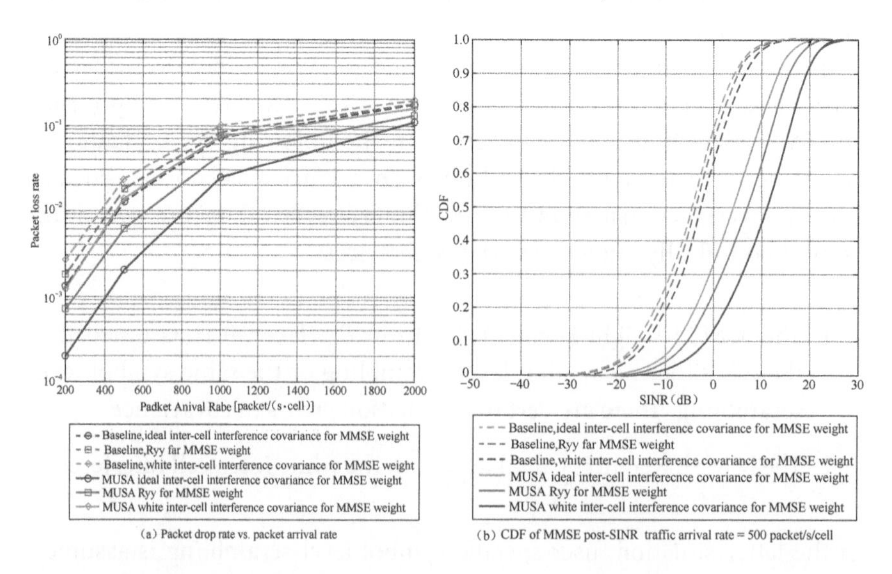

FIGURE 7.26 System simulation results for Case 3 with different assumptions of inter-cell interference estimation. (a) PDR vs. PAR and (b) CDF of MMSE post-SINR (traffic arrival rate = 500 packet/s/cell).

MMSE-PIC receiver. It can be seen that for the target PDR = 1%, MUSA can support twice of system load of the baseline with the MMSE-IRC receiver and about 1.5 system load of the baseline with the MMSE-PIC receiver.

Figure 7.27b shows the maximum number of users on a resource. It can be observed that the maximum number of users increases with the system loading. The maximum number of users in MUSA is much higher than that of the baseline because a spreading length of 4 is used, meaning that a user in MUSA occupies more resources. At PDR = 1% and 2,000 packet/s/cell system loading, the maximum number of users can reach 16 for MUSA.

Figure 7.27c shows the CDFs of the number of users on each resource for MUSA. It is found that in many times, the number of users per resource is not very large.

Figure 7.27d shows the RU of the baseline and MUSA. It is seen that the RU increases with the system loading. The RU in MU/SA is much higher than that in the baseline. The reason is similar as above, e.g., each user occupies more resource in MUSA than in the baseline.

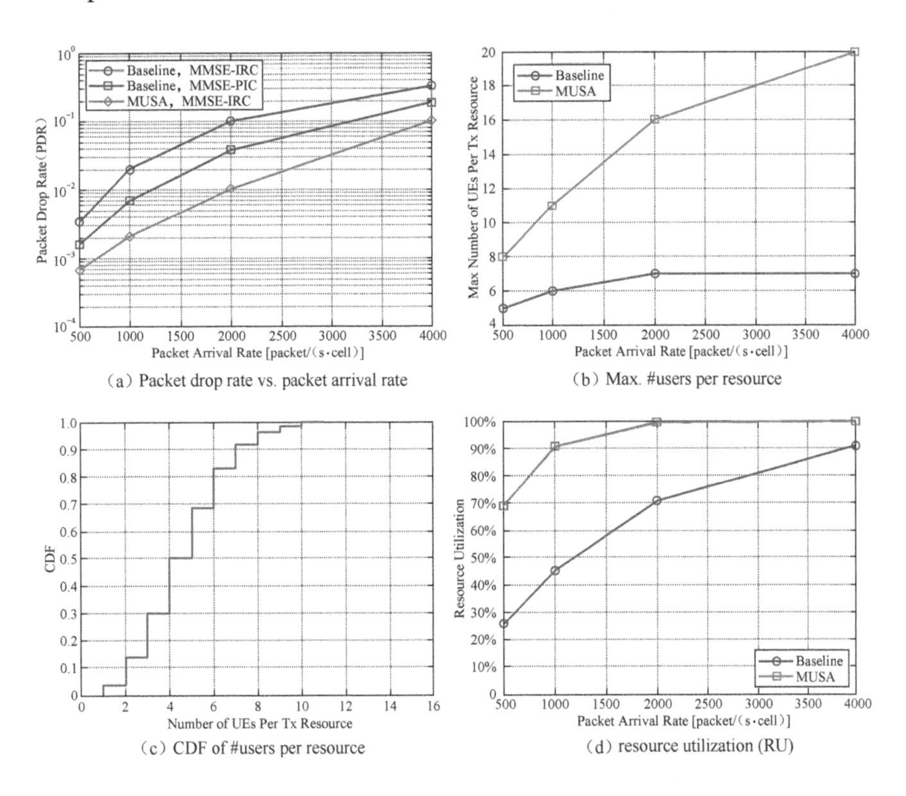

FIGURE 7.27 System simulation results for Case 1 of eMBB small data. (a) PDR vs. PAR, (b) max. #users per resource, (c) CDF of #users per resource and (d) RU.

7.3.2.2 Case 2: Each User Occupies 12 PRBs + 1 ms Time and Frequency Resource in Both the Baseline and MUSA

Since a user occupies the entire bandwidth of the same time duration for the baseline and MUSA, the code rate of the baseline is lower. The baseline in this case is LCRS.

Figure 7.28 shows the system simulation results of Case 2 for eMBB small data. Among them, Figure 7.28a shows the PDR vs. PAR for the baseline with MMSE-IRC receiver and MMSE-PIC receiver, as well as MUSA with the MMSE-PIC receiver. It is observed that at PDR = 1%, MUSA can support 3 times system loading of the baseline with the MMSE-IRC receiver and 2 times the system loading of the baseline with the MMSE-PIC receiver.

Figure 7.28b shows the maximum number of users in a resource. It is seen that due to the same resource usage, e.g., 12 PRBs + 1 ms, the maximum number of users is similar in the baseline and MUSA. Figure 7.28c shows the CDF of the number of users in a resource for MUSA. Figure 7.28d shows the RU. It is found that the baseline and MUSA have a similar RU.

Figure 7.28e shows the CDFs of MMSE post-SINR of the baseline and MUSA. A significant difference is observed between the two. In the simulation, the code rate of the baseline is 1/6, whereas the code rate of MUSA is 2/3. Even with the difference in the threshold for detection, MUSA still outperforms the baseline, which can be explained by the inter-cell interference suppression capability by joint spatial and spreading domain MMSE detection for MUSA. Here, the covariance of inter-cell interference is assumed to be structured (e.g., colored) and ideally known.

7.3.3 uRLLC Scenario

For the uRLLC scenario, two cases are evaluated:

7.3.3.1 Case 1: Each User Occupies 3 PRBs + 0.25 ms Time and Frequency Resource in the Baseline and 12 PRBs + 0.25 ms Resource in MUSA

In this case, resources in the baseline are divided for different users. For MUSA, each user occupies the entire simulation bandwidth (e.g., 12 PRBs). DMRS is pre-configured.

Figure 7.29 shows the percentages of users satisfying the reliability and delay requirement as a function of PAR for Case 1 when the TBS is 60 bytes. It is observed that at the target percentage of 95%, even with the MMSE-IRC receiver, MUSA can support 3 times the system loading of

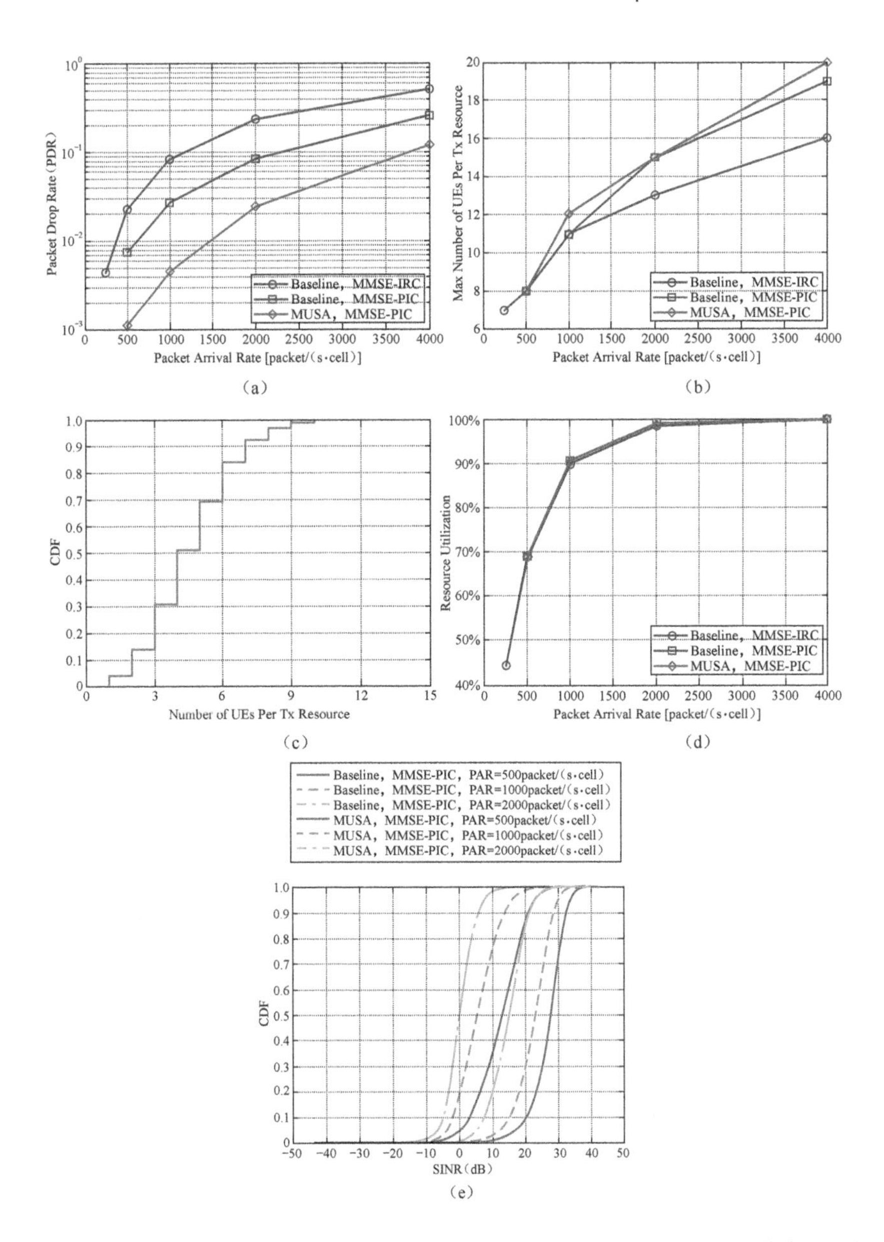

FIGURE 7.28 System-level simulation results for Case 2 of eMBB small data. (a) PDR vs. PAR, (b) max #users per resource, (c) CDF of #users per resource, (d) RU and (e) CDF of MMSE post-SINR.

the baseline. When the MMSE-PIC receiver is used, the PDR can be further reduced, and the percentage of users satisfying the reliability and delay requirement is increased in the baseline. However, the PDR of MUSA

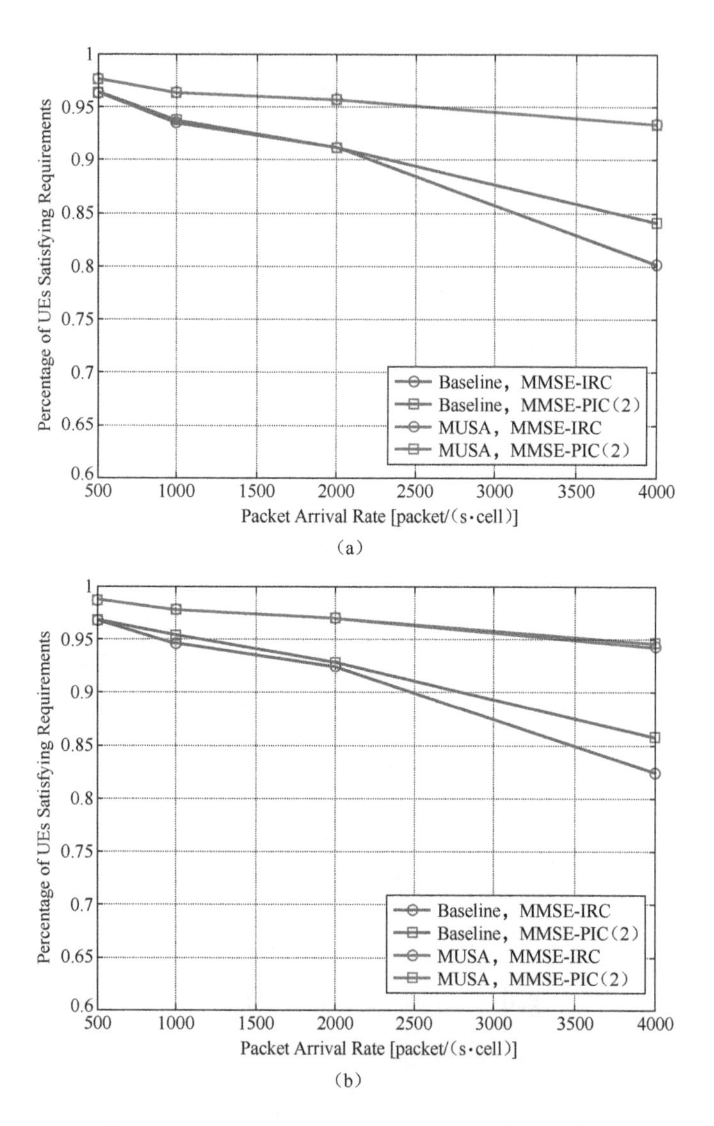

FIGURE 7.29 Percentages of users satisfying the reliability and delay requirement for Case 1 uRLLC, TBS = 60 bytes. (a) 4 GHz + 200 m and (b) 700 MHz + 500 m.

is already very low for the MMSE-IRC receiver. The performance gain of MUSA over the baseline is not very significant when MMSE-PIC is used for both the baseline and MUSA.

Figure 7.30 shows the percentage of users satisfying the reliability and delay requirement for Case 1 when TBS = 200 bytes. The observation is similar to TBS = 60 bytes, except that the overall performance is slightly inferior to that of TBS = 60 bytes.

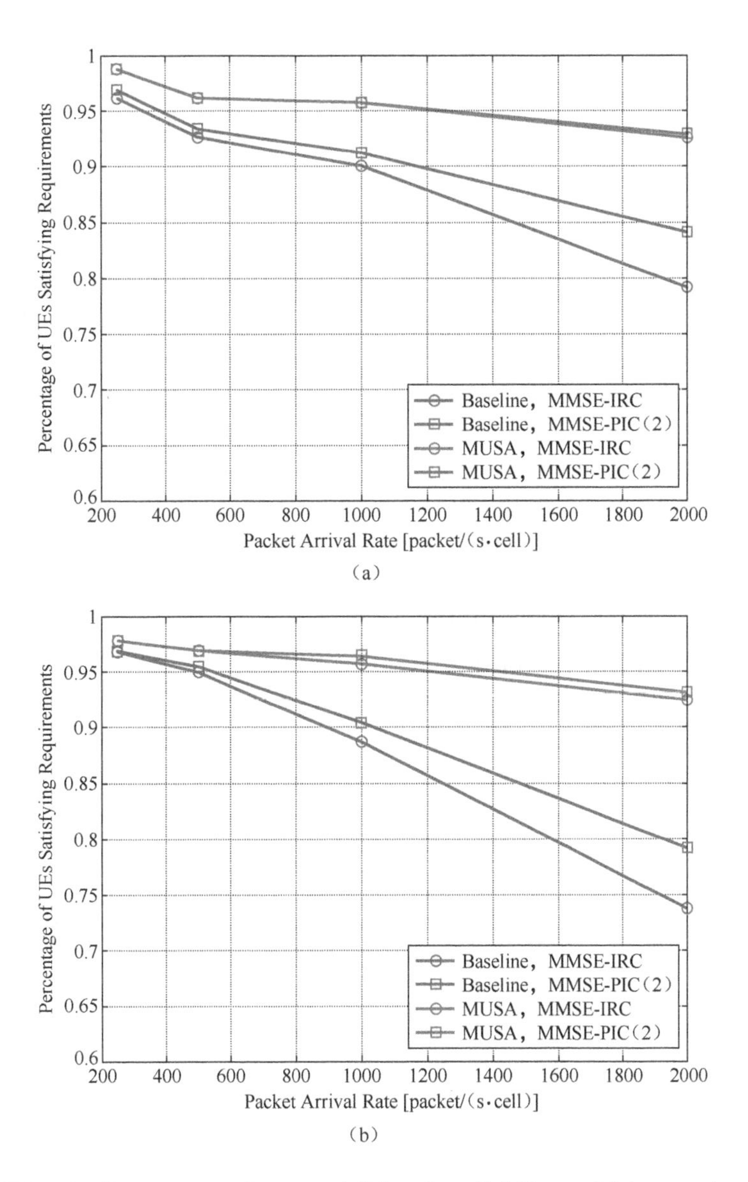

FIGURE 7.30 Percentages of users satisfying the reliability and delay requirement for Case 1 uRLLC, TBS = 200 bytes. (a) 4 GHz + 200 m and (b) 700 MHz + 500 m.

7.3.3.2 Case 2: Each User Occupies 12 PRBs + 0.25 ms Time and Frequency Resource in the Baseline and MUSA

In this case, since a user occupies the entire simulation bandwidth over the same time duration, the code rate of the baseline is lower than that of MUSA. Here, the baseline scheme is LCRS.

Figure 7.31a shows the percentages of users satisfying the reliability and delay requirement when the carrier frequency is 4 GHz, the site-to-site distance is 200 m and the TBS is 60 bytes. It is found that at the target percentage of 95%, MUSA can support 2,000 packets/s/cell of system load, whereas the baseline can only support about 700–800 packets/s/cell system load.

Figure 7.31b shows the CDFs of MMSE post-SINR of the baseline and MUSA. About 10 dB difference is observed. In the simulation, the effective code rate of the baseline is 0.1435. The code rate of MUSA is 0.574. There is about a 7 dB difference in the threshold for detection. However, even with this discount, MUSA still outperforms the baseline because of the inter-cell interference suppression capability of the joint spatial and spreading domain MMSE detection. Here, the covariance matrix of the inter-cell interference is structured (e.g., colored) and ideally known.

7.4 PEAK-TO-AVERAGE POWER RATIO

7.4.1 CP-OFDM Waveform

For NOMA, due to the introduction of user differentiation mechanisms to the transmitter, the PAPR of the waveform is affected to some extent. The level of the impact depends on the specific NOMA solution:

- For IDMA, LCRS type of schemes, the processing is limited to the bit level and can be considered as using a lower code rate to transmit. Hence, the PAPR is almost the same as in the case of OMA
- For MUSA, the non-orthogonal code access (NOCA) type of symbol-level spreading schemes, as illustrated in Figure 7.32, depending on whether the spreading is done in the time domain or frequency domain, the impact on PAPR would be different
    - If the spreading is done in time domain, as seen in Figure 7.33, the processing is equivalent to weighting and repletion of the OFDM symbol, without changing the PAPR
    - If the spreading is in the frequency domain, as shown in Figure 7.34, the spreading would bring certain correlation to the signal and hence increase the PAPR. Symbol-level scrambling can randomize the signal between adjacent subcarriers and reduce the PAPR. In [6], the constant amplitude zero auto-correlation (CAZAC) sequence is proposed to scramble the signal at the

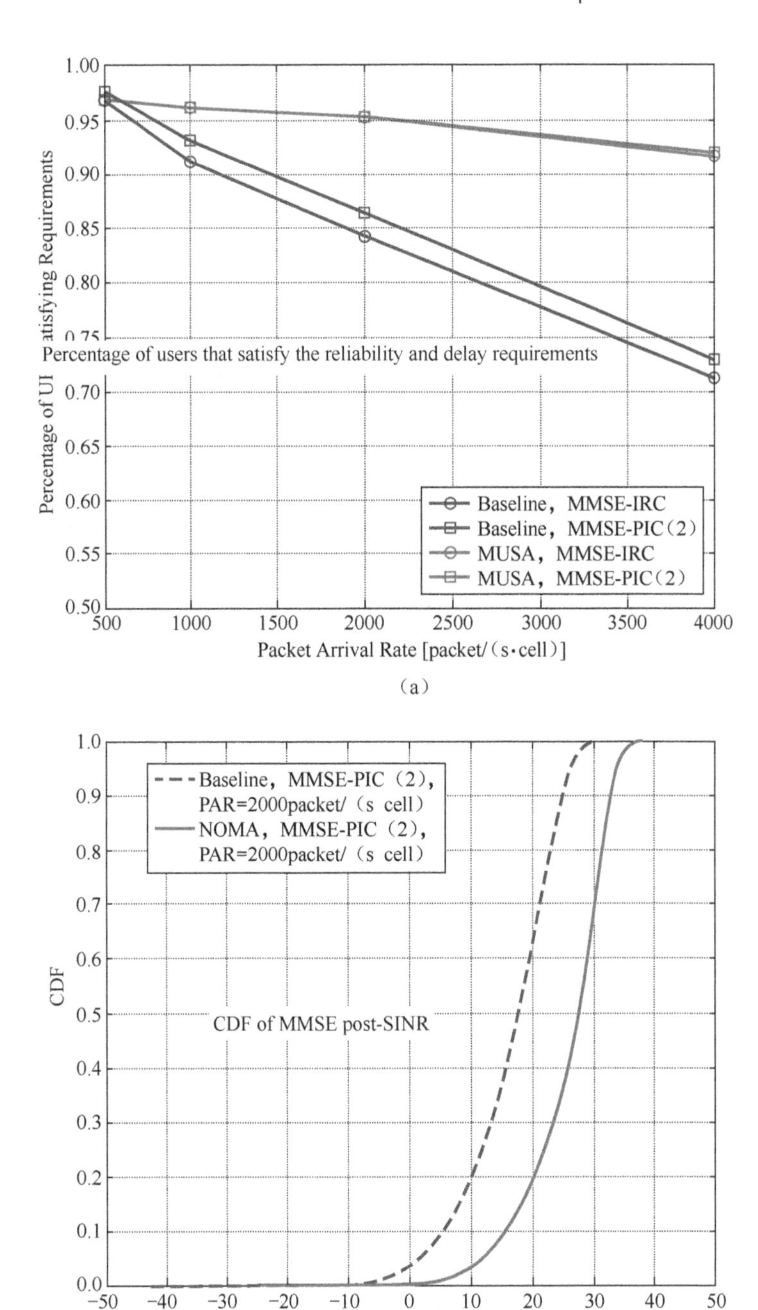

FIGURE 7.31 System-level simulation results for Case 2 of uRLLC, 4GHz carrier, ISD = 200 m, TBS = 60 bytes. (a) Percentage of users that satisfy the reliability and delay requirements and (b) CDF of MMSE post-SINR.

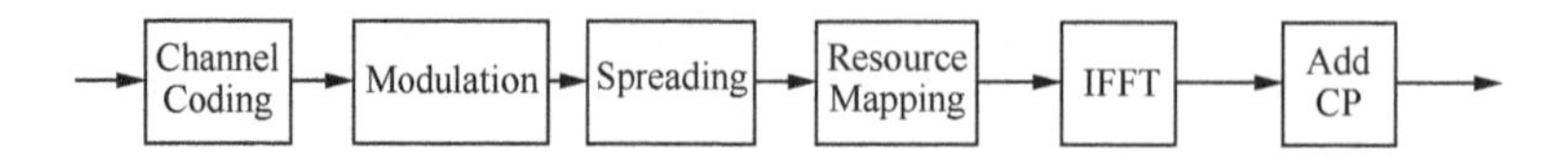

FIGURE 7.32 Transmitter side processing of NOMA in the case of the CP-OFDM waveform.

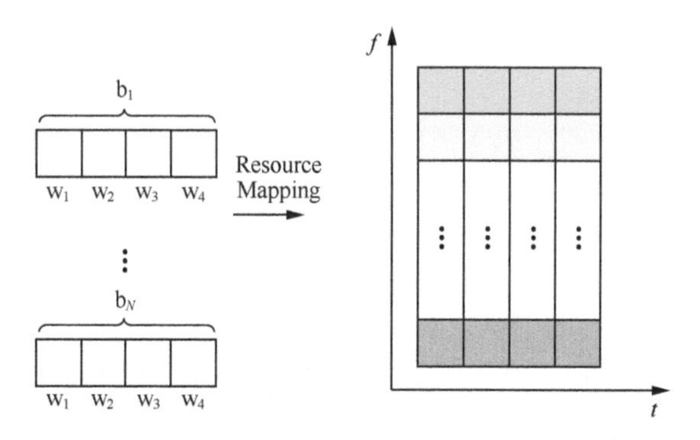

FIGURE 7.33 Time-domain spreading in the case of the CP-OFDM waveform.

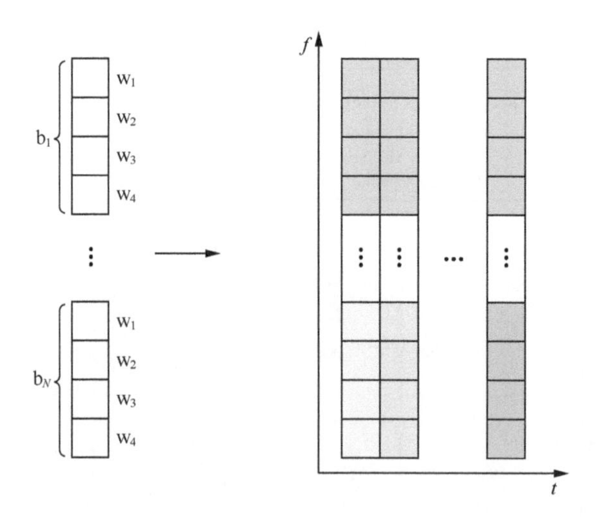

FIGURE 7.34 Frequency domain spreading in the case of the CP-OFDM waveform.

symbol level, which can reduce the PAPR to a level that is even lower than OMA.

- For frequency domain spreading, the PAPR also depends on the specific spreading sequences. If a spreading sequence tends to have the time-domain signals of different subcarriers constructively superimposed, the increase of PAPR would be significant. If certain optimization can be done to avoid such constructive superposition, like RSMA [1] shown below, the increase of PAPR compared to OMA would be smaller

$$s_k(l)=\frac{1}{\sqrt{L}}\exp\left(j\pi\left(\frac{(k+l)^2}{K}\right)\right)w(l) \tag{7.9}$$

where $w(l)$ is an all-1 sequence or a periodic sequence satisfying $\sum_{l=1}^{L} w(l)w^*(l+m)=L\delta(m)$.

- For SCMA and IGMA, their PAPR is slightly higher than that of OMA

Figure 7.35 shows the PAPRs of several NOMA schemes. For the symbol-level linear spreading type of schemes, MUSA and RSMA are listed here as the representative of other spreading schemes whose PAPRs are similar. The PAPR of LCRS is the same as OMA.

It is observed that the PAPR of IDMA and time-domain spreading is the same as OMA (which uses CP-OFDM waveform). The PAPR is slightly higher than that of OMA. The PAPR of frequency domain spreading of MUSA is the highest. Due to the special design of its spreading sequence, the PAPR of frequency domain spreading of RSMA is lower than that of MUSA. After symbol-level scrambling, the PAPR of the frequency domain spreading is significantly reduced, sometimes even lower than that of OMA (with CP-OFDM).

7.4.2 DFT-S-OFDM Waveform

For the scenarios that are very sensitive to PAPR, the DFT-s-OFDM waveform is often used. Hence, it is of more practical value to investigate the PAPR of NOMA with the DFT-s-OFDM waveform.

Similar to the case of CP-OFDM waveform, when NOMA is combined with the DFT-s-OFDM waveform,

- The operation of IDMA, LCRS, etc. is only at the bit level, which is equivalent to using a lower code rate. Its PAPR is the same as that of OMA.
- In the case of DFT-s-OFDM, for symbol-level linear spreading schemes such as MUSA and NOCA, the spreading can be in the time

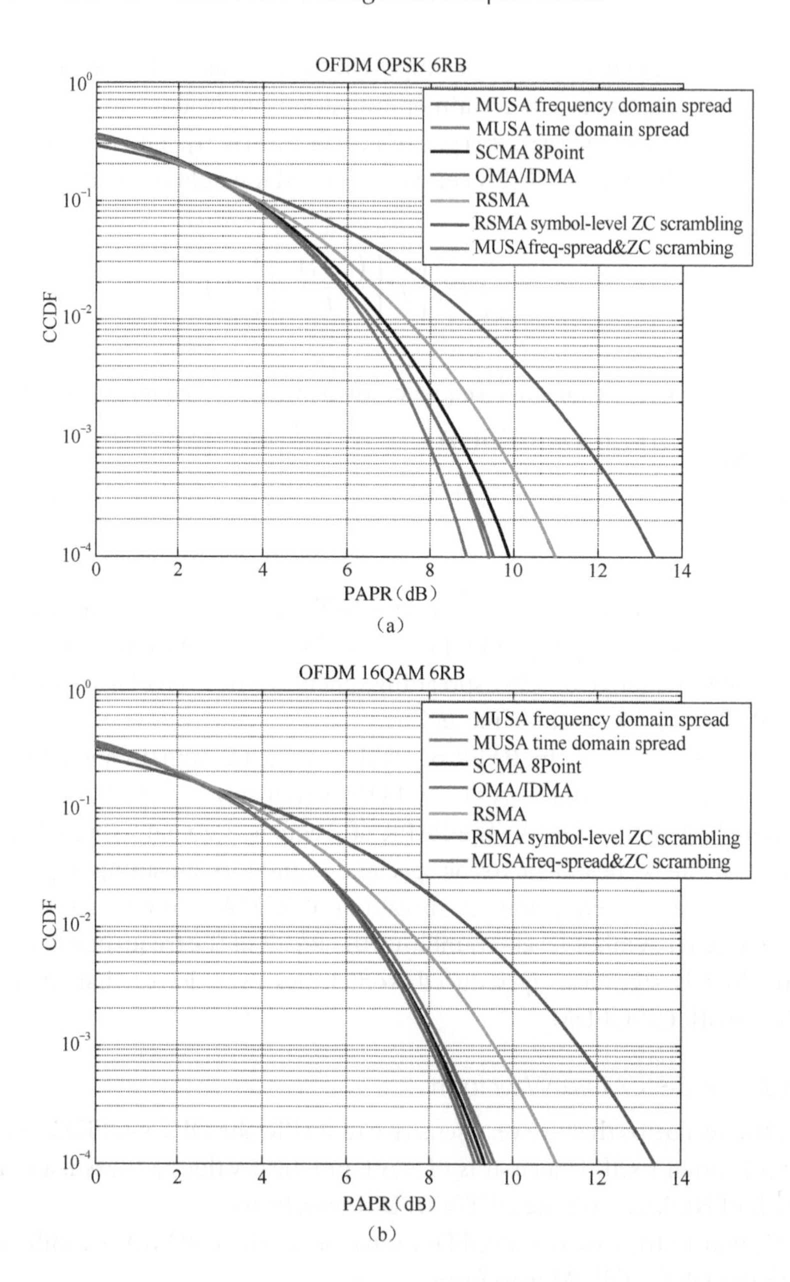

FIGURE 7.35 Distributions of PAPR for different schemes for the CP-OFDM waveform.

domain, frequency domain or transform domain which can have different impacts on the PAPR. For transform domain and time-domain spreading, the transmitter side block diagram is shown in Figure 7.36 where the size of DFT is the same as the number of occupied subcarriers. The transmitter side block diagram for the frequency domain spreading is shown in Figure 7.37. The size of DFT is the number of occupied subcarriers divided by the spreading factor, as shown in Figure 7.38.

- If the spreading is in the time domain, this is equivalent to weighting and repetition of the DFT-s-OFDM symbol and would not increase the PAPR.
- If the spreading is in the frequency domain, certain correlation is introduced which would increase PAPR. Symbol-level spreading can be used to randomize the signals in adjacent subcarriers, therefore reducing the PAPR. CAZAC-chu sequence is proposed

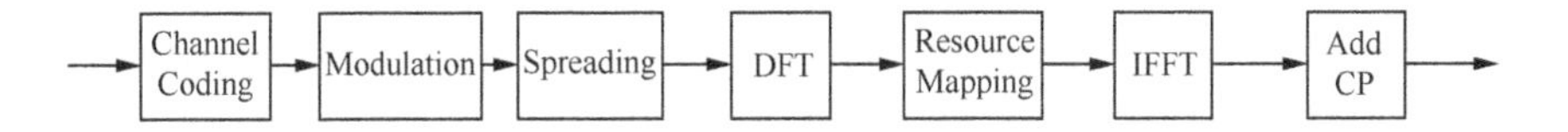

FIGURE 7.36 Transmitter side block diagram of transform-domain and time-domain spreading for the DFT-s-OFDM waveform.

FIGURE 7.37 Transmitter side block diagram of frequency domain spreading for the DFT-s-OFDM waveform.

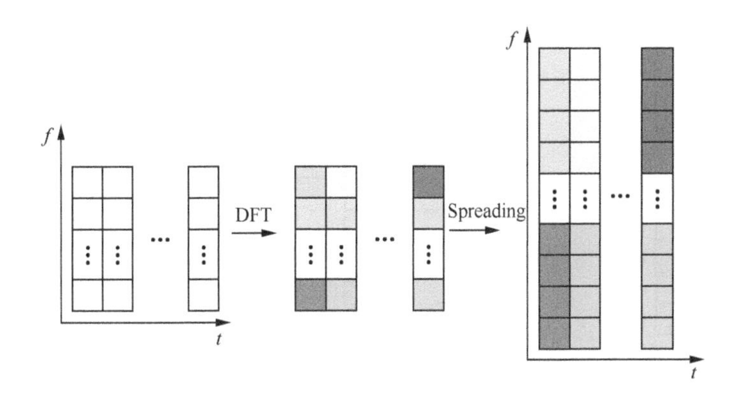

FIGURE 7.38 An illustration of signal generation of frequency domain spreading for the DFT-s-OFDM waveform.

for symbol-level spreading [6] which would reduce the PAPR to a level even below that of OMA (using DFT-s-OFDM).

- If the spreading is in the transform domain, as shown in Figure 7.39, the PAPR is slightly higher than OMA. The PAPR can be further reduced with symbol-level scrambling. However, at the receiver, it is not possible to perform the joint spreading domain and spatial-domain MMSE, which would affect the performance to some extent.

- For SCMA and IGMA, the PAPR is slightly higher than that of OMA. In [7], a 4-point constellation is proposed whose basic principle is similar to π/2 -BPSK, that is, the phase transition limited to π/2, leading to smoother amplitude variation and reduced PAPR.

Figure 7.40 shows PAPR of several NOMA schemes in the case of the DFT-s-OFDM waveform. MUSA is shown as a representative to symbol-level linear spreading schemes since the PAPR of linear spreading family is very similar. LCRS does not introduce any new processing and its PAPR is the same as OMA (using DFT-s-OFDM).

It is observed that IDMA and time-domain spreading have the similar PAPR as OMA (using DFT-s-OFDM waveform). The PAPR of SCMA is higher than OMA. Frequency domain spreading has the highest PAPR. After the symbol-level ZC sequence scrambling, the PAPR of frequency domain spreading is significantly reduced but still higher than that of OMA. The PAPR of transform domain spreading is slightly higher

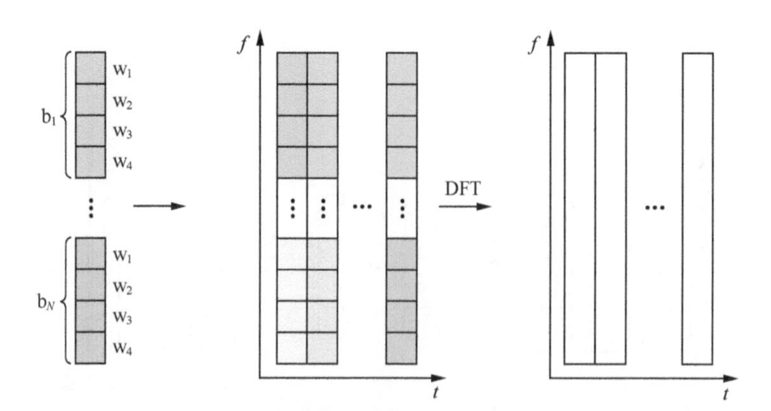

FIGURE 7.39 An illustration of signal generation of transform domain spreading for the DFT-s-OFDM waveform.

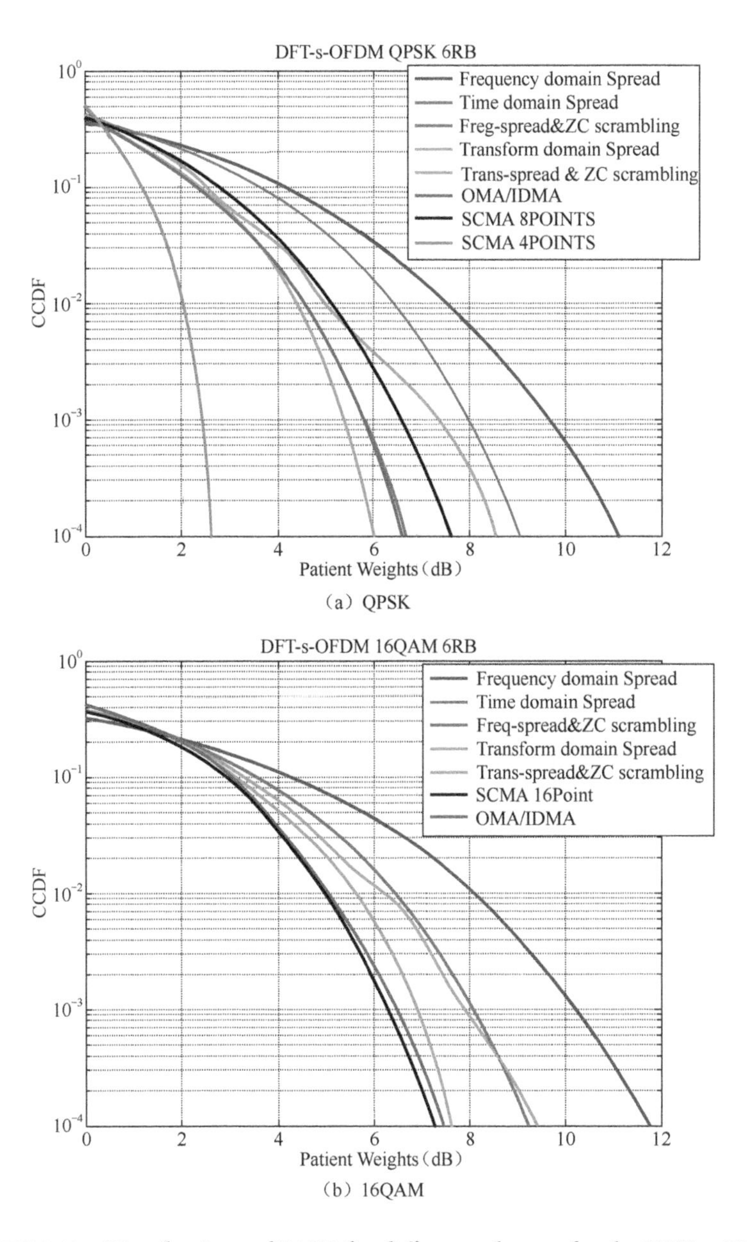

FIGURE 7.40 Distributions of PAPR for different schemes for the DFT-s-OFDM waveform.

than that of OMA. It can be further reduced by symbol-level CAZAC sequence scrambling.

In the simulation of PAPR for SCMA, the constellation of the 4-point spreading sequences is listed in Table 7.5.

TABLE 7.5 Four-Point Spreading Sequences of SCMA

Bit Values	**00**	**01**	**10**	**11**
Corresponding symbol sequence	$\begin{bmatrix} 1 \\ j \\ -1 \\ -j \end{bmatrix}$	$\begin{bmatrix} 1 \\ -j \\ -1 \\ j \end{bmatrix}$	$\begin{bmatrix} -1 \\ j \\ 1 \\ -j \end{bmatrix}$	$\begin{bmatrix} -1 \\ -j \\ 1 \\ j \end{bmatrix}$

REFERENCES

1. 3GPP, TR 38.802, Study on new radio access technology, Physical layer aspects.
2. B. Dai, Y. Yuan, and Y. Yu, *Narrow-Band (NB-IoT) Standards and Key Technologies*. People's Telecom Press, Beijing, China, 2016.
3. 3GPP, TR 38.812, Study on non-orthogonal multiple access (NOMA) for NR.
4. 3GPP, TR 38.901, Channel model for 5G NR.
5. 3GPP, RP-182630, NOMA link and system level performance, ZTE, RAN#82, December 2018, Sorrento, Italy.
6. 3GPP, R1-1811243, Transmitter side signal processing schemes for NOMA, Qualcomm, RAN1#95, November 2018, Spokane, USA.
7. 3GPP, R1-1812187, Discussion on the design of NOMA transmitter, Huawei, RAN#95, November 2018, Spokane, USA.

CHAPTER 8

System Design and Performance Evaluation of Contention-based Grant-free NOMA Transmissions

Nan Zhang, Wei Cao, Zhifeng Yuan, Jianqiang Dai, Ziyang Li, Hong Tang, Weimin Li, Jian Li and Yihua Ma

8.1 PROCEDURE OF CONTENTION-BASED GRANT-FREE ACCESS

To support contention-based grant-free access, the interaction between the base station and the user terminal is illustrated in Figure 8.1, with each step elaborated.

- Step 1: configuration set announcement for uplink transmission. This information is to be used to indicate the time/frequency resources, non-orthogonal signature pool, reference signal configuration, etc. for potential uplink transmission. In contention-based access, this information cannot be sent in the way for grant-based transmission,

DOI: 10.1201/9781003336167-8

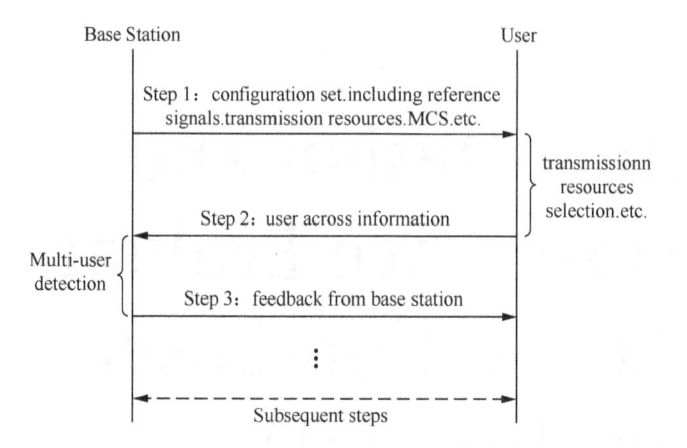

FIGURE 8.1 Procedure of contention-based grant-free access.

e.g., unicast is not possible. Instead, the broadcast mode is normally used, for instance, via system information to be announced to all the users within the coverage.

- Step 2: When the user receives the configuration set, it would autonomously select the resources, non-orthogonal signature, etc. Either the "preamble+data" structure to be discussed in Section 8.3 or the "data-only" structure to be discussed in Section 8.4 can be used. In the case of the "preamble+data" structure, the data part can be transmitted in a non-orthogonal manner as discussed in Section 8.6, in order to improve the system capacity. If a "data-only" structure is used, normally only the short spreading-based non-orthogonal multiple access (NOMA) schemes can be used because their blind detection is less complex. Some control information can be embedded in the payload, for instance, the user ID, in addition to the application layer data.
- Step 3: once the base station receives the uplink signals from multiple users, it can perform multi-user detection and user identification based on the reference signals and data. Then based on the outcome of the detection and whether there is a requirement for scheduling, the base station would send the corresponding feedback to the users:
    - When multiple users share resources and there are detection failures, the base station can send common feedback information to users. This resembles the common downlink control information (DCI) scrambled by RA-RNTI.

  - When a specific user is successfully detected and there is subsequent grant-based transmission needed, the base station can feedback the user-specific information, like DCI scrambled by C-RNTI.
- Subsequent steps: when the user receives the feedback from the base station via Step 3, it would interact with the base station according to the information contained in the feedback, for instance, to repeat the transmission, or transmit based on the scheduling grant.

In the meantime, to avoid too many users concentrating on a certain part of the resources, the base station can indicate some additional information, for example, the statistics usage of reference signals, so that users can more efficiently utilize the idle resources.

8.2 PREAMBLE + DATA CHANNEL STRUCTURE

8.2.1 Candidate Channel Structure

To support contention-based grant-free access, the behavior of the users would resemble the traditional random access procedure. The base station would broadcast the system information to all the users about the time and frequency resources for grant-free access. Each user autonomously selects a non-orthogonal signature and generates the signal to be transmitted. Intuitively, the channel structure of contention-based grant-free access would be a preamble for random access, followed by a data block. The preamble part itself is contention-based and primarily contains some forms of signature. In the data block, non-orthogonal signatures can be used, which has a certain relationship with the signature in the preamble.

In Figure 8.2, five candidate channel structures are shown with different shading corresponding to different constituent parts of each channel structure. These structures are designed by various combinations of the ingredient blocks in the channel structures already supported by 5G New Radio (NR) standards. More variants may be possible when NOMA is to be used for practical systems.

- The first candidate, "preamble + data" is the most straightforward. The preamble and the data have their independent time-frequency resources. The preamble is at the beginning of the structure. Hence in the receiver, the processing can be in two steps. First, the receiver tries to find the user via preamble detection. Then, the receiver carries out the detection in the data part.

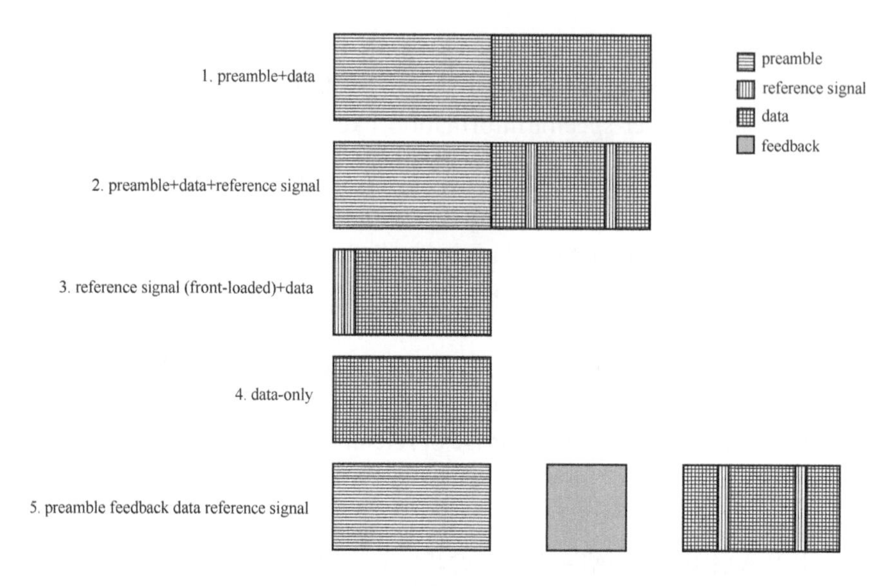

FIGURE 8.2 Candidate channel structures for (preamble + data).

- The second candidate, "preamble + data + reference signal", is featured as the reference signal that can be used for channel estimation for the equalization in the data part. Certainly, the reference signal can carry the NOMA signature or part of the signature.
- The third candidate, "reference signal (front-loaded) + data", can be considered as one special case of "preamble + data". DMRS like in long-term evolution (LTE) PUSCH can be front-loaded so that the receiver can first conduct the user detection and then the data decoding. Figure 8.3 shows an example where the legacy position of DMRS in LTE PUSCH is illustrated in Figure 8.3a. By moving the DMRS to the front of the subframe, we can get the front-loaded DMRS as seen in Figure 8.3b and 8.3c.
- In the fourth candidate "data-only", there is no preamble, nor DMRS. Only the NOMA signature will be used to generate the signal. On the receiver side, a blind receiver would be used to decode the data. For the successfully received data, the NOMA signature, as well as the user indicator embedded in the data payload, can be used as the outcome of the detection.
- The fifth candidate, "preamble + feedback + data + reference signal", is quite similar to the legacy random access procedure. A user randomly selects a preamble for accessing, the base station will send the feedback based on the outcome of the preamble detection.

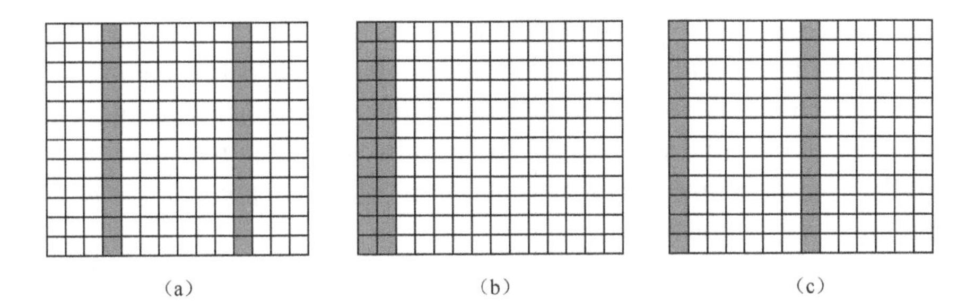

FIGURE 8.3 Three possible locations of DMRS in the time domain. (a) Legacy DMRS locations, (b) Localized front-loaded DMRS locations and (c) Distributed front-loaded DMRS locations.

When the user has received the feedback, it will generate the data based on the NOMA signature, embed the reference signal to the data, and then send it to the base station. Compared to the legacy random access, the contention resolution may be finished earlier. Alternatively, the contention resolution can be solved during the data detection.

8.2.2 Function Description

A user sends a message via contention-based grant-free access, the base station would receive this message with two goals: (1) to know which user is sending this message and (2) to know the information bits in the payload. Hence, for contention-based access, the most basic functions are user detection and data detection.

8.2.2.1 User Detection

There are two cases of contention-based grant-free access: pre-configuration-based, and pure random access.

In the pre-configuration-based access, users are already registered to the base station. The base station would pre-configure the users according to their traffic characteristics. This is suitable for periodically transmitted messages with similar packet sizes, for instance, VoIP traffic. To efficiently utilize the resources, the base station can allocate multiple users in the same time-frequency resources. For each of the pre-configured resources, the user may mute or transmit data, depending on the buffer status. Hence, the user detection carried out at the base station is essentially the detection of active users. More specifically, a list of all the potential active users is known prior. User detection is just to find out which users are actually transmitting. Both preamble signatures and reference signals can be used for active user detection. For instance, when preamble signatures/

resources are one-to-one mapped to the users, the list of active users can be obtained by preamble detection. It is seen that the scenarios suitable for such pre-configuration are rather limited and require maintaining a big table for all potential active users. There is also some requirement for user traffic. Hence, for massive internet of things (IoT), such pre-configuration may not be very practical.

In pure random access, there is no registration beforehand to the base station. The base station may need to configure the appropriate size of the resource pool, according to the characteristics of user traffic. This situation reflects a more general operation. However, it raises the bar for user detection. First of all, the user identification should be carried in the channel structure and can be a preamble signature, reference signal or data payload. If the resources of the preamble are large enough, a one-to-one mapping may be feasible between the preamble signatures and the users. However, this assumption may be too idealistic because the resources of the preamble are limited, however, a legacy identification of a user would require about dozens of bits, which can hardly be used. Similarly, a reference signal is even more limited in time and frequency resources and can hardly be mapped to user identifications. One possible way is to let the preamble and reference signal carry part of the user's identification. The complete user identification will be carried jointly by the data part as well as the preamble/reference signal. In another word, user identification would not be completed until the data payload part is decoded successfully.

8.2.2.2 Data Detection

Data detection is directly related to demodulation and decoding of the data part. Considering that preamble and reference signal may carry part of user identification information, preamble and reference detection may be part of data detection. The functions of data detection may not be limited to channel estimation, timing/frequency offset estimation, etc.

In general, both preamble and reference signal can be used for channel estimation. Depending on time and frequency locations between preamble/reference signal and data part, a receiver can use them selectively. If data only is used for the transmission, the base station can perform blind detection for the data itself. That is, the channel estimation should rely on the traffic channel.

Timing/frequency offset estimation can also be carried out by using a preamble and/or reference signal. Among them, the estimated timing offset can be used for scheduling-based multi-user access. Similarly, in the case of the data-only channel structure, the base station can literally use all the data symbols as the reference signal to estimate the frequency offset for later use.

8.2.3 Basic Design Aspects

Without losing the generality, preambles are used in this section to explain the fundamental issues with the channels in the case of contention-based grant-free access.

8.2.3.1 Time and Frequency Resource Allocation

The time/frequency resource relationship is part of the basic design for channel structure. The relative positions of time-frequency resources between preamble, reference signals and payload data determine the functionalities of each part. Normally, reference signal and payload data are time-division multiplexed (TDM), occupying the same frequency portion, so that the reference signal can be used for the channel estimation of the data payload. A preamble is primarily for user detection and its time/frequency locations with respect to the data payload can have more freedom, as illustrated in Figure 8.4.

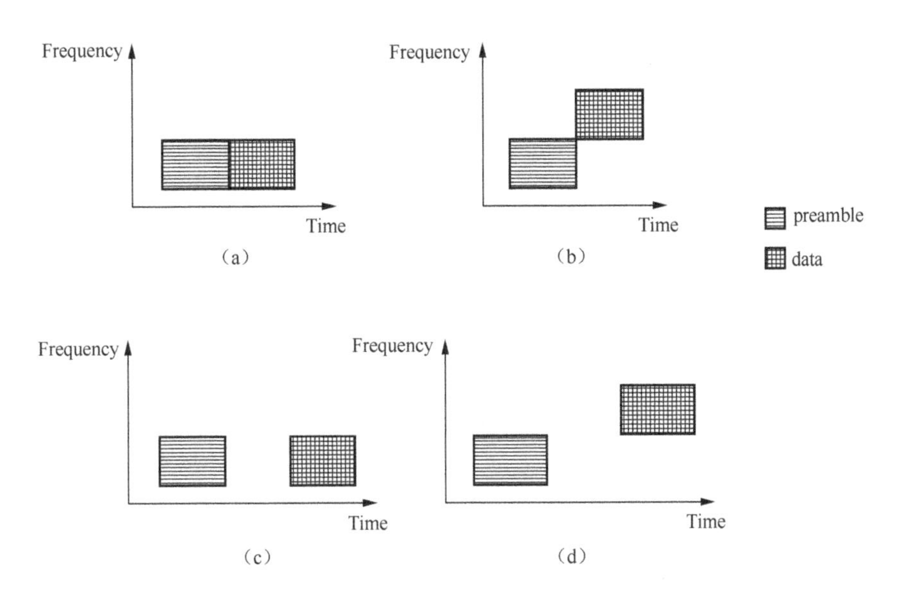

FIGURE 8.4 Time and frequency domain relationship between preamble and data payload.

a. The preamble and data payload are placed side by side in the time domain without a gap. They occupy the same frequency of resources. If the channel response is considered coherent within this amount of time, the channel estimate via the preamble can be used to demodulate the data payload.

b. There is no overlap in frequency between the resources for the preamble and the resources for the data payload. If the channel coherence holds within this amount of time, channel estimate via preamble can be used for demodulation of the data payload. With this basic relation, multiple data payloads can be mapped to a single frequency block in the frequency resources of the preamble, thus achieving a certain mixture of orthogonal transmission and non-orthogonal transmissions for the data payload.

c. The preamble and data payload are TDM but not contiguous in time. They occupy the same frequency of resources. If the channel response is still considered constant within this amount of time, the channel estimate via the preamble can be used for the demodulation of the data payload. The time gap between the preamble and the data payload would reduce the burden of the base station receiver to process the data in time and also provide certain flexibility in resource configuration of the preamble and data.

d. The time and frequency resources of the preamble and data payload are neither overlapped nor contiguous. If the channel response is still considered coherent within this time and frequency, channel estimate via preamble can be used for the demodulation of the data payload. This layout represents the most flexible resource configuration of preamble and data.

8.2.3.2 Sequences

As mentioned before, since the reference signal is often tightly coupled with the data payload, its sequence length is quite limited by the frequency domain resources of the payload, as well as the number of non-orthogonally multiplexed users. By contrast, the preamble, as a rather separate part, can have quite different subcarrier spacing compared to the subcarrier spacing for the data payload. Normally, the subcarrier spacing of the preamble is narrower, resulting in a longer symbol duration of the preamble than of the data payload. Such a setting has the following advantages:

a. A longer symbol duration means a longer cyclic prefix (when the overhead of the cyclic prefix is unchanged) and thus more robust to the timing offset in the case of asynchronous transmission.

b. Very long sequences can be used for the preamble, allowing a very large pool of preambles for users to choose from. This can effectively reduce the collision probability when users randomly select preambles.

c. Preambles have narrower subcarrier spacing and span over many subcarriers. This ensures very accurate timing offset estimation over a wide range. If preambles take certain responsibility in channel estimation, it would be very beneficial to the detection/decoding of the subsequent data payload.

Considering the widely used Zadoff-Chu (ZC) sequence for LTE and its variants (for instance, PRACH, SRS and DMRS in MU-MIMO) which have a very good property for detection, a ZC sequence may also be used for channel estimation over its occupied bandwidth, and thus can be a candidate. Similar to PRACH, the preamble can be defined by its root and cyclic shift. One key design parameter is the resource pool size. The length of ZC sequence and the spacing of adjacent cyclic shifts determine the number of opportunities for access per root. If the number of opportunities per root is not enough, multiple roots would be needed. Considering the mathematic properties of the ZC sequence: perfect auto-correlation (between different cyclic shifts of the same root) and constant modulo of cross-correlation (between different roots), it is beneficial to the detection at the receiver, if as many as possible opportunities of the same root can be provided.

The base station notifies the users in its cell about the resource pool of the preambles, via a broadcast channel. Each user autonomously chooses a preamble to generate the signal. If the preamble can carry part of the user identification, the user would select the preamble based on its identification. If there is a certain mapping between the preamble and the time/frequency resources of the subsequent reference signal or data payload, once the preamble is chosen by a user, the corresponding information about the resources of the reference signal and data payload would be determined as well.

Another important design aspect of the preamble is the overhead of time/frequency resources. Using a longer ZC sequence is helpful in increasing the pool size of the preamble. When the resource overhead is given, a longer ZC sequence means that the subcarrier spacing has to be narrower. Similarly, given a specific subcarrier spacing, a longer ZC sequence means more subcarriers and more frequency domain resource overhead. In addition, a longer ZC sequence and more opportunities mean more complicated processing for cross-correlation detection. All these factors should be weighed in when designing the channel structure for contention-based grant-free access.

8.3 DATA-ONLY SOLUTION

The above-mentioned "preamble+data" solution can simplify the preamble-based blind detection. However, such a solution faces a challenging issue for contention-based grant-free access:

In contention-based access, all the resources for transmission of a user, including the preamble, are autonomously chosen by each user. It is essentially contention oriented. Hence, it is inevitable that different users choose the same preamble, causing collision. With more users simultaneously accessing a system, the collision probability increases quickly and would throttle the performance of multi-user detection once the collision probability reaches a certain level. When collisions occur, it would not only cause miss detection, but also severe degradation of channel estimation. These two effects would significantly impact the performance of multi-user detection.

For a given number of simultaneously transmitting users, the only way to reduce the collision probability is to increase the number of preambles. However, such practice can cause two problems:

a. Increasing the number of preambles usually means a longer preamble sequence, therefore more time/frequency resources for the preamble. This would invade the resources for data payload, and degrade the performance of data payload reception. For small packet transmission, a long preamble may require more time/frequency resources than for the data payload itself.

b. The increased detection complexity is associated with a longer preamble.

Hence, in practical systems, the number of preambles normally would not be very large. Under this constraint, when the "preamble+data" solution is used for contention-based grant-free access, the number of simultaneously accessing users or the performance of multi-user detection is limited by the collision of the preamble.

In order to avoid the issue of preamble collision, the data-only solution can be considered where the transmitted signal is solely made up of data symbols (not spread, or spread) without a reference signal. The receiver would perform fully blind multi-user detection based on the data symbols' structure and the geometric characteristics of the constellation of low-order modulations such as binary phase-shift keying (BPSK).

In order to improve the performance of multi-user detection in contention-based grant-free access, short spreading sequences discussed in Section 6.1 can be used in a data-only solution. By doing so, the receiver can exploit the structure of the spreading sequence, together with the geometric characteristics of the constellations of low-order modulations. For the high-spectral-efficiency scenario, the data-only solution can be used alone without symbol-level spreading.

8.3.1 Channel Structure

The data-only solution is illustrated as the fourth structure in Figure 8.2. Furthermore, symbol-level linear spreading of short length, as shown in Figure 6.1, is preferable in order to improve the performance of contention-based grant-free access. Discussion of codebook design for linear spreading can be found in Section 6.1. In this section, we focus on a data-only solution based on the multi-user shared access (MUSA) sequence, especially on the design of advanced receivers for data only.

First of all, the main characteristic of short-length spreading for data only is that the L symbols after length-L spreading are placed into contiguous resource elements. For instance, if the user u is to be spread with the sequence $\mathbf{c}_u = \left[c_{u,1}, c_{u,2}, \ldots, c_{u,L}\right]^T$ and the j-th modulation symbol is $s_{u,j}$, after spreading we get L symbols: $\mathbf{c}_u \cdot s_{u,j} = \left[c_{u,1}s_{u,j}, c_{u,2}s_{u,j}, \ldots, c_{u,L}s_{u,j}\right]^T$, placed side by side in time or frequency. For example, they can be contiguous in time, in order to fully exploit the channel coherence in time to simplify blind detection.

Secondly, there is a requirement for the transmitter: in addition to the user ID information, the data should include information about the spreading sequence. Such a measure is to handle the situation where, for instance, a user's data is successfully decoded with the hypothesis that spreading sequence A is used, however, in fact the data was spread with sequence B. To ensure the correctness of the reconstructed signal (e.g., to be spread with sequence B), this information is needed. Certainly, the information about the spreading code can be indicated by the information bits, to avoid the extra overhead. One typical way is to use UE_ID to indicate the spreading code.

8.3.2 Receiver Algorithm

In the data-only solution, the processing at the transmitter is relatively simple, and the collision/overhead of the reference signal/preamble can be avoided, which is quite attractive. However, the burden mainly rests on the receiver where full blind detection is needed, which can be very challenging.

Without the channel estimation, the advanced receiver for data-only can fully rely on the discrimination between different users from the spatial domain, code domain, power domain and constellation domain to improve the performance. The receiver also needs to reduce the complexity of blind detection by utilizing the characteristic of the data, for instance, to conduct the blind detection according to the structure of the spread signal and to take advantage of the geometric property of low-order modulation for the purpose of blind estimation of the channel. In addition, due to the contention-based access, there exist non-controllable collision and interference between users and the near-far effect. Considering these factors, a successive interference cancellation (SIC)-based receiver seems more suitable. Hence, the data-only receiver discussed in this section combines the advanced technology of blind detection and blind estimation and SIC-based multi-user detection.

Before describing the algorithm, let us clarify the setup and streamline the notations. Assuming that each user has only one transmit antenna and the base station has R receiver antennas. The modulation order of all users' data is the same. After modulation, there are J modulation symbols. The modulation symbols of user u can be expressed as $\left\{s_{u,j}\right\}, j=1,\ldots,J$. After the spreading, the j-th modulation symbol of user u becomes L symbols $\mathbf{c}_u \cdot s_{u,j}=\left[c_{u,1}s_{u,j}, c_{u,2}s_{u,j},\ldots,c_{u,L}s_{u,j}\right]^T$ which are contiguous in time and/or frequency. At the r-th antenna of the base station, the channel vector is $\left[g_{r,u,j,1}, g_{r,u,j,2},\ldots,g_{r,u,j,L}\right]^T$. Then, the L spread symbols corresponding to the j-th modulation symbol of User u on the r-th receiver antenna can be represented as

$$\mathbf{y}_{r,u,j}=\left[g_{r,u,j,1}, g_{r,u,j,2},\ldots,g_{r,u,j,L}\right]^T \odot \left[c_{u,1}s_{u,j}, c_{u,2}s_{u,j},\ldots,c_{u,L}s_{u,j}\right]^T \quad (8.1)$$

The notation $\odot$ here denotes the element-by-element multiplication between two vectors or two matrices.

Since L symbols are placed contiguously, for instance, in the same subcarrier, and L is relatively small, the channel coefficients are strongly cor related and can be considered as the same, e.g., $g_{r,u,j,1} \approx g_{r,u,j,2}, \ldots \approx g_{r,u,j,L}$. Using the average value $g_{r,u,j} = \frac{1}{L}\sum_{l=1}^{L} g_{r,u,j,l}$, Eq. (8.1) can be simplified as

$$\mathbf{y}_{r,u,j} = g_{r,u,j}\left[c_{u,1}s_{u,j}, c_{u,2}s_{u,j}, \ldots, c_{u,L}s_{u,j}\right]^T = \left[c_{u,1}, c_{u,2}, \ldots, c_{u,L}\right]^T g_{r,u,j}s_{u,j} \quad (8.2)$$

This equation can be written in the following two forms:

$$\mathbf{y}_{r,u,j} = \mathbf{h}_{r,u,j}s_{u,j} \quad (8.3)$$

or

$$\mathbf{y}_{r,u,j} = \mathbf{c}_u s_{r,u,j}' \quad (8.4)$$

where $\mathbf{h}_{r,u,j} = \left[c_{u,1}, c_{u,2}, \ldots, c_{u,L}\right]^T g_{r,u,j} = \mathbf{c}_u g_{r,u,j}$ in Eq. (8.3) is the effective channel from the modulation symbol $s_{u,j}$ to the r-th receiver antenna, including the spreading operation. In Eq. (8.4), $s_{r,u,j}' = g_{r,u,j}s_{u,j}$ is the modulation symbol weighted by the channel. $\mathbf{c}_u s_{r,u,j}'$ can be seen as the spreading operation for the weighted modulation symbol $s_{r,u,j}'$.

Assuming that U users are accessing, the j-th modulation symbol of each user is spread into L symbols. The received signal on the r-th antenna is $\mathbf{y}_{r,j} = \sum_{u=1}^{U} \mathbf{y}_{r,u,j} + \mathbf{n}_r$, which can be written into the following two forms, respectively:

$$\mathbf{y}_{r,j} = \sum_{u=1}^{U} \mathbf{h}_{r,u,j}s_{u,j} + \mathbf{n}_r \quad (8.5)$$

or

$$\mathbf{y}_{r,j} = \sum_{u=1}^{U} \mathbf{c}_u s_{r,u,j}' + \mathbf{n}_r \quad (8.6)$$

In fact, Eq. (8.5) is from the traditional perspective. It essentially reveals that as long as we know the effective channel $\mathbf{h}_{r,u,j}$ of each user, the modulation symbol of each user $s_{u,j}$ can be detected by the multi-user detection algorithm, and then decoded. In another word, the objective of traditional multi-user detection is the modulation symbol $s_{u,j}$ itself. However, in contention-based grant-free, it is difficult to know the accurate value of $\mathbf{h}_{r,u,j}$. The traditional receiver relies on each user's reference signal for the activation detection, and then to estimate the channel weight vector $g_{r,u,j}$. Then via the mapping relationship between the reference signal and spreading sequence to determine the spreading sequence $\mathbf{c}_u$ been used, and then the effective channel $\mathbf{h}_{r,u,j} = \mathbf{c}_u g_{r,u,j}$. However, under the contention-based grant-free, the collision of reference signals would cause severe distortion of the effective channel, therefore the significant degradation of the data detection.

Equation (8.6) considers this issue from another perspective: blind multi-user detection. Here, the objective of the multi-user detection is the weighted modulation symbol $s_{r,u,j}'$. In this way, it appears that there is only a modulation symbol $s_{r,u,j}'$ as well as the spreading sequence in the received signal $\mathbf{y}_{r,j}$, e.g., no channel weighting is applied. Via this formulation, the receiver can first carry out the de-spreading to suppress multi-user interference. Then, to carry out the blind estimation of the channel based on the modulation symbol $s_{r,u,j}'$ with a higher SINR after the multi-user interference suppression. Then equalization and decoding are performed. This is the key idea of a data-only solution which can be combined with the SIC mechanism to achieve blind multi-user detection.

To facilitate the explanation, let us first consider the data only for a single receiver antenna and then later on for multiple receiver antennas. To simplify the representation, we start from a flat fading channel first, without the time and frequency offset.

For more complicated channels, a similar principle can be used to carry out a blind estimation of channels via modulation symbols $s_{r,u,j}'$ with a higher SINR after multi-user interference suppression, albeit requiring more complicated processing to estimate frequency selective fading and time-frequency offsets. It is noted that for massive infrequent small packets scenario, each transmission would not occupy a large amount of resources, which may relieve some of the difficulties of blind estimation of channels.

8.3.2.1 Blind Detection for the Data-Only Solution of Single Receiver Antennas

In the case of a single receiver antenna, e.g., $R = 1$, Eqs. (8.5) and (8.6) can be simplified as

$$\mathbf{y}_j = \sum_{u=1}^{U} \mathbf{h}_{u,j} s_{u,j} + \mathbf{n} \tag{8.7}$$

$$\mathbf{y}_j = \sum_{u=1}^{U} \mathbf{c}_u s_{u,j}' + \mathbf{n} \tag{8.8}$$

where $\mathbf{h}_{u,j} = \mathbf{c}_u g_{u,j}$, $s_{u,j}' = g_{u,j} s_{u,j}$. These two equations are essentially the simplified form of Eqs. (8.5) and (8.6) without the subscript r for antenna index. If the channel is frequency flat, $g_{u,j} \equiv g_u$. Then $\mathbf{h}_{u,j} \equiv \mathbf{h}_u$, $\mathbf{h}_u = \mathbf{c}_u g_u$, $s_{u,j}' = g_u s_{u,j}$.

In the following, a very simple blind detection is described assuming a single receiver antenna.

Let us consider a set $\mathbf{S}_N$ that is comprised of N spreading sequences $\{\mathbf{c}_k\}$, $k = 1, \ldots, N$. Each user is modulated with BPSK. The basic processing in the data-only blind detection is as follows (without using reference signal for channel estimation):

1. First to blindly de-spread, and try all N spreading sequences for de-spreading, then get N symbol streams:

$$\hat{s}_{k,j} = \mathbf{c}_k^H \mathbf{y}_j, k = 1, \ldots, N, j = 1, \ldots, J$$

2. To treat N symbol streams as the potential BPSK symbol streams, perform the blind estimation of the channel based on the geometric property of the BPSK constellation, and then do the equalization.

3. To calculate the SINR of N equalized BPSK symbol streams, pick the top F symbol streams with the highest SINRs and do the decoding. To avoid an excessive number of decoding, F should be set small, e.g., $F = 4$.

4. To reconstruct (including encoding, modulation and spreading) the users' signal from the successfully decoded bits. The reconstructed users' symbols can be used as a "reference signal" to improve the channel estimation.

5. To remove channel-weighted reconstructed symbols from the received signal and get a cleaner received signal, and repeat the process from Step 1.

It should be pointed out that in order to improve the performance of contention-based grant-free access, it is better to design as many as possible the spreading sequences $\mathbf{S}_N$ with low cross-correlations. Take the MUSA codebook in Table 6.5 as an example, there are $N = 64$ spreading sequences, which means that the de-spreading operation needs to loop through 64 sequences and carry out a blind estimation of the channel and equalization 64 times, which requires a lot of processing. After the equalization, SINR ordering would be performed and only the F out of

64 symbol streams with the highest SINRs would be input to the decoder. The rest of the 64-F streams would not be further processed, i.e., the de-spreading, blind estimation of channel and equalization are somewhat wasted, as illustrated in Figure 8.5.

Hence, the above single-antenna data-only blind receiver just shows the principle. However, it does not fully utilize the characteristic of the data to optimize multi-user detection. Its implementation is not simple either.

The above naïve form of data-only blind detection has the following major disadvantages:

1. De-spreading, blind estimation of channel and equalization have to be performed for N times. N is usually not small.
2. De-spreading implemented is of matched filter type which is not optimal for short spreading.

8.3.2.1.1 Blind Sequence Detection

To reduce the collision probability, the codebook size of the spreading sequences $\mathbf{S}_N$ is normally very big. However, a large number of spreading sequences would result in non-decodable de-spread symbol streams. Hence, the operation of the de-spreading and the subsequent blind estimation/equalization would be wasteful.

In order to reduce the complexity of blind detection, it is preferable to discard as early as possible those spreading sequences that are not like to

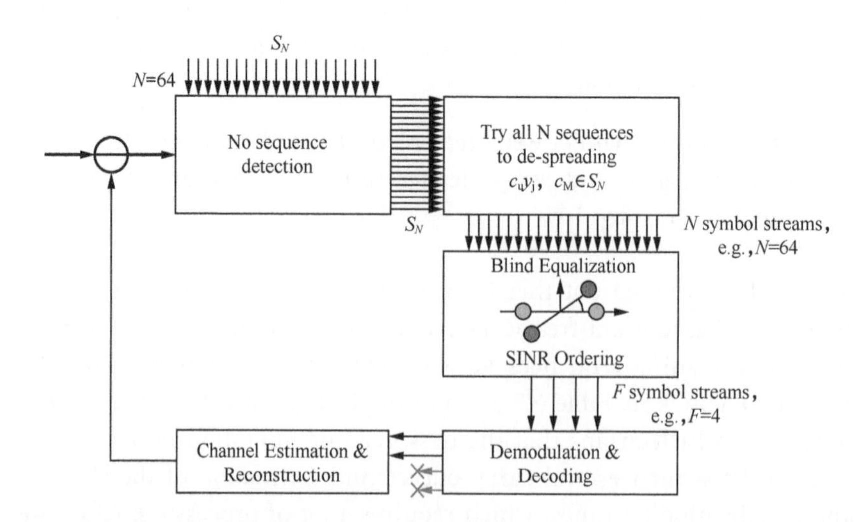

FIGURE 8.5 Naïve form of data-only blind detection for single Rx antenna.

produce decodable streams. We can rely on "detection of active sequence". That is, to pick M spreading sequences from $\mathbf{S}_N$ and form $\mathbf{S}_M$, usually, for each iteration, M is much smaller than N. Due to the lack of a reference signal to determine which user's spreading sequence, the characteristic of the received spread signal can be used for the detection of active sequences. Such reference signal-free activity detection is called blind detection of the sequences.

In fact, in the received spread signal $\mathbf{y}_j = \sum_{u=1}^{U} \mathbf{h}_{u,j} s_{u,j} + \mathbf{n}$, e.g., Eq. (8.7), there is an important statistic, that is, the covariance matrix $\mathbf{R}_y = \frac{1}{J}\sum_{j=1}^{J} \mathbf{y}_j \mathbf{y}_j^*$. For the flat channel, $\mathbf{h}_{u,j} \equiv \mathbf{h}_u$. Hence, Eq. (8.7) can be simplified as

$$\mathbf{y}_j = \sum_{u=1}^{U} \mathbf{h}_u s_{u,j} + \mathbf{n} \tag{8.9}$$

When J is large, the covariance matrix can be approximated as

$$\mathbf{R}_y = \frac{1}{J}\sum_{j=1}^{J} \mathbf{y}_j \mathbf{y}_j^* \approx \sigma^2 \mathbf{I} + \sum_{u=1}^{U} \mathbf{h}_u \mathbf{h}_u^* \tag{8.10}$$

By utilizing this covariance matrix, the receiver for data only can evaluate the SINR_k of each symbol stream corresponding to the spreading sequence $\mathbf{c}_k$, without the knowledge of channel coefficients.

$$A_k = \mathbf{c}_k^* \mathbf{R}_y^{-1} \mathbf{c}_k \tag{8.11}$$

When SINR_k is large, the metric A_k is usually small. Hence, the detection of the active sequence can be based on this metric, e.g., to select M spreading sequences from $\mathbf{S}_N$ whose A_k are the smallest, and form the set $\mathbf{S}_M$. This operation can be represented mathematically as

$$\mathbf{S}_M = \arg\min_{(M)} \mathbf{c}_k^* \mathbf{R}_y^{-1} \mathbf{c}_k, \mathbf{c}_k \in \mathbf{S}_N \tag{8.12}$$

where $\arg\min_{(M)}(\cdot)$ denotes selecting M elements whose values are the smallest in a full set.

It should be noted that there is no vigorous proof so far for the criterion above "when SINR_k is large, the metric A_k is usually small," although [1] provides a quasi-proof. Further data and simulations are provided in [1,2] which verify that quite high accurate estimate can be achieved in flat fading channels, and thus maintain a fairly good performance with the reduced complexity for blind detection. In fact, even with the presence of timing and frequency errors, as long as they are not very large, the covariance matrix $\mathbf{R}_y = \frac{1}{J}\sum_{j=1}^{J}\mathbf{y}_j\mathbf{y}_j^*$ would not change much. Hence, the impact of timing/frequency errors on the accuracy of blind detection based on Eq. (8.12) would not be significant.

8.3.2.1.2 Blind MMSE De-Spreading

After $\mathbf{S}_M$ is selected via blind detection of sequences, the subsequent de-spreading will be performed on these M sequences. The previously mentioned matched-filter-based de-spreading does not consider the cross-user interference, and the SINR of the de-spread symbol stream is not optimal. In contrast, MMSE-based de-spreading is the optimal linear operation which is to be discussed below.

Based on the formula of the received signal $\mathbf{y}_j = \sum_{u=1}^{U}\mathbf{h}_{u,j}s_{u,j} + \mathbf{n}$, e.g., Eq. (8.7), in the case of ideal channel estimation, the effective channel of each user $\mathbf{h}_{u,j}$ is known. Hence, it is straightforward to perform MMSE estimation for $s_{u,j}$, that is, $\tilde{s}_{u,j} = \mathbf{w}_{u,j}\mathbf{y}_j$, where $\mathbf{w}_{u,j} = \mathbf{h}_{u,j}^*\left(\sigma^2\mathbf{I} + \sum_{u=1}^{U}\mathbf{h}_{u,j}\mathbf{h}_{u,j}^*\right)^{-1}$ is the MMSE weight. In the context of spreading, the MMSE weight is also called the MMSE de-spreading vector. However, for the receiver of data-only, the effective channel $\mathbf{h}_{u,j}$ is not known, and there is no way to calculate the exact value of $\mathbf{w}_{u,j}$. If the channel is flat, the above MMSE estimation can be simplified as $\tilde{s}_{u,j} = \mathbf{w}_u\mathbf{y}_j$, that is, all the symbols would share the same MMSE de-spreading vector $\mathbf{w}_u = \mathbf{h}_u^*\left(\sigma^2\mathbf{I} + \sum_{u=1}^{U}\mathbf{h}_u\mathbf{h}_u^*\right)^{-1}$. Furthermore, from $\mathbf{h}_u = \mathbf{c}_u g_u$ and $\mathbf{R}_y = \frac{1}{J}\sum_{j=1}^{J}\mathbf{y}_j\mathbf{y}_j^* \approx \sigma^2\mathbf{I} + \sum_{u=1}^{U}\mathbf{h}_u\mathbf{h}_u^*$, e.g., Eq. (8.10), we can obtain $\mathbf{w}_u \approx g_u^*\mathbf{c}_u^*\mathbf{R}_y^{-1}$. Hence, $\tilde{s}_{u,j} = \mathbf{w}_u\mathbf{y}_j$ can be converted into:

$$\frac{\tilde{s}_{u,j}}{g_u^*} \approx \mathbf{c}_u^*\mathbf{R}_y^{-1}\mathbf{y}_j \tag{8.13}$$

Equation (8.13) is the blind MMSE estimate of the modulation symbol $s_{u,j}$. This estimate does not require the information about the channel while being able to get the MMSE estimate of a variant of $s_{u,j}$ which is $c\frac{s_{u,j}}{g_u^*}$.

Then $\frac{s_{u,j}}{g_u^*}$ can be used for blind estimation of the channel to get $\frac{1}{g_u^*}$. After that, this scalar weight is removed by equalization to get the actual MMSE estimate $\tilde{s}_{u,j}$.

Let us look at Eq. (8.13) from a different angle and compare the two MMSE-based optimization problems according to the basic definition of MMSE.

1. Classic way for MMSE weight calculation essentially boils down to the optimization for $\underset{\mathbf{x}}{\arg\min} E\left\|\mathbf{x}\mathbf{y}_j - s_{u,j}\right\|^2$. Obviously, the optimal solution of $\mathbf{x}$ is $\mathbf{x} = \mathbf{w}_u = \mathbf{h}_u^*\left(\sigma^2\mathbf{I} + \sum_{u=1}^{U} \mathbf{h}_u\mathbf{h}_u^*\right)^{-1}$.
2. The determination of blind MMSE weight is essentially the optimization problem of $\underset{\mathbf{x}}{\arg\min} E\left\|\mathbf{x}\mathbf{y}_j - \frac{s_{u,j}}{g_u^*}\right\|^2$. According to Eq. (8.10), that is, $\mathbf{x} \approx \mathbf{c}_u^*\mathbf{R}_y^{-1}$. Then we get the blind MMSE estimation as Eq. (8.13). Hence, the blind MMSE treats the channel-weighted modulation symbol $\frac{s_{u,j}}{g_u^*}$ (e.g., $\frac{1}{\left|g_u^*\right|^2} g_u s_{u,j}$) as the target for estimation, with the hope that the estimate $\frac{\tilde{s}_{u,j}}{g_u^*}$ is of the MMSE to $\frac{s_{u,j}}{g_u^*}$. A similar notion can also be found in Eq. (8.6). That is, to incorporate channel weight into the modulation symbol and form $g_u s_{u,j}$ to be treated as the target for estimation. By doing so, multi-user interference suppression can be carried out without the knowledge of the channel gain. Then, blind estimation of the channel is performed by using the channel-weighted modulation symbols.

8.3.2.1.3 Blind Estimation of Channel and Equalization

Via MMSE de-spreading operation $\mathbf{c}_u^*\mathbf{R}_y^{-1}\mathbf{y}_j$, we just obtain $\frac{1}{\left|g_u^*\right|^2} g_u s_{u,j}$ which is the symbol stream weighted by $\frac{1}{\left|g_u^*\right|^2} g_u$. This weight would cause the rotation of the modulation constellation. For signals with low-order modulations, the rotated constellation still exhibits strong geometrical characteristics which can be exploited for blind estimation of channel and equalization. After the rotation factor $\frac{1}{\left|g_u^*\right|^2} g_u$ is estimated, the rotated constellation can be counter-rotated to restore to the original position to accomplish the equalization. In the following, BPSK is used as an example to illustrate blind estimation. Figure 8.6 shows the rotated BPSK constellation after blind MMSE and the restored constellation.

Clustering-based algorithms can be used to differentiate the constellation points and then restore the constellation based on the information of the constellation points. However, clustering-based algorithms are quite complex. Take the k-means algorithm for instance; it requires multiple iterations to reach convergence which involve many computations of Euclidean distance. In fact, for constellations of low-order modulations, the "center mass" of each cluster of scattering points shows a quite regular property. The optimal boundary is normally a straight line between the clusters. With this property, a very simple Partition Matching algorithm [2] can be used to calculate the rotation factor of the BPSK constellation. Here $\left\{s_j\right\}, j=1,\ldots,J$ denotes J-rotated BPSK modulation symbols. ρ_s is the estimated rotation factor for J BPSK symbols.

Partition Matching

$$\rho_x = \sum_{j\in\left\{\Re(s_j)>0\right\}} s_j - \sum_{j\in\left\{\Re(s_j)\le 0\right\}} s_j$$

$$\rho_y = \sum_{j\in\left\{\Im(s_j)>0\right\}} s_j - \sum_{j\in\left\{\Im(s_j)\le 0\right\}} s_j$$

$$\rho_{y=x} = \sum_{j\in\left\{\Re(s_j)>\Im(s_j)\right\}} s_j - \sum_{j\in\left\{\Re(s_j)\le\Im(s_j)\right\}} s_j$$

$$\rho_{y=-x} = \sum_{j\in\left\{\Re(s_j)>-\Im(s_j)\right\}} s_j - \sum_{j\in\left\{\Re(s_j)\le-\Im(s_j)\right\}} s_j$$

$$\rho_s = \frac{1}{J}\arg\max_{\rho\in\{\rho_x,\rho_y,\rho_{y=x},\rho_{y=-x}\}}\left|\rho\right|^2$$

The above equations can be explained as follows:

Step 1: Initial partition. Assuming four initial partition lines: $x = 0, y = 0, y = x, y = -x$ which can be used to determine whether the in-phase component and the quadrature component are positive or negative, as well as the relative amplitude between the in-phase component and the quadrature component.

Step 2: To pick the optimal from the initial partitions. There is at least one partition that is the closest to the right partition and the closer the initial partition to the right one, the larger the amplitude of the "mass centers" from this initial partition, this can be based on the second norm of the "mass centers". Step 3: Equalization based on the "mass center".

From the above discussion, we can see that "partition matching" is quite effective, e.g., only summation operation is involved in order to estimate the rotation factor, which is much simpler than clustering-based algorithms. "Partition matching" can be directly extended to quadrature phase-shift keying (QPSK).

Note that for BPSK, the estimation target is $g_u s_{u,j}$ (here the amplitude $\frac{1}{|g_u^*|^2}$ is skipped since it can be easily normalized). There would be two possible outcomes with equal opportunities: g_u or $-g_u$ (e.g., g_u rotated by 180°). In the case of QPSK, there would be four possible outcomes, corresponding to g_u rotated by 0°, 90°, 180°, and 270°. This means that the "partition matching" algorithm would cause phase ambiguity problems. Certainly, clustering-based algorithms cannot completely solve this issue. Phase ambiguity cannot be solved solely by the modulation symbol alone. For BPSK, there are only two hypotheses: g_u and $-g_u$. We can carry out the equalization for both hypotheses and then try to decode both. Apparently, only the correctly equalized symbol stream can be successfully decoded. Extending this to QPSK, four hypotheses are needed for equalization and channel decoding, meaning that more processing is required to resolve the phase ambiguity. This is the price the blind detection-based receiver has to pay.

8.3.2.1.4 Channel Estimation Based on the Reconstructed Signal

The blindly equalized signal goes through demodulation and channel decoding. If the cyclic redundancy check (CRC) passes, the receiver can reconstruct the clean signal based on the correctly decoded bits, and then estimate the channel based on the reconstructed signal. At this time, the purpose of channel estimation is to generate the clean signal at the receiver to be extracted. This can improve the demodulation performance of the weaker users. Although the correlation between symbols of different users is low, in order to improve the accuracy of channel estimation, joint channel estimation based on the least square (LS) criterion between reconstructed signals of multiple users can be considered. The joint estimation for k users can be represented as

$$\mathbf{G}=[g_1,\ldots,g_k]^T=(\mathbf{S}*\mathbf{S})^{-1}\mathbf{S}*\mathbf{y} \tag{8.14}$$

where $\mathbf{S}$ is the symbol matrix of reconstructed signals of k users already successfully decoded, e.g., $\mathbf{S}=[\mathbf{s}_1,\ldots,\mathbf{s}_k],\mathbf{s}_k=[s_{k,1},\ldots,s_{k,J}]^T$.

When using the joint estimation of channels, as more users are successfully decoded in the course of SIC, the channel estimation of each user gets more accurate. As Figure 8.7 shows, with more users decoded successfully, the channel estimate of a user gets closer to its true value. In turn, the residual error of SIC becomes smaller and smaller, which is very beneficial to the weaker users' demodulation. Such estimation is crucial for improving the overload capability of grant-free access.

In summary, with the above-discussed blind sequence detection, blind MMSE de-spreading, blind equalization and joint channel estimation, the architecture of the advanced SIC for multi-user blind detection

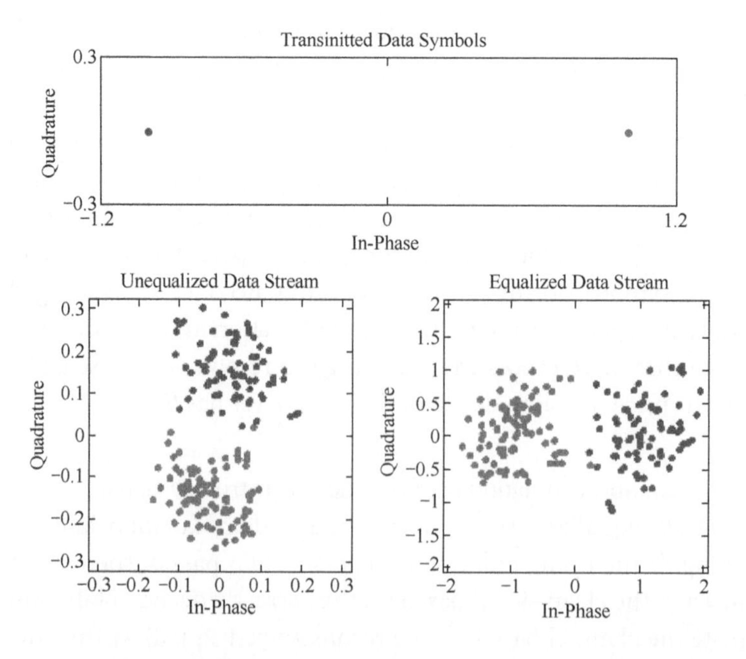

FIGURE 8.6 BPSK constellations at the transmitter, before and after equalization.

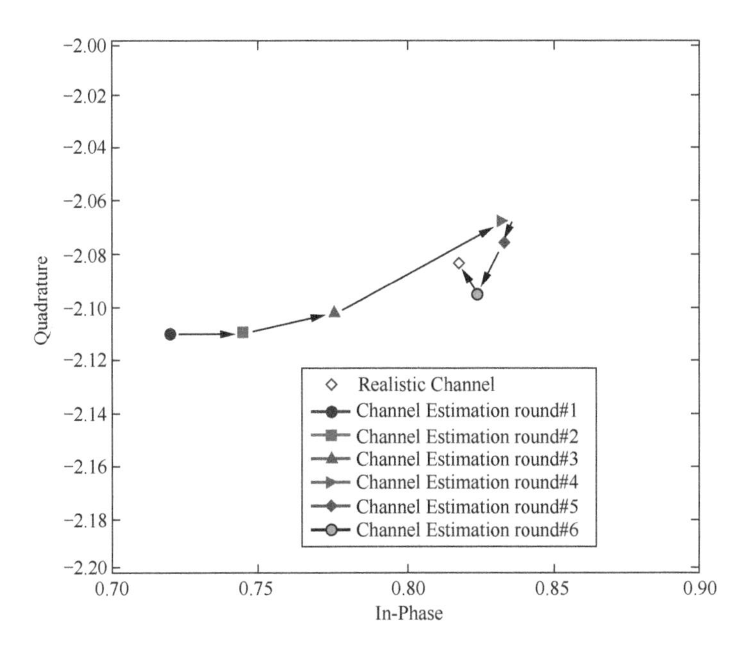

FIGURE 8.7 Channel estimate in each iteration of SIC and the comparison with the true channel.

is illustrated in Figure 8.8. Take the MUSA codebook $\mathbf{S}_N$ in Table 6.5 as an example, there are $N = 64$ spreading sequences with low cross-correlations. In either SIC iteration, to pick $M = 8$ sequences (to form $\mathbf{S}_M$) and perform MMSE de-spreading, blind channel estimation and equalization for these eight streams. Then calculate the SINR of these eight equalized modulation symbol streams and pick four of them with the highest SINRs for further demodulation and channel decoding. Due to the 180° phase ambiguity of blind channel estimation on the BPSK symbol, two hypotheses of each of the four symbol streams would be decoded "positively" and "negatively", respectively. Then the joint channel estimation is carried out based on the reconstructed signal. In the end, the channel-weighted reconstructed signal (clean received signal) is canceled.

Compared to the most basic (e.g., naïve) blind receiver for a single antenna illustrated in Figure 8.5, by implementing blind sequence detection, the receiver's complexity is significantly reduced. In addition, in contrast to the matched-filter-based de-spreading in Figure 8.5, MMSE-based de-spreading is used in Figure 8.8 which has better performance.

Simulations based on the advanced blind receiver in Figure 8.8 are carried out for a single receiver antenna and the performance is shown in Figure 8.9. The detailed simulation configuration is as follows.

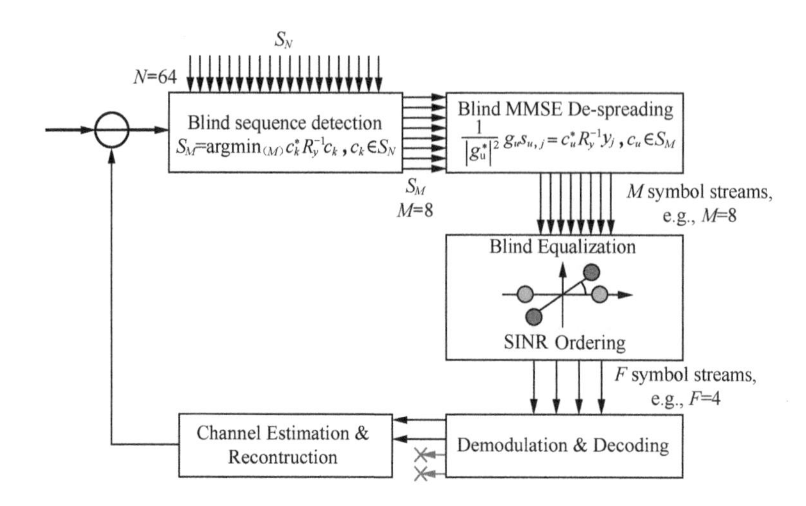

FIGURE 8.8 Advanced blind receiver for data-only, single receiver antenna.

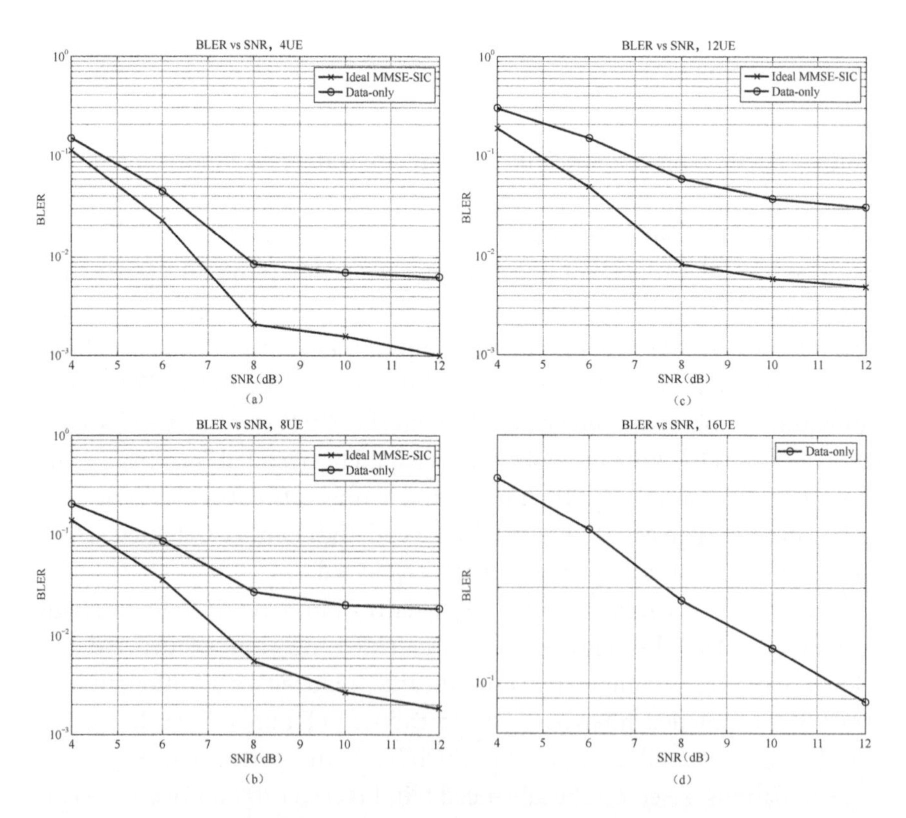

FIGURE 8.9 Link-level performance of advanced blind receiver for the data-only solution, single receiver antenna. (a) 4 UEs, (b) 8 UEs, (c) 12 UEs and (d) 12 UEs.

Each user has a single transmit antenna. The information bits go through the LTE Turbo encoder with rate = 1/2. The modulation is BPSK and the spreading length is 4. After the spreading, the symbol stream goes through CP-OFDM with 15 kHz subcarrier spacing. More specifically, the spreading symbols are mapped into a resource of 180 kHz in frequency and 4 ms in time. MUSA sequences (total of 64) as shown in Table 6.5 are used. Each user randomly selects one of the 64 sequences for spreading. Each BPSK modulation symbol is spread into four symbols and placed to four consecutive resource elements in the time domain. To mimic the near-far effect due to open-loop power control in contention-based grant-free access, the transmit power of each user is not the same, rather it follows a uniform distribution with the center SNR (corresponding to the values in the horizontal axis of link curves) as the mean value, ranging between [–8, 8] dB. A narrow band, e.g., 180 kHz, is simulated. The channel is flat Rayleigh fading.

From the simulation results shown in Figure 8.9, it is seen that even with a single receiver antenna under contention-based grant-free access, fairly high system loading can be achieved by the advanced blind receiver for a data-only solution using MUSA spreading sequences, without reference signal. For instance, at the operating point of SNR = 12 dB, e.g., each user's transmit SNR uniformly distributed over [4, 20] dB, 400% overloading, e.g., 16 users, can be supported with BLER = 10%. It is worth pointing out that even with 12 users (300% overloading), the performance between blind receivers for the data-only solution is not far from the MMSE-SIC receiver with ideal channel estimation. In fact, for reference signal-based channel estimation, when the number of users is increased to six or more, the potential collision of the reference signals due to contention-based access would severely degrade the channel estimation and result in a big performance gap to the ideal channel estimation case [4].

8.3.2.2 Blind Receiver for Data-Only Solution under Multiple Receiver Antennas

As mentioned earlier, the received symbol on the r-th antenna can be represented as either $Y_{r,j} = \sum_{u=1}^{U} h_{r,u,j} s_{u,j} + n_r$ (e.g., Eq. (8.5)) or $Y_{r,j} = \sum_{u=1}^{U} c_u s_{r,u,j} + n_r$ (e.g., Eq. (8.6)). For flat fading, $g_{r,u,j} \equiv g_{r,u}, \mathbf{h}_{r,u,j} \equiv \mathbf{h}_{r,u}$. Then $\mathbf{h}_{r,u} = \mathbf{c}_u g_{r,u}, s_{r,u,j}' = g_{r,u} s_{u,j}$.

In contention-based grant-free access, cross-user interference is quite severe, including the collision of reference signals (if used), as well as the signature collision in the data part. While the larger codebook size of non-orthogonal sequences can reduce the collision probabilities to some extent, collisions are still inevitable, especially in high overloading cases.

It is noted that the spatial channels $\mathbf{g}_u = [g_{1,u},\ldots,g_{R,u}]^T$ are independent between different users. Such independence would provide spatial discrimination to discern the data of multiple users. For example, if the spreading code of UE1 and UE2 collides, there is no way to differentiate these two users in the spreading code domain. However, their spatial channels $\mathbf{g}_1$, $\mathbf{g}_2$ are independent. The cross-correlation $\mathbf{g}_1^*\mathbf{g}_2$ has a high probability of low correlation. As the number of receiver antennas increases, the probability of low correlation grows. In the case of ideal channel estimation, the base station can perfectly know the spatial channels of the two users and form the appropriate spatial receiving weights $\mathbf{v}_1$, $\mathbf{v}_2$ via for instance MMSE criterion. If the spatial correlation between UE1 and UE2 is low, MMSE type of combining can quite well separate the two users' signals. In another word, with the ideal channel estimation, MMSE for multiple receiver antennas can provide both the diversity gain and interference rejection.

Multi-user spatial discrimination can also be explained by receiver beamforming. A spatial combining is essentially a receiver beam. If the spatial correlation is low, it means that the two users are in two beams well separated from each other, as illustrated in Figure 8.10a. However, in the realistic channel estimation, if the same preamble is chosen (collision), the receiver has no idea that there are two users trying to access, and may only estimate the combined channel which is $\mathbf{g}_1 + \mathbf{g}_2$. With such a combined channel, the receiver can only get a highly distorted beam, as shown in Figure 8.10b. Obviously, this inaccurate beam estimation would only result in the overlapped signals of users, e.g., would neither align the target user nor reject the interference of other users. Because of the preamble collision, the discrimination in the spatial domain with multiple receiver antennas loses significantly.

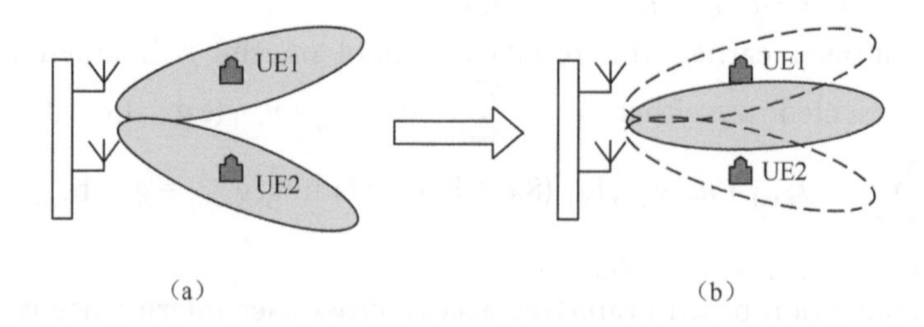

FIGURE 8.10 Beam separation capability under ideal channel estimation and RCE for contention-based grant-free access.

Such understanding has been verified by many simulations in the wireless industry. When the preamble-based traditional transceiver scheme is used for contention-based grant-free access, the consequence is: (1) the performance with realistic channel estimation is significantly worse than that of ideal channel estimation and (2) the performance with realistic channel estimation deteriorates rapidly as the number of users increases. In order to improve the performance in contention-based grant-free access, we would consider not using the traditional reference signal-based receiving algorithm as shown in Figure 8.11a where the channel is first estimated via reference signal, followed by inter-user interference suppression and channel equalization that are often carried out jointly. The collision of the reference signal would severely distort the channel estimation and thus affect the performance of multi-user detection. The traditional receiving algorithm can be reversed: first to carry out spatial and code domain multi-user interference suppression, followed by blind channel estimation and the subsequent equalization, demodulation and decoding for modulation symbol streams of higher SINR, as shown in Figure 8.11b. It can be considered as the extension of a blind receiver for a single receiver antenna to multiple receiver antennas.

Without the channel estimation via reference signal, the spatial domain multi-user interference suppression would be blind spatial combining or blind beamforming. If the receiver antennas have a certain special structure to make the channel exhibit certain characteristics, the blind spatial combining may follow a certain rule. However, for independent antennas, for instance, a low-frequency band of mMTC scenario, the spatial

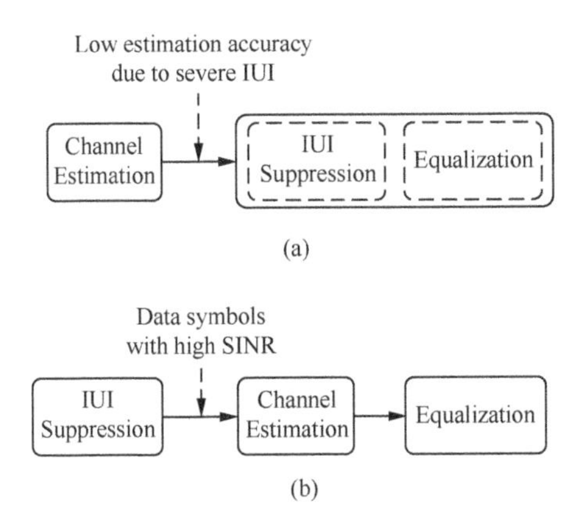

FIGURE 8.11 Comparison between the reference signal-based receiver and data-only-based blind receiver. (a) Traditional reference signal based receiving procedure and (b) data-only receiving procedure.

channels of different antennas are almost independent, requiring fully blind combining. Fortunately, the number of antennas is not large, e.g., the number of hypotheses of different combinations is limited. For two receiver antennas, six hypotheses can be considered for the blind combining:

$$\{[1,0],[0,1],[1,1],[1,-1],[1,j],[1,-j]\}$$

In the case of four receiver antennas, there are more combinations. Here two sets of combining hypotheses are provided, as shown in Tables 8.1 and 8.2 with 24 entries and 16 entries, respectively. With these combining weights, blind spatial combining can be carried out to achieve the spatial domain multi-user interference suppression, without the channel estimation.

By now, the discussion of the key modules for data-only receivers for multiple antennas is complete. Next, we show a specific design for two receiver antennas. As shown in Figure 8.12, first carry out spatial domain blind combining over the two receiver antennas. There are a total of six combining vectors (hypotheses). Each combining vector corresponds to a combining "channel". The processing in each combining "channel" is the same, e.g., blind receiver for data-only of single receiver antenna. More specifically, for each combining channel, in each SIC iteration, first to carry out blind sequence detection, pick M active sequences. Then perform blind MMSE de-spreading and blind channel estimation/equalization to get the equalized potential constellations. Then calculate the SINRs of all equalized potential constellations over six combining channels, and

TABLE 8.1 Spatial Combining Weights for 4 Rx Antennas, 24 Hypotheses

Index	Combining Weights	Index	Combining Weights
1	$[1, 1, 0, 0]$	13	$[0, 1, 1, 0]$
2	$[1, -1, 0, 0]$	14	$[0, 1, -1, 0]$
3	$[1, j, 0, 0]$	15	$[0, 1, j, 0]$
4	$[1, -j, 0, 0]$	16	$[0, 1, -j, 0]$
5	$[1, 0, 1, 0]$	17	$[0, 1, 0, 1]$
6	$[1, 0, -1, 0]$	18	$[0, 1, 0, -1]$
7	$[1, 0, j, 0]$	19	$[0, 1, 0, j]$
8	$[1, 0, -j, 0]$	20	$[0, 1, 0, -j]$
9	$[1, 0, 0, 1]$	21	$[0, 0, 1, 1]$
10	$[1, 0, 0, -1]$	22	$[0, 0, 1, -1]$
11	$[1, 0, 0, j]$	23	$[0, 0, 1, j]$
12	$[1, 0, 0, -j]$	24	$[0, 0, 1, -j]$

TABLE 8.2 Spatial Combining Weights for 4 Rx Antennas, 16 Hypotheses

Index	Combining Weights	Index	Combining Weights
1	$[1, 1, 1, 1]$	9	$[1, -1, -j, -j]$
2	$[1, -1, 1, -1]$	10	$[1, 1, -j, j]$
3	$[1, 1, -1, -1]$	11	$[1, -1, j, j]$
4	$[1, -1, -1, 1]$	12	$[1, 1, j, -j]$
5	$[1, -j, -j, 1]$	13	$[1, j, -1, j]$
6	$[1, j, j, -1]$	14	$[1, -j, -1, -j]$
7	$[1, -j, -j, -1]$	15	$[1, j, 1, -j]$
8	$[1, j, -j, 1]$	16	$[1, -j, 1, j]$

sort the SINRs. Pick F symbol streams with the highest SINRs for further demodulation and channel decoding. Once the CRC check of one of the streams succeeds, signal reconstruction can be carried out with the detected spreading sequence and decoded information bits. The reconstructed signal can be used for channel estimation, after which this user's signal can be extracted from the received signal.

Using the blind receiver as shown in Figure 8.12, the performance is shown in Figure 8.13. More detailed simulation assumptions are as follows.

Each user has a single transmit antenna. LTE Turbo code of 1/2 code rate and BPSK modulation are used. The spreading length is 4. After the spreading, the signal goes through LTE's CP-OFDM with a subcarrier spacing of 15 kHz. Then the symbols are mapped to a resource of 180 kHz wide in frequency and 4 ms in the time domain. Each user would randomly select one of 64 MUSA spreading sequences (spreading codebook can be found in Table 6.5 in Chapter 6). The four spread symbols are placed in consecutive resource elements in time domain. The difference between the single receiver antenna is that each user's transmit power is

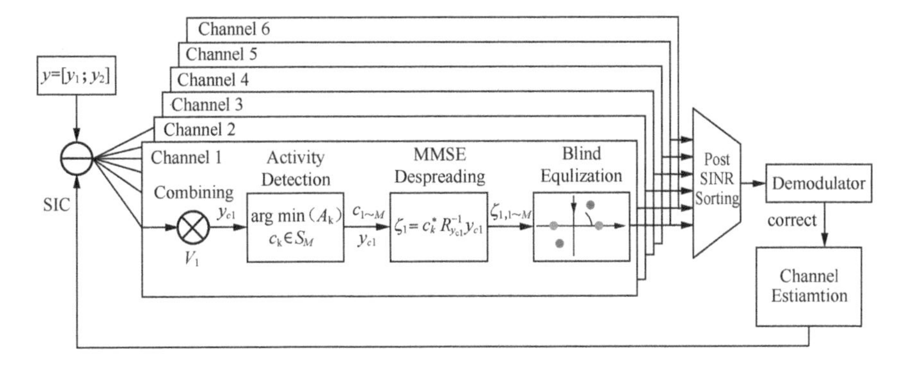

FIGURE 8.12 Data-only blind receiver for two antennas.

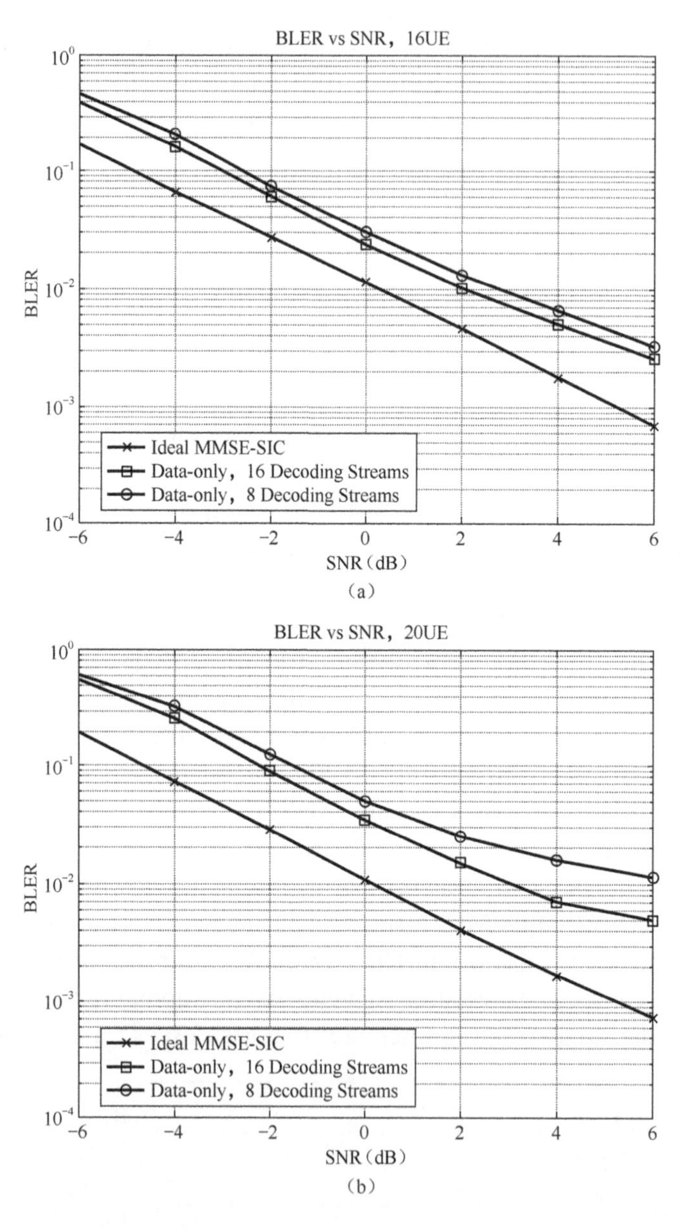

FIGURE 8.13 Link-level performance of data-only blind receiver for 2Rx antennas. (a) BLER vs SNR for data-only-based transmission of 16UEs and (b) BLER vs SNR for data-only-based transmission of 20UEs.

the same in two receiver antennas. In this case, the near-far effect is only reflected in the fast fading. For simplicity without losing generality, flat Rayleigh fading is assumed.

The simulation setting for the blind receiver is as follows:

1. Using above six combining vectors {[1, 0], [0, 1], [1, 1], [1, −1], [1, j], [1, $-j$]}. In each SIC iteration, six spatially combined data streams are generated.
2. In each SIC iteration, for each spatially combined data stream, blind sequence detection is carried out. Out of 64 spreading sequences, pick eight sequences with the highest SINR and perform MMSE blind de-spreading and blind channel estimation/equalization for the eight corresponding symbol streams.
3. After the blind equalization, there would be a total of $6 * 8 = 48$ BPSK symbol streams. By examining the SINR of 48 equalized constellations, pick 8 or 16 streams with the highest SINRs and perform demodulation and channel decoding.

It is seen from Figure 8.13 that with two receiver antennas at the base station and MUSA spreading sequence, data-only receiver with advanced blind detection can achieve quite high overloading capability, e.g., 500% corresponding to 20 users simultaneously transmitting, in contention-based grant-free access, without reference signal. Even with 20 users, the performance of the data-only advanced receiver is not far from the performance of MMSE-SIC with ideal channel estimation. In contrast, for reference signal-based solutions, when the number of users increases, e.g., six to eight users, the high probability of collision under contention-based grant-free would severely degrade the quality of channel estimation and thus the performance [4].

8.4 DMRS ENHANCEMENTS

A demodulation reference signal (DMRS), as a reference signal for data demodulation, is very important to the performance. In the current specification of NR, Type 1 (based on frequency domain comb+time/frequency domain length-2 code division multiplexing) and Type 2 (based on frequency division multiplexing+time/frequency domain length-2 code division multiplexing), as illustrated in Figure 8.14. The total number of orthogonal ports is 8 and 12, respectively.

8.4.1 Enhanced Designs

In order to support a larger number of users and reduce the collision probabilities of DMRS ports, several enhancements of DMRS are proposed [4].

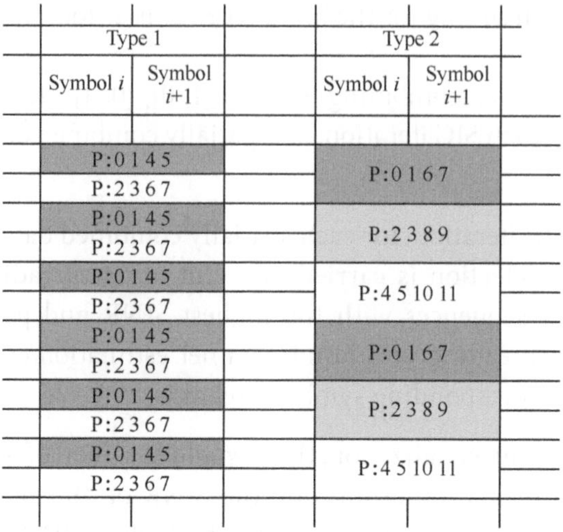

Type 1		Type 2	
Symbol i	Symbol i+1	Symbol i	Symbol i+1
P:0 1 4 5		P:0 1 6 7	
P:2 3 6 7			
P:0 1 4 5		P:2 3 8 9	
P:2 3 6 7			
P:0 1 4 5		P:4 5 10 11	
P:2 3 6 7			
P:0 1 4 5		P:0 1 6 7	
P:2 3 6 7			
P:0 1 4 5		P:2 3 8 9	
P:2 3 6 7			
P:0 1 4 5		P:4 5 10 11	
P:2 3 6 7			

FIGURE 8.14 DMRS design in NR specification.

1. To use longer orthogonal code in the time/frequency domain: this family of schemes is based on Type 2 DMRS, including the following flavors:

    a. Length-4 orthogonal cover codes (OCC) are in the frequency domain while Length-2 codes are in the time domain. As shown in Figure 8.15, each orthogonal code division multiplexing (CDM) group (corresponding to a type of shade) can support up to eight DMRS ports. In each physical resource block (PRB), three CDM groups can be configured, bringing a total of 24 DMRS ports.

    b. Length-2 OCC in the frequency domain and length-4 in the time domain, as shown in Figure 8.16. In this solution, resources of both front-loaded DMRS and additional DMRS are used. Each CDM group is shown with its unique shade and can support 8 DMRS ports. There are a total of three CDM groups, e.g., 24 DMRS ports in a PRB. Compared to Figure 8.15, the solution in Figure 8.16 has a higher density in the frequency domain for more accurate channel estimation and is more suitable for the low Doppler shift case.

    c. Length-4 OCC is applied to both time and frequency domains, as shown in Figure 8.17, which can be considered as the combination of the above two designs and be able to support more DMRS ports, e.g., 48. However, such design is at the cost of performance for high-frequency selectivity and high Doppler channels.

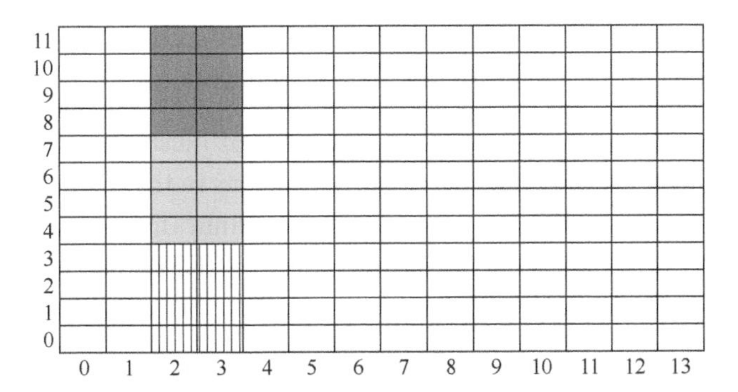

FIGURE 8.15 DMRS enhancements (frequency domain OCC4+time domain OCC2) to support 24 DMRS ports.

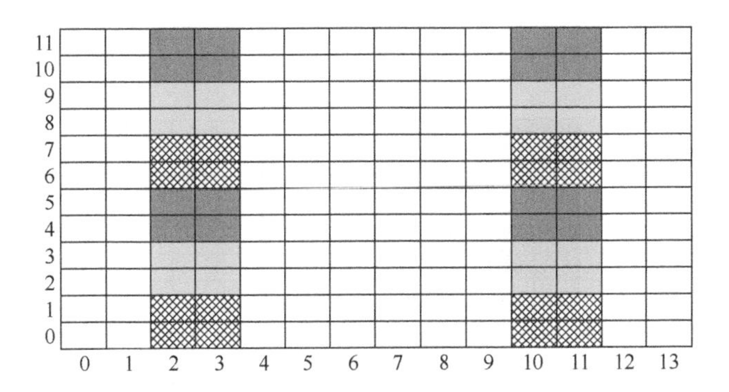

FIGURE 8.16 DMRS enhancements (frequency domain OCC2+time domain OCC4) to support 24 DMRS ports.

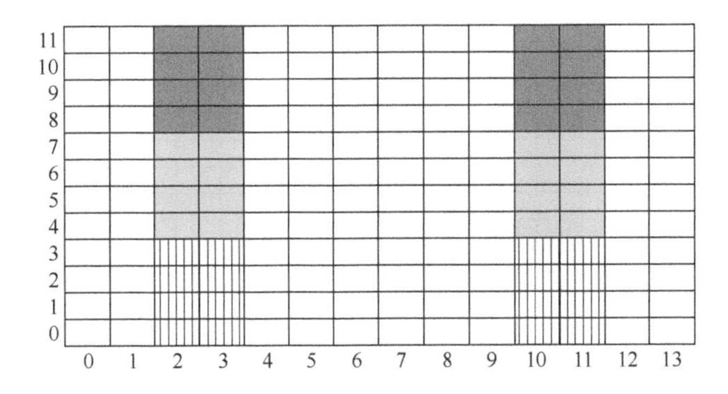

FIGURE 8.17 DRMS enhancement (OCC4 for both time and frequency domain) to support 48 DMRS ports.

2. Sparser density in the frequency domain, suitable for Type 1 and Type 2 DMRS, including the following flavors:

    a. Based on the design of Type 2 DMRS, as illustrated in Figure 8.18. Each port occupies two resource elements in the frequency domain (density cut in half). Length-2 OCC continues to be used in the time/frequency domain. Each time/frequency CDM group can support 4 DMRS ports. In total, 24 ports can be supported in a PRB.

    b. Based on Type 1 DMRS design, as seen in Figure 8.19. A comb factor of 4 is used in the frequency domain. Reuse cyclic shift (CS = 2) and length-2 OCC as in current NR spec. Each comb can support 4 DMRS, with the total being 16 in a PRB.

    c. Based on Type 1 DMRS design, as shown in Figure 8.20. Comb factor of 6 in frequency domain and reuse frequency

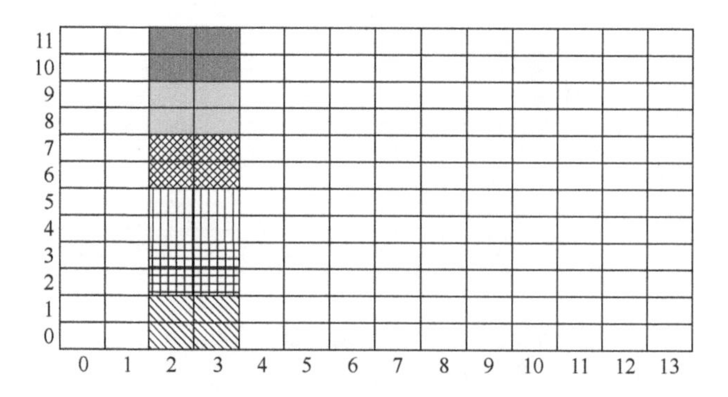

FIGURE 8.18 DMRS enhancement (each port in frequency occupies only two resource elements), supporting 24 ports.

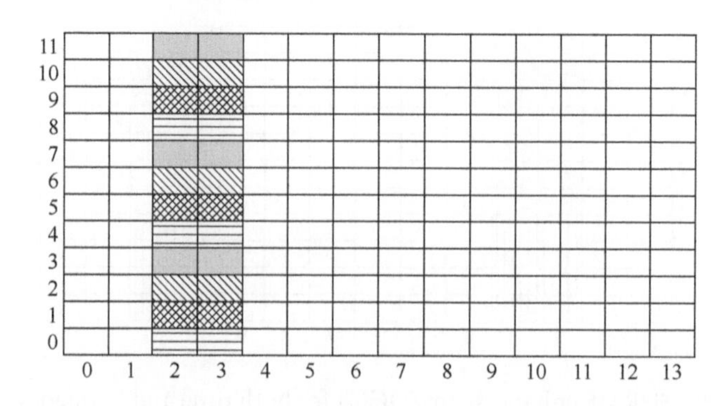

FIGURE 8.19 DMRS enhancement (comb factor of 4), supporting 16 DMRS ports.

domain cyclic shift (CS = 2) and time domain length-2 OCC of NR spec. Each comb supports 4 DMRS ports, with the total being 24 in a PRB.

d. Reuse Type 1 DMRS pattern. To use larger cyclic shift (CS = 6) and length-2 OCC in time domain. Each comb can accommodate 12 DMRS ports, with the total being 24, as illustrated in Figure 8.21.

3. To use ZC sequence with configurable subcarrier spacing. To adjust the target number of users and the number of DMRS ports via adjusting the number of root sequences and the number of cyclic shifts.

4. To use more resources for DMRS, as illustrated in Figure 8.22, additional resources are used in addition to the front-loaded resources. Length-2 OCC is still used in the time domain. Each CDM group can support four ports. There are a total of 24 ports in a PRB.

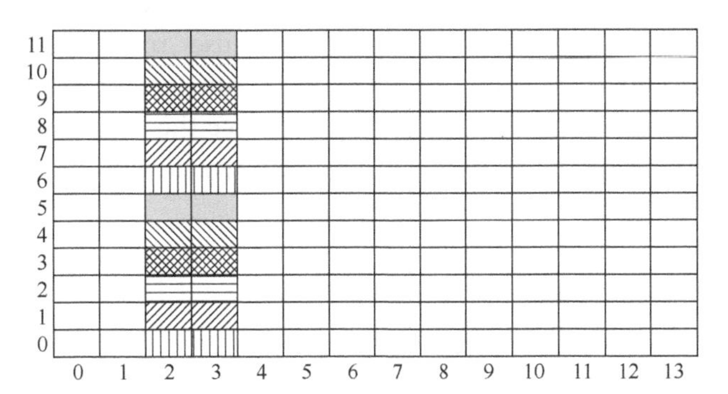

FIGURE 8.20 DMRS enhancement (comb factor of 6), supporting 24 DMRS ports.

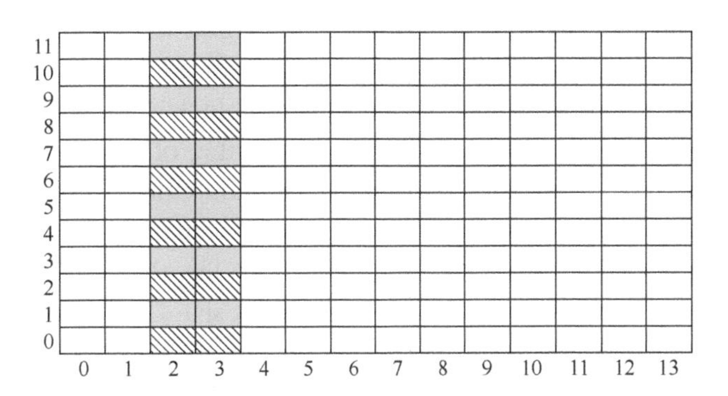

FIGURE 8.21 DMRS enhancement (comb factor = 2, CS = 6).

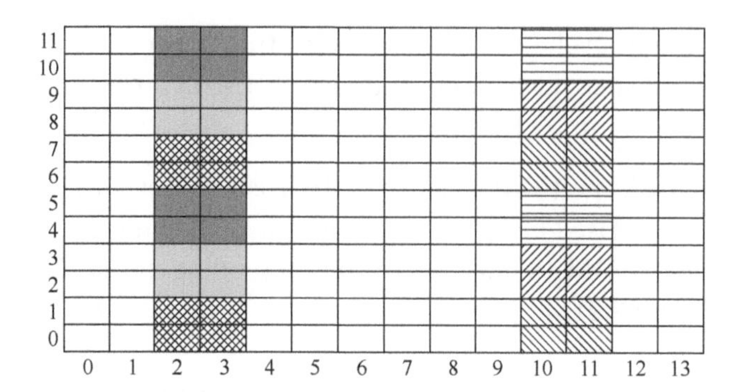

FIGURE 8.22 DMRS enhancement (additional resources to be used) to support 24 DMRS ports.

8.4.1.1 Configuration Signaling

As seen in Section 8.1, in order to support contention-based grant-free access, time/frequency resources and the corresponding configurations for reference signals, e.g., DMRS, are normally signaled via the broadcast channel, such as system information. Once a user receives the signal, it would randomly select a DMRS port. The parameters for DMRS depend on the above enhancement of DMRS. To reduce the complexity of blind detection, the mapping relationship, e.g., between the reference signal and time/frequency resources, may be signaled.

8.5 PERFORMANCE EVALUATION AND METHODOLOGY

8.5.1 Line-Level Simulation Parameters

Table 8.3 lists the simulation cases for contention-based grant-free in the 3GPP NOMA study. It is seen that contention-based transmission is mainly used for mMTC and eMBB scenarios. In the mMTC scenario, the typical band is 700 MHz. The number of antennas is 2. Both CP-OFDM waveform and DFT-s-OFDM waveform can be used. Due to contention-based access, the number of bits in a transport block is not large, e.g., only 10 and 20 bytes need to be simulated. The number of simultaneous users is not very large either, e.g., four should be mandatorily simulated while six can be optionally simulated. For the eMBB scenario, the typical band is 4 GHz. The number of receiver antennas is 4. Only the CP-OFDM waveform needs to be simulated. The transport block size (TBS) can be 40 or 80 bytes due to the double resources and double receiver antennas compared to the mMTC scenario. The number of users can be 4 or 6.

TABLE 8.3 Link-Level Simulation Cases for Contention-Based Grant-Free (Mandatory)

Case Index	Scenario	Carrier Frequency (Hz)	# Antennas	Average SNR	Waveform	MA Signature Allocation	Channel Model	TBS (Bytes)	# Users	TO/FO
10	mMTC	700M	2	Unequal	CP-OFDM	Random	TDL-C	10	4	>0
11	mMTC	700M	2	Unequal	CP-OFDM	Random	TDL-A	20	4	>0
12	mMTC	700M	2	Unequal	DFT-S	Random	TDL-A	10	4	>0
13	mMTC	700M	2	Unequal	DFT-S	Random	TDL-A	20	4	>0
24	eMBB	4GHz	4	Unequal	CP-OFDM	Random	TDL-A	40	4	>0
25	eMBB	4GHz	4	Unequal	CP-OFDM	Random	TDL-A	80	4	>0

8.5.2 Link to System Mapping (PHY Abstraction)

8.5.2.1 Preamble or Reference Signal-Based

For link to system mapping with the potential collision of the preamble/reference signal, the focus in this section is how to capture the impact of collision on user identification and channel estimation. All other aspects would be similar to that without preamble/reference signal collision. Since two or more users select the same DMRS, the base station can identify only one user, and the estimated channel would be the superposition of all those channels with DMRS collision. Assuming that UE1 and UE2 choose the same DMRS, the estimated channel can be modeled as

$$H_R = H_{I1} + H_{I2} + H_e \tag{8.15}$$

where H_{I1} and H_{I2} represent the true channel coefficients of UE1and UE2, respectively. The channel estimation error H_e can be modeled with the same method in Section 7.1.2.

Due to the impact of collision, the channel estimation based on DMRS may be not accurate, so LS channel estimation based on the reconstructed symbols of all successfully decoded users can be considered to refine the channel estimation results for interference cancellation. As the number of decoded users increases, the LS channel estimation results would be more accurate. Here assuming UE 1 has been decoded correctly, the LS channel estimation error of UE 1 can be modeled approximately as

$$\frac{|h_e|^2}{|h_1|^2} \approx \frac{\sum \mathrm{SNR}_{\mathrm{int}} + 1}{\left(x_1^H x_1\right)\mathrm{SNR}_1} \tag{8.16}$$

where h_1 is the channel coefficient in the frequency domain for UE1. h_e is the channel estimation error of UE1 using LS-based estimation. x_1 is the symbol sent by UE1. SNR_1 is the ideal SNR of UE1. $\mathrm{SNR}_{\mathrm{int}}$ includes the ideal SNR of other not-yet successfully decoded users.

When there is DMRS collision, missed detection would occur. Hence, when a user is successfully decoded and its contribution to the interference is canceled, we can repeat the entire process as described in Figure 7.2 to redo the user identification. This can help to find the missed-detected users and improve the performance.

8.5.2.2 Validation of LS Channel Estimation

TDL-A 30 ns channel model is assumed in the validation. Take two users as an example, assuming that one user's data is successfully decoded. To perform LS channel estimation based on the reconstructed data. Then to compare the error from the model and the statistics of the actual estimation error. Figure 8.23 shows the variance of the normalized realistic channel estimation (RCE) error and the variance of the normalized error from the model. They are very close.

8.5.2.3 Validation of Link to System Mapping

For the mMTC scenario, assuming the CP-OFDM waveform, TDL-A 30 ns fading channel model, and equal or unequal average SNR, simulations are carried out under different loading of users. The results of the validation are shown in Figure 8.24.

8.5.2.4 Data-Only-Based

With a data-only solution, the issue of preamble or reference signal collision due to the limited number of preamble/reference signals can be avoided. The data-only solution is accompanied by blind detection, and its physical layer abstraction should be able to model the key processing modules in the receiver which are illustrated in Figure 8.25, including the following steps:

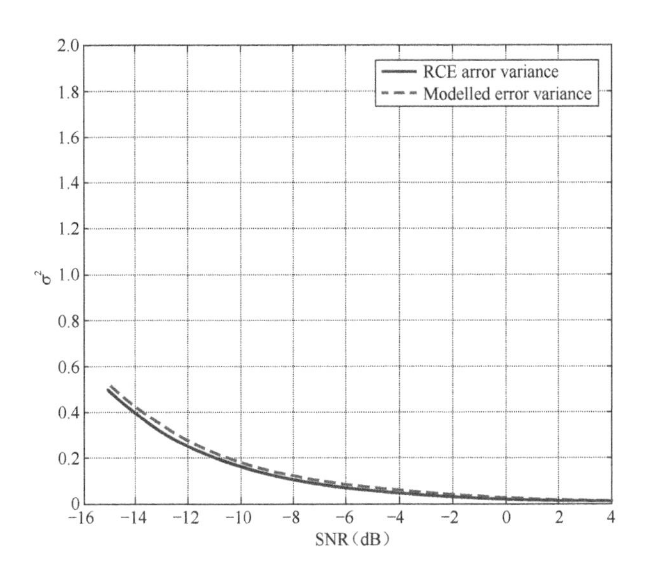

FIGURE 8.23 Variance of normalized channel estimation errors, of true estimate vs. model.

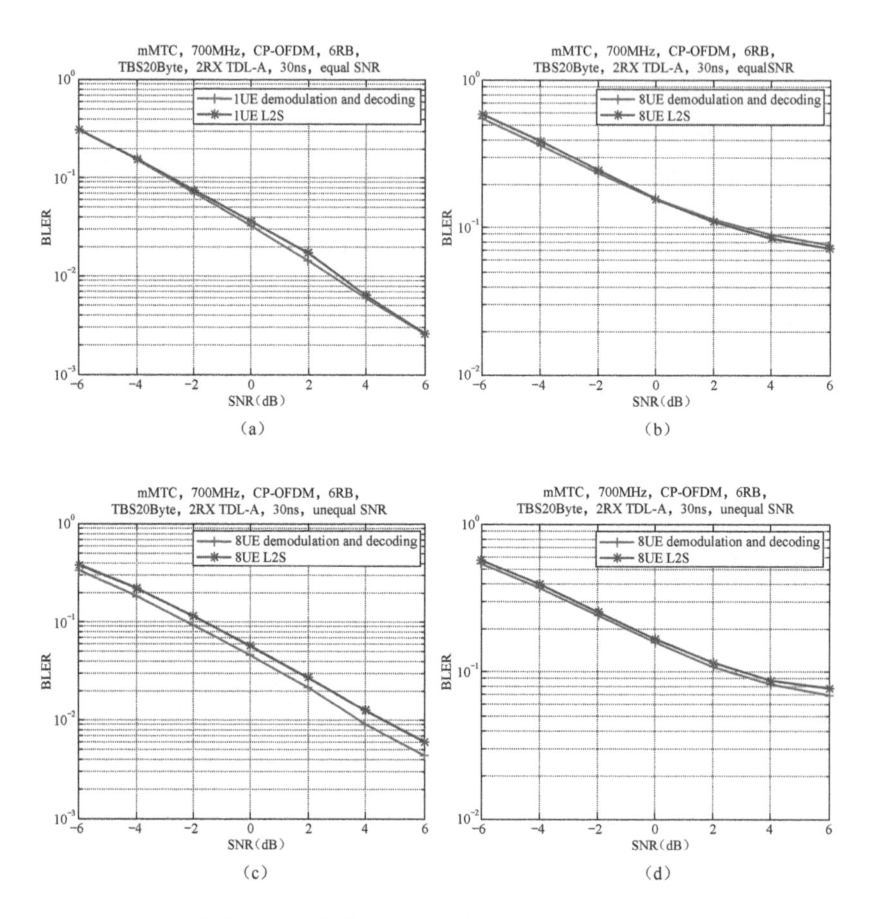

FIGURE 8.24 Validation of link to system mapping. (a) mMTC, CP-OFDM, 1 UEs, 6 RBs, 1 ms, TDL-A 30 ns, equal SNR, (b) mMTC, CP-OFDM,8 UEs, 6 RBs, 1 ms, TDL-A 30 ns, equal SNR, (c) mMTC, CP-OFDM, 1 UEs, 6 RBs, 1 ms, TDL-A 30 ns, unequal SNR, uniform [–3, 3] and (d) mMTC, CP-OFDM, 8 UEs, 6 RBs, 1 ms, TDL-A 30 ns, unequal SNR, uniform [–3, 3].

1. To calculate the effective channel from the ideal channel estimation

   Assuming there are K users of single transmit antenna and M spreading sequences. The number of receiver antennas at the base station is N. The true effective channel of the k-th user on the i-th receiver antenna is

$$\hat{h}_{i,k} = h_{i,k} c_k \tag{8.17}$$

   where $h_{i,k}$ is the true channel of the k-th user on the i-th receiver antenna and c_k is the spreading sequence selected by the k-th user.

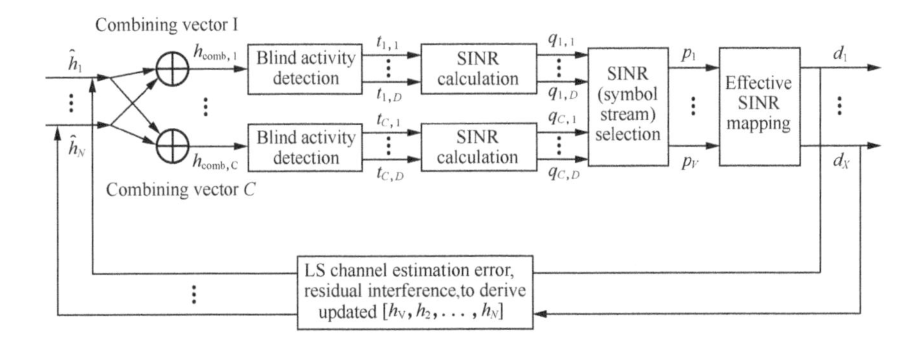

FIGURE 8.25 Blind MMSE-SIC for the data-only solution.

2. Antenna combining and blind identification of users

Using the pre-defined combining vectors to combine the effective channels to get the j-th combined signal of the k-th user:

$$h_{\text{comb},j,k} = c_{\text{comb},j}\hat{h}_k \tag{8.18}$$

where $c_{\text{comb},j}$ is the j-th combining vector. $\hat{h}_k = \left(\hat{h}_{1,k}{}^T, \hat{h}_{2,k}{}^T, \ldots, \hat{h}_{N,k}{}^T\right)^T$. For two receiver antennas, six combining vectors are defined: {(1, 0), (0, 1), $(1/\sqrt{2}, 1/\sqrt{2})$, $(1/\sqrt{2}, -1/\sqrt{2})$, $(1/\sqrt{2}, j/\sqrt{2})$, $(1/\sqrt{2}, -j/\sqrt{2})$}. Then calculate the covariance matrix R by using the combined signal h_{comb}.

$$R = E[YY^*] = \frac{1}{S}\sum_{k=1}^{K} h_{\text{comb},k}h^*_{\text{comb},k} + R_{nn} \tag{8.19}$$

where S is the number of modulation symbols transmitted by each user. R_{nn} is the additive white Gaussian noise. Then, with the information about the covariance matrix R and the set of spreading sequences, based on the following criterion, pick D spreading sequences from M sequences whose metrics are the smallest.

$$\arg\min_{(D)} c_m^H R^{-1} c_m, m = 1, \ldots, M \tag{8.20}$$

3. Calculation of the SINR and selection of data streams

For the k-th user, if the used spreading sequence is identified, the SINR is calculated based on $h_{\text{comb}, k}$:

$$\text{SINR}_k = h^*_{\text{comb},k} \left(\sum_{i=1, i \neq k}^{K} h_{\text{comb},i} h^*_{\text{comb},i} + \sigma^2 I \right)^{-1} h_{\text{comb},k} \tag{8.21}$$

where σ^2 is the variance of additive white Gaussian noise. I is the identity matrix. Then to sort the derived SINRs of users and pick V streams with the highest SINR for decoding.

4. Mapping of the effective SINR

   To obtain the effective SINRs of V streams, based on the RBIR-SINR mapping formula. Then using the effective SINRs to look up BLER – SNR curves in additional white Gaussian noise (AWGN) channel to figure out the BLER. To compare this BLER with a random variable uniformly distributed over [0, 1]. If less than the random variable, this stream is considered decoded successfully. Otherwise, the decoding fails.

5. LS channel estimation, interference cancellation and update

   From the successfully decoded bits, we can get the user identification information and the spreading sequence this user has selected. Then to reconstruct the signal, followed by LS channel estimation and interference cancellation. Here LS channel estimation error and residual interference can be modeled. The procedure (2) – (5) can be repeated until no users can be successfully decoded.

In the mMTC scenario, assuming CP-OFDM waveform and TDL-C 300ns fading model, frequency domain or time domain spreading, and equal or unequal average SNR, the validation results under the different user loading are shown in Figure 8.26.

8.6 PERFORMANCE EVALUATIONS

8.6.1 Link-Level Simulation Results

For contention-based grant-free access, not many companies in the 3GPP NOMA study item have participated in the corresponding simulations. Even when some of them do, their understanding of grant-free is not fully aligned, which results in different simulation assumptions. Hence it is not

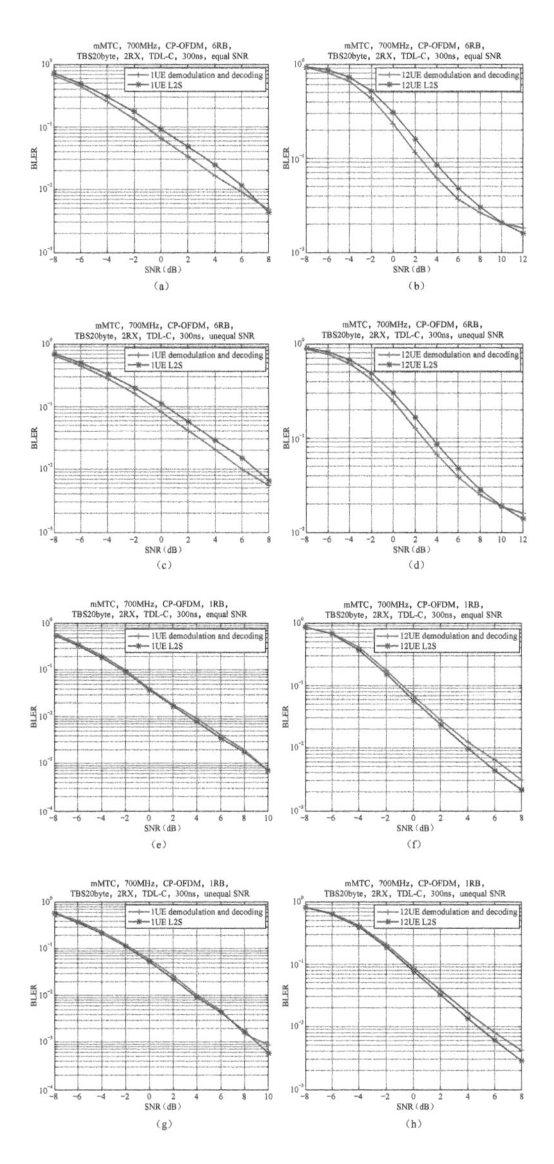

FIGURE 8.26 Validation of link to system mapping for the data-only solution with a blind receiver. (a) mMTC, CP-OFDM, 1 UE, 6 RBs, 1 ms , TDL-C 30 ns, equal SNR, (b) mMTC, CP-OFDM, 12 UE, 6 RBs, 1 ms , TDL-C 30 ns, equal SNR, (c) mMTC, CP-OFDM, 1 UE, 6 RBs, 1 ms, TDL-C 30 ns, unequal SNR, uniform [–3, 3], and (d) mMTC, CP-OFDM, 12UE, 6 RBs, 1 ms, TDL-C 30 ns, unequal SNR, uniform [–3, 3], (e) mMTC, CP-OFDM, 1 UEs, 1 RBs, 6 ms, TDL-C 300 ns, equal SNR, (f) mMTC, CP-OFDM, 12 UEs, 1 RBs, 6 ms, TDL-C 300 ns, equal SNR, (g) mMTC, CP-OFDM, 1 UEs, 1 RBs, 6 ms, TDL-C 300 ns, equal SNR, uniform [–3, 3] and (h) mMTC, CP-OFDM, 12 UEs, 1 RBs, 6 ms, TDL-C 300 ns, equal SNR, uniform [–3, 3].

meaningful to compare all the results. Instead, the results for simulation Case 11 of 4 or 6 users may be compared.

There are two understandings of grant-free transmission: random selection vs. random activation. Random selection is contention-based where different users randomly select their preambles or DMRS sequences, resulting in potential collisions. Since the pool size of preamble or DMRS is fixed, as the number of users increases, collision probability grows. This is the primary reason for significant performance degradation in contention-based scenarios. Random activation is essentially a content-free scenario where the potential number of users is less than the pool size of preamble or DMRS. Each potential user is allocated a unique sequence. When a user is activated, it just uses the pre-allocated preamble or DMRS sequence without worrying about the collision.

Obviously, due to collision-free nature, the performance of random activation is better than that of random selection, and this difference becomes more significant as the number of users increases. However, in real systems, random selection is more common because the pool size of preamble or DMRS cannot be excessive, yet the potential number of users can keep increasing. When the number of potential users is larger than that of preamble or DMRS, random activation may not be very valid.

In Figure 8.27, MUSA and SCMA2 are of contention-based access, while interleaved grid multiple access (IGMA) and SCMA1 are of contention-free access. Here our focus is contention-base. Comparing the two contention-based solutions, the performance difference is mainly due to the pool size of the preamble or DMRS. For MUSA, the TBS is 20 bytes. It is based on the realistic user detection algorithm. The transmission duration is two-slot where 50% overhead is spent on preamble for the purpose of the reference signal (no dedicated DMRS). Preamble and data have the same bandwidth. The pool size of the preamble is 64. In the case of four users, BLER = 10% can be achieved at SRN = −1.7 dB. For six users, SNR has to be increased to 1.2 dB to get BLER = 10%. For SCMA2, the TBS is 20 bytes. The transmission duration is 1 slot and the overhead of DMRS is 2/7. The pool size of the SCMA codebook is 24. It is observed that BLER = 10% cannot be achieved for 4 or 6 users, no matter how high the SNR is.

In addition to the different understanding of grant-free, the timing error in the MUSA simulation of Figure 8.27 is assumed to be uniformly distributed within [0, 1.5*NCP], whereas the timing error in SCMA and IGMA is assumed to be uniformly distributed within [0, 0.5*NCP].

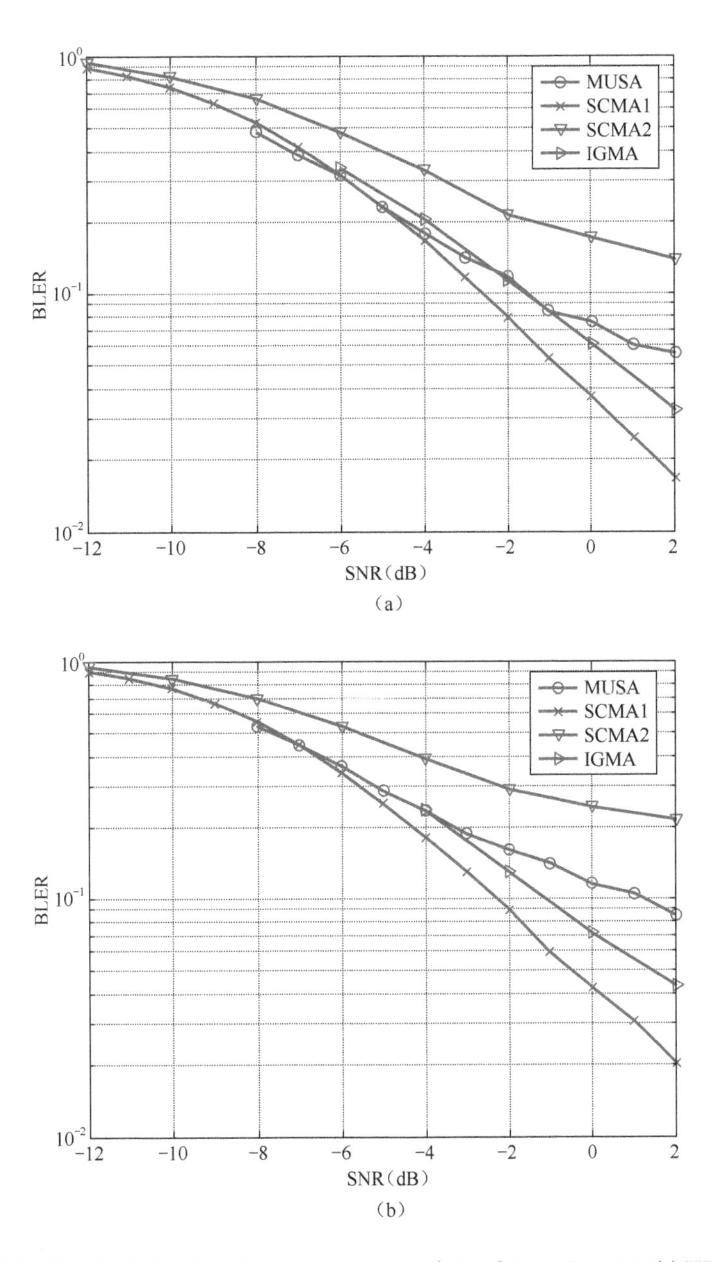

FIGURE 8.27 Link-level performance curves of simulation Case 11. (a) TBS = 20 bytes, 4 UEs and (b) TBS = 20 bytes, 6 UEs.

Next, we consider different deployment scenarios, different waveforms and different pool sizes to see their impact on the performance of contention-based grant-free access. Most of these results are from ZTE's

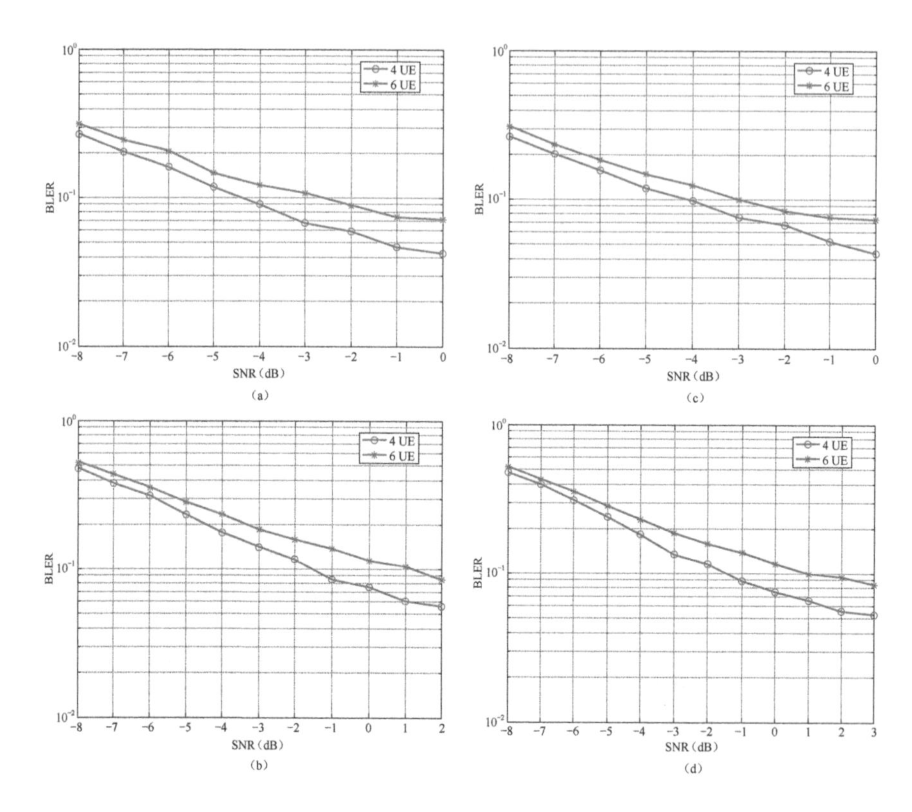

FIGURE 8.28 Link-level performance of contention-based grant-free access in mMTC scenario. (a) TBS = 10 bytes, CP-OFDM waveform, (b) TBS = 20 bytes, CP-OFDM waveform, (c) TBS = 10 bytes, DFT-S-OFDM waveform, and (d) TBS = 20 bytes, DFT-S-OFDM waveform.

contributions to 3GPP. Two scenarios: mMTC and eMBB are simulated, as well as CP-OFDM and DFT-s-OFDM waveforms for the mMTC scenario, and pool sizes of 64 and 96. The timing error is modeled as uniformly distributed within [0, 1.5*NCP]. The simulation results for the mMTC scenario are shown in Figure 8.28.

It is observed that in the mMTC scenario, the link-level performance of contention-based grant-free access is not much different between using CP-OFDM and DFT-s-OFDM waveforms. In both waveforms, up to six users, each having TBS of 20 bytes, can be supported.

In the eMBB scenario, a random selection of MA signatures is assumed. The results are shown in Figure 8.29. It is observed that up to eight users, each having TBS of 80 bytes, can be supported in contention-based grant-free access.

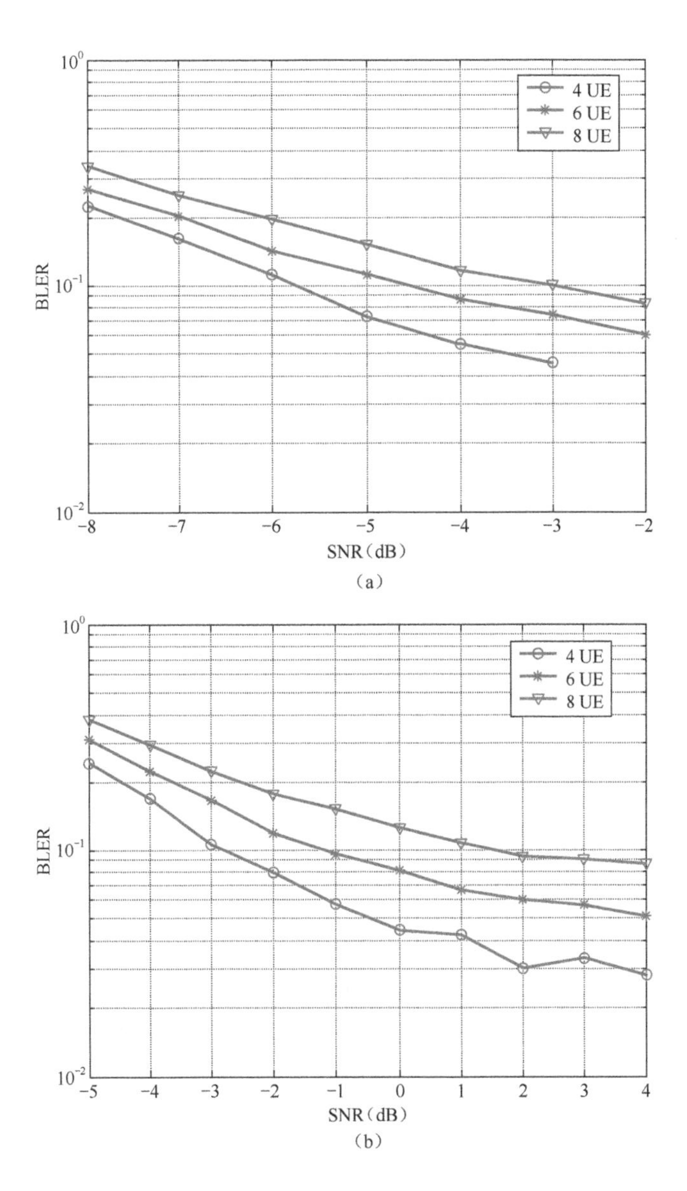

FIGURE 8.29 Link-level performance of contention-based grant-free access, in eMBB scenario. (a) TBS = 10 bytes, CP-OFDM waveform, and (b) TBS = 20 bytes, CP-OFDM waveform.

It is specified in NR standards that the pool size of preambles is 64. Apparently, the collision probability would decrease as the pool size increases. The corresponding performance evaluation is shown in Figure 8.30 where the pool sizes of 64 and 96 are compared. Each spreading sequence is one-to-one mapped to each preamble sequence.

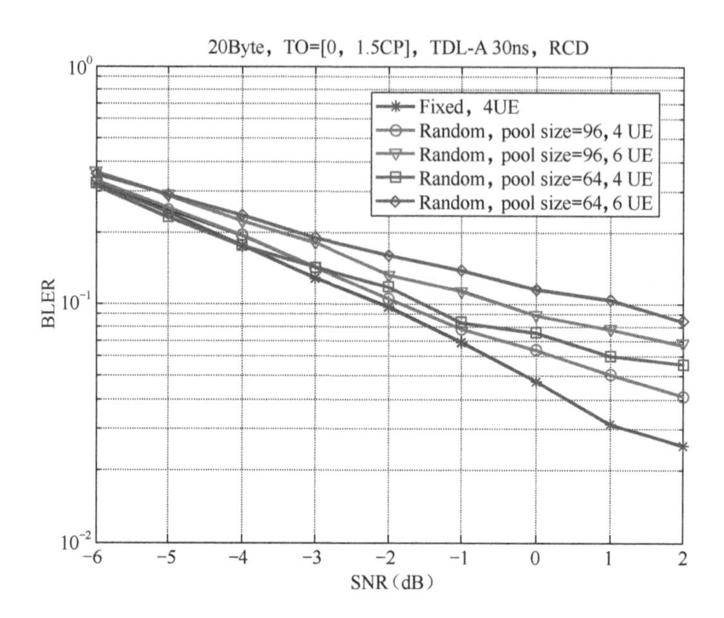

FIGURE 8.30 Simulation results for different pool sizes, TO uniformly distributed within [0, 1.5*NCP].

"Fixed" means pre-allocation without collision. "Random" means random selection with potential collision.

Assuming that there are two windows, e.g., [0, NCP] and [NCP, 2 * NCP], for preamble detection at the receiver, MMSE-SIC would be carried out for each window. It is found that

- When the pool size is 96, the performance difference between contention-based and contention-free grant-free access is quite small, e.g., $\Delta = 0.2$ dB @10% BLER when there are four users, each having TBS of 20 bytes. When the number of users increases to 6, the performance difference is around 1.5 dB.
- When the pool size is 64, the performance difference between contention-based and contention-free grant-free access is not big, e.g., $\Delta = 0.5$ dB @10% BLER when there are four users, each having TBS of 20 bytes. When the number of users increases to 6, the performance difference is around 3 dB.

Hence it can be concluded that when the overloading factor is not high, e.g., four users, the performance degradation due to the random collision of MA signatures is acceptable. However, as more users (e.g., 6) try to access, the pool size of the preambles/MA signatures has to be increased. to reduce the collision probability.

8.6.2 System-Level Simulation Results

In this section, simulations are carried out to evaluate the system-level performance of uplink NOMA that is based on contention-based grant-free access. Here, MUSA is used as an example where both data-only solution and preamble+data solution are evaluated. The deployment scenario is mMTC.

8.6.2.1 Data-Only Solution

In this simulation, for the baseline solution, there are 24 DMRS, and the time/frequency resources and DMRS sequences are randomly selected by each UE. For DMRS-based MUSA, there are 64 DMRS sequences and 64 spreading sequences of length 4, with one-to-one mapping for each. Each user randomly selects time/frequency resources and DMRS sequences. While in MUSA with data-only solutions, only the 64 spreading sequences are used, i.e., no overhead for DMRS.

To ensure the coverage, users in a cell are put into three groups according to their wideband downlink SNR. Each group of users occupy one frequency location (e.g., resource pool) not overlapped with other groups. Each user randomly selects one PRB within a resource pool. For the

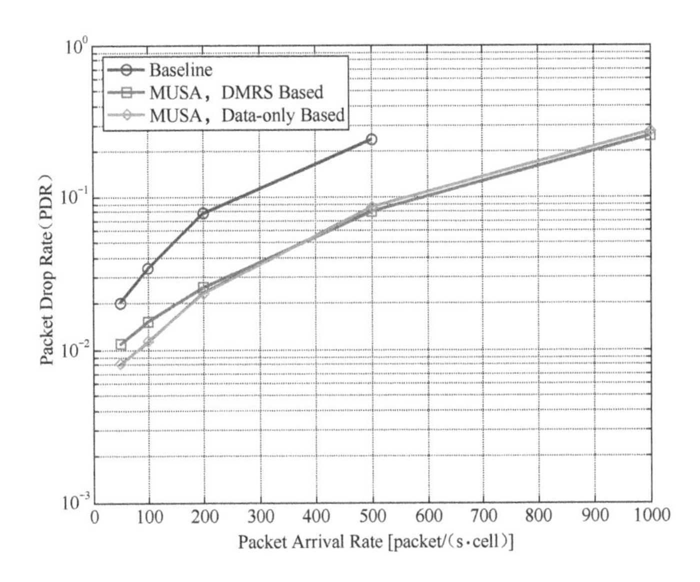

FIGURE 8.31 System-level performance comparison of the baseline and MUSA with DMRS and data-only for contention-based grant-free access.

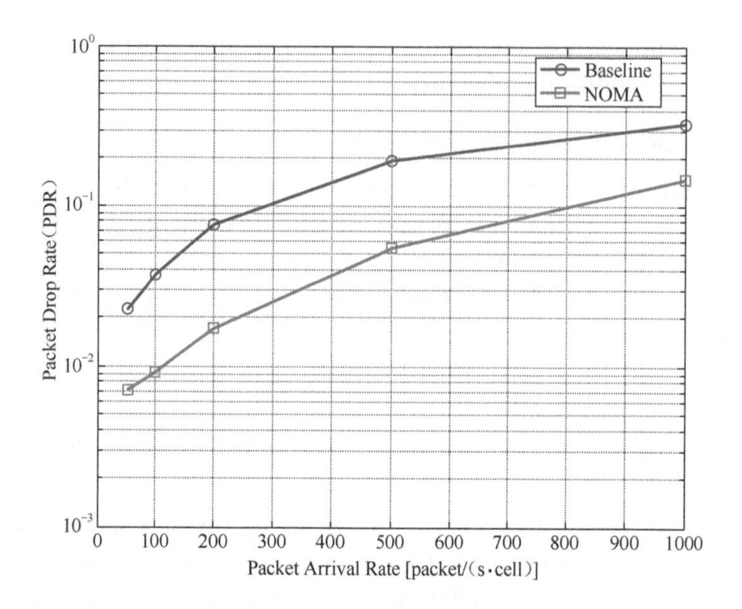

FIGURE 8.32 System-level performance comparison of the baseline and MUSA with (preamble + data) for contention-based grant-free access.

baseline solution, 1, 4 and 16 repetitions (1 means no repetition) are carried out for three groups, respectively. Repeated transmissions can improve the coverage. For the MUSA solution, 1, 1 and 4 repetitions are carried out, in addition to length 4 spreading. A blind MMSE-IC receiver algorithm is used for both the baseline solution and MUSA solution.

In Figure 8.31, system performances of the baseline, DMRS-based MUSA and data-only-based MUSA are compared in terms of packet drop rate. It is found that DMRS-based MUSA and data-only-based MUSA significantly outperform the baseline solution. Part of the reason is the low probabilities of collisions in MUSA.

8.6.2.2 (Preamble + Data) Solution

In this simulation, each UE occupies 6 PRBs. In the baseline, there are 64 preambles, randomly selected by each user. A similar setting is for the NOMA solution in the sense that 64 preambles are used with random selection by each user. The data part of the NOMA solution is spread by length = 4 spreading sequences. There are 64 spreading sequences, one-to-one mapped to preambles. To ensure coverage, four repetitions are applied.

Figure 8.32 shows the performances of (preamble + data) of the baseline and NOMA solution, in terms of packet drop rate (PDR) vs. packet arrival rate. At PDR = 10%, NOMA can support 800 packets/s/cell whereas the baseline can only support 300 packets/s/cell. The gain is about 150%.

REFERENCES

1. Z. Yuan, Y. Hu, W. Li, and J. Dai, "Blind multi-user detection for autonomous grant-free high-overloading MA without reference signal," arXiv:1712.02601 cs.IT], 2017.
2. Z. Yuan, Y. Hu, W. Li, and J. Dai, "Blind multi-user detection for autonomous grant-free high-overloading multiple-access without reference signal," in *IEEE 87th Vehicular Technology Spring Conference RAMAT*, Porto, Portugal, 2018.
3. 3GPP, TS 38.211, NR physical channels and modulation (Release 15).
4. 3GPP, TR 38.812, Study on non-orthogonal multiple access (NOMA) for NR.